Omics and Plant Abiotic Stress Tolerance

Edited by

Narendra Tuteja

Sarvajeet Singh Gill

&

Renu Tuteja

eBooks End User License Agreement

CONTENTS

FOREWORD

Ten-thousand years of selection have turned wild species into established useful plants. A few hundred years of breeding with evermore sophistication have lead to the generation our extant crops. All of this progress has been achieved by phenotype observation, increasingly augmented by measurements. The concepts of evolutionary change, of genetics and genes, and finally the reality of DNA have been added during the last century. Plant biology and the crop-centered biological sciences as well, profited from the advances, but at the same time many biological disciplines lost sight of the organismic dimension in the pursuit of detail. A large number of plant species with intriguing evolutionary adaptations that had previously been studied were abandoned and replaced by focus on a handful of model species with fervent supporters and fierce detractors. To be sure, the focus on models provided immense new insights into the functioning of plants but the new insights became possible only because of a paradigm change. The steady and incremental advance in knowledge has been replaced by a view of plants as an integrated system of many processes and pathways. The sea change in conceptual approach and high-throughput experimentation has been accomplished by the availability of new technologies. These summarily termed 'omics' concepts have opened a way for evolutionary specialties to be studied and understood in previously unimaginable complexity.

The volume edited by Drs. Narendra Tuteja, Sarvajeet Singh Gill and Renu Tuteja places plant stress tolerance behaviour in this 'omics' context and does it well. The book succeeds in presenting a large variety of concepts, models and viewpoints. The book presents a wealth of excellent articles, both broad overviews as well as detailed accounts that discuss genomics, transcriptomics, proteomics and metabolomics principles aimed at advancing our understanding of plant abiotic stress tolerance phenomena. The chapters, written by experts in their respective fields, cover a large array of topics and interpret our recently dramatically enlarged view of the genetic basis of stress-affected plant development, biochemistry and physiology. The various contributions integrate the stress topic from the view of crop species as well as from the vantage of established models. This comprehensiveness should make this volume equally valuable not only to basic investigators and application-oriented plant scientists but also for teachers and students entering this field of plant biology.

Hans J. Bohnert, Ph.D.
University of Illinois at Urbana-Champaign
Urbana, IL
USA

PREFACE

"Take care of the earth and she will take care of you"
Anonymous

World population is increasing at an alarming rate and is expected to reach more than nine billion by the end of 2050, whereas, plant productivity is being seriously limited by various abiotic stresses all over the world. Global climatic pattern is becoming more unpredictable with increased occurrence of drought, flood, storm, heat waves, and sea water intrusion. It has been estimated that abiotic stresses are the principal cause for decreasing the average yield of major crops by more than 50%, which causes losses worth hundreds of millions of dollars each year. Therefore, to feed the world population maintaining crop productivity even under unfavourable environment is a major area of concern for all nations. Developing crop plants with ability to tolerate abiotic stresses is need of the day which demands modern novel strategies for thorough understanding of plant's response to abiotic stresses. In particular, an array of new innovative "omics" tools, comprising of genomics, proteomics and metabolomics, are evolving at rapid pace, which is empowering the scientists to systematically analyze the genome at various levels and their effect on phenotypic variability. Omics tools particularly genomics allowed the use of important strategies like genome-wide expression profiling which is useful to identify genes associated with stress response. Furthermore, proteomics helped a lot to analyze the relationship between gene expression (transcriptomics) and metabolism (metabolomics). Metabolomic studies, thus along with transcriptomics and proteomics, and their integration with systems biology, will lead to strategies to alter cellular metabolism for adaptation to abiotic stress conditions.

This volume deals with up-to-date information on omics approaches for crop improvement and abiotic stress tolerance. The first and second chapters describe the importance of omics and its benefits for crop improvement and related issues. Third chapter will deal with up-to-date literature on major abiotic stresses like cold, drought, salt and heavy metals, their effect on plant performance and possible measures to counteract it. The following chapters from four to seven will deal with the use of omics approaches like genomics, proteomics, metabolomics, transcriptomics and many more for crop tolerance to various abiotic stresses. Chapter eight uncovers the epigenetic studies in reference to abiotic stress tolerance. Chapter nine deals with rizotoxic ions and importance of omics approaches has been uncovered in detail. Chapter ten and eleven will discuss the role of nitric oxide, S-nitrosoproteome and abscissic acid in abiotic stress signaling in plants. Chapter twelve is summarizing the importance of RNA silencing in plants following various stresses. Finally, in the chapter thirteen, system biology approaches has been taken into consideration for abiotic stress tolerance.

The editors and contributing authors hope that this book will include a practical update on our knowledge of abiotic stress tolerance and lead to new discussions and efforts to the use of omics tools for the improvement of crop plants in the era of global climatic change.

We would like to thank Prof. Hans Bohnert for writing the foreword and Bentham Science Publishers, particularly Manager Bushra Siddiqui and Salma, for their support and efforts.

Narendra Tuteja, Sarvajeet Singh Gill and Renu Tuteja
International Centre for Genetic Engineering and Biotechnology
India

CONTRIBUTORS

Antonio F. Tiburcio
Unitat de Fisiologia Vegetal,
Facultat de Farmàcia, Universitat de Barcelona, Diagonal 643, 08028 Barcelona, Spain

Aryadeep Roychoudhury
Department of Botany,
Plant Molecular Biology and Biotechnology Laboratory, University of Calcutta, 35,
Ballygunge Circular Road, Kolkata-700 019

Atle M. Bones
Department of Biology,
Norwegian University of Science and Technology, Høgskoleringen 5, NO-7491 Trondheim, Norway

Cheng-Ri Zhao
Applied Plant Science,
Faculty of Applied Biological Sciences,
*Gifu University, Gifu, 501-1193, Gifu, Japa*n

Cristina Bortolotti
Unitat de Fisiologia Vegctal,
Facultat de Farmàcia, Universitat de Barcelona,
Diagonal 643, 08028 Barcelona, Spain

Faïçal Brini
Plant Molecular Genetic Laboratory,
Centre of Biotechnology of Sfax (CBS) Route Sidi Mansour Km 6,
B.P ''1177'' 3018 Sfax –Tunisia

Habib Khoudi
Plant Molecular Genetic Laboratory,
Centre of Biotechnology of Sfax (CBS) Route Sidi Mansour Km 6,
B.P ''1177'' 3018 Sfax –Tunisia

Henry I. Miller
The Hoover Institution,
Stanford University, Stanford,
California 94305–6010, U.S.A.

Hiroyuki Koyama
Applied Plant Science,
Faculty of Applied Biological Sciences,
Gifu University, Gifu, 501-1193, Gifu, Japan

Jasmeet Kaur Abat
Plant Molecular Physiology,
Biochemistry and Proteomics Laboratory,
Department of Botany, University of Delhi,
Delhi-110007, India

John Einset
Department of Plant and Environmental Sciences, P. O. Box 5003, Norwegian University of Life Sciences (UMB),
1432 Aas, Norway

Juan C. Cuevas
Unitat de Fisiologia Vegetal,
Facultat de Farmàcia, Universitat de Barcelona, Diagonal 643, 08028 Barcelona, Spain

Kailash C. Bansal
National Research Centre on Plant Biotechnology, Indian Agricultural Research Institute, New Delhi – 110 012, India

Kaouther Feki
Plant Molecular Genetic Laboratory,
Centre of Biotechnology of Sfax (CBS) Route Sidi Mansour Km 6, B.P ''1177'' 3018 Sfax –Tunisia

Karabi Datta
Department of Botany,
Plant Molecular Biology and Biotechnology Laboratory, University of Calcutta, 35,
Ballygunge Circular Road, Kolkata-700 019

Khaled Masmoudi
Plant Molecular Genetic Laboratory,
Centre of Biotechnology of Sfax (CBS) Route Sidi Mansour Km 6, B.P ''1177'' 3018 Sfax –Tunisia

Konika Chawla
Department of Biology,
Norwegian University of Science and Technology, Høgskoleringen 5, NO-7491 Trondheim, Norway

M.E. González
IIB-INTECH, Camino Circ. Laguna km 6,
(B7130IWA) Chascomús, Buenos Aires, Argentina

Martin Kuiper
Department of Biology,
Norwegian University of Science and Technology, Høgskoleringen 5, NO-7491 Trondheim, Norway

Meenu Kapoor
University School of Biotechnology,
Guru Gobind Singh Indraprastha University,
Delhi, India

Ming-Bo Wang
CSIRO Plant Industry,
GPO Box 1600, Canberra,
ACT 2601, Australia

Moez Hanin
Plant Molecular Genetic Laboratory,
Centre of Biotechnology of Sfax (CBS) Route Sidi Mansour Km 6, B.P ''1177'' 3018 Sfax –Tunisia

Narendra Tuteja
Plant Molecular Biology Group,
International Centre for Genetic Engineering and Biotechnology, Aruna Asaf Ali Marg, New Delhi – 110 067, India

Ngoc Tuan Le
CSIRO Plant Industry,
GPO Box 1600, Canberra,
ACT 2601, Australia

Oscar A. Ruiz
IIB-INTECH, Camino Circ. Laguna km 6,
B7130IWA, Chascomús, Buenos Aires, Argentina

Pankaj Barah
Department of Biology,
Norwegian University of Science and Technology, Høgskoleringen 5, NO-7491 Trondheim, Norway

Pedro Carrasco
Departament de Bioquímica i Biologia Molecular, Universitat de València, Facultat de Ciències Biològiques, Dr. Moliner 50, 46100 Burjassot, València, Spain

Prasanta K. Subudhi
School of Plant, Environmental, & Soil Sciences, 215 M.B. Sturgis Hall, Louisiana State University Agricultural Center,
Baton Rouge, LA 70803, USA

Renu Deswal
Plant Molecular Physiology,
Biochemistry and Proteomics Laboratory,
Department of Botany, University of Delhi,
Delhi-110007, India

Renu Tuteja
International Centre for Genetic Engineering and Biotechnology, Aruna Asaf Ali Marg, New Delhi – 110 067, India

Ruben Alcázar
IIB-INTECH, Camino Circ. Laguna km 6,
(B7130IWA) Chascomús, Buenos Aires, Argentina

Sangram K. Lenka
National Research Centre on Plant Biotechnology, Indian Agricultural Research Institute, New Delhi – 110 012, India

Sanjay Kapoor
Department of Plant Molecular Biology,
University of Delhi South Campus Benito Juarez road, New Delhi, India

Sarvajeet S. Gill
Plant Molecular Biology Group,
International Centre for Genetic Engineering and Biotechnology, Aruna Asaf Ali Marg, New Delhi – 110 067, India

Swapn K. Datta
Department of Botany,
Plant Molecular Biology and Biotechnology Laboratory, University of Calcutta, 35,
Ballygunge Circular Road, Kolkata-700 019

Teresa Altabella
Unitat de Fisiologia Vegetal,
Facultat de Farmàcia, Universitat de Barcelona, Diagonal 643, 08028 Barcelona, Spain

Yoshiharu Y. Yamamoto
Applied Plant Science,
Faculty of Applied Biological Sciences,
Gifu University, Gifu, 501-1193, Gifu, Japan

CHAPTER 1

"Omics": Will Regulation and Activism Stifle Its Potential?

Henry I. Miller*

The Hoover Institution, Stanford University, Stanford, California 94305–6010, U.S.A.

Abstract: National and international regulation of recombinant DNA–modified, or "genetically engineered" (also referred to as "genetically modified," or GM), plants, including those with enhanced abiotic stress tolerance, is unscientific and illogical, a lamentable illustration of the maxim that bad science makes bad law. Instead of regulatory scrutiny that is proportional to risk – a fundamental principle of regulatory oversight of any product or activity – the degree of oversight of genetically engineered organisms is actually inversely proportional to risk. The current approach to regulation, which captures organisms to be field tested or commercialized according to the techniques used to construct them rather than their properties, flies in the face of scientific consensus. This approach has been costly in terms of economic losses and human suffering. The poorest of the poor have suffered the most because the hugely inflated development costs of genetically engineered plants and food have prevented robust development of new plant varieties with enhanced abiotic stress tolerance and other desirable characteristics. An approach to the regulation of field trials, known as the "Stanford Model," is designed to assess risks of new agricultural introductions, whether or not the organisms are genetically engineered, and independent of the genetic modification techniques employed. It offers a scientific, rational, risk–based basis for field trial regulations. Using this sort of model for regulatory review would not only better protect human health and the environment but would also permit more expeditious development and more widespread use of innovative, new plants and seeds.

"OMICS": THE ESSENTIAL PIECES OF THE PUZZLE OF LIFE

Biotechnology's powerful research tools have enabled researchers to dissect cellular and genetic processes to enhance our understanding of biological systems at their most fundamental – at the molecular level. But reductionism has its limits: Biological organisms are more than the sum of their inert, disparate components, and much current research focuses on the way that the parts and networks and systems interact with one another. The quest for such knowledge has given rise to research approaches that have been dubbed "omics" – genomics, proteomics, metabolomics, immunomics, and transcriptomics – which attempt to elucidate how various systems and levels of macromolecular synthesis and regulation interact, and to unravel the mysteries of living cells – for example how it is that "the genome is largely constant, irrespective of cell type and age, but the proteome varies from one cell type to the next, from one year to the next, and even from moment to moment"[1]. The understanding of such phenomena will be the basis for the genetic engineering of the future, but only if stakeholders can rationalize government regulation.

"GENETIC ENGINEERING" IS NOT NEW

Over many millennia, there has been a virtually seamless continuum of genetic improvement of crops with increasingly sophisticated techniques [2]. Recombinant DNA modification, which I will refer to henceforth as "genetic engineering," was introduced as part of this progression of technologies during the 1970s. Because genetic modification, or improvement, has been with us for centuries, "genetically modified organism," and its abbreviation, "GMO," are unfortunate choices of terminology. Defined arbitrarily as organisms that genes transferred across species lines – but only when accomplished by recombinant DNA techniques – the use of the terms "GM" or "GMO" ignores the fact that genetic modification is achieved using many techniques and technologies.

Millions of new genetic variants of plants field tested each year are derived from "wide–cross hybridizations," in which genes have been moved across species or genus barriers [3]. Wide crosses have been performed for almost a century, and thousands of such "non–molecular transgenic varieties" (as they might be called) are in commerce

---

*Address correspondence to: **Dr. Henry I. Miller**, Research Fellow, The Hoover Institution, Stanford University, Stanford, California 94305–6010, U.S.A., Phone: +1 (650) 725–0185; E–mail: henry.miller@stanford.edu

Narendra Tuteja, Sarvajeet Singh Gill and Renu Tuteja [Eds.]

around the world. Examples include:

> *Triticum agropyrotriticum*, a man–made 'species' that resulted from combining genes from bread wheat and a grass called quackgrass or couchgrass, that contains all the chromosomes of wheat and one extra whole genome from the quackgrass.

- Triticale, also a man–made grain, a wheat–rye hybrid.

- Pluots and apriums, plum–apricot hybrids.

- *Triticum agropyrotriticum* is particularly illustrative. Throughout development, from field testing through scaling up and commercialization to being fed to animals and humans, neither regulators nor activists were concerned with whether the tens of thousands of genes from quackgrass would make *T. agropyrotriticum* more weedy or whether any of the expression products of the introduced genes were toxigenic or allergenic. Nor has the pluot elicited any resistance from activists or scrutiny from regulators.

By contrast, if someone was to move a *single* gene from quackgrass into Triticum or from plum to apricot using recombinant DNA techniques, the resulting plants would be subject to expansive, extensive, lengthy and debilitating regulatory regimes.

As discussed by Fedoroff in this volume [4], most agricultural crops are the products of hundreds, if not thousands, of years of genetic improvement. Maize, for example, has undergone drastic, gradual modification, from the original grass–like plant with rock–hard seeds into modern maize, with many rows of firmly attached soft kernels full of carbohydrate, oil and protein [5].

A more recent example of the irrationality of current conceptions of "natural" versus "genetically engineered" is Golden Rice, several varieties biofortified with beta–carotene, the precursor of vitamin A. An examination of the "pedigree" of the immediate precursor of Golden Rice, IR64 – a strain of rice widely used in many parts of the world – and the addition of two genes that convert IR64 into Golden Rice is revealing [6]. What is astonishing about this construction is that for regulatory purposes, all of the complex genetic changes, including mutations, recombinations, deletions and translocations leading to IR64 are somehow considered "natural" – and therefore elicit no concern or review – while the insertion of two well–characterized genes that enable the plant to synthesize beta–carotene (which is converted to vitamin A in vivo) elicits an extraordinary burden of regulatory costs and delays. Although "GMOs" (or variations on the theme) are not a genuine, meaningful category, in most regulatory regimes around the world, merely the use of recombinant DNA techniques to modify an organism – that is, the creation of and intent to field test a "GMO" – is the trigger for draconian, dilatory and expensive regulatory regimes.

BENEFITS AND OBSTACLES

Genetically engineered plants have persuasively demonstrated extraordinary benefits:

- Increased yield, which permits conservation of cultivated land and avoidance of expanded upslope farming.

- Decreased use of chemical pesticides, which leads to less runoff and fewer poisonings. For example in China, the use of Bt cotton has substantially reduced poisoning incidents by pesticides amongst farmers and their families [7].

- Reduced water requirements with drought resistant or saline tolerant varieties may in the near future become the most important applications world–wide. With recurrent droughts over southern Europe, Australia, parts of the United States and much of sub–Saharan Africa, small improvements in water requirements can make a large difference in yields and cost–effectiveness of farming.

- Shifts in herbicide usage lead to more environmental friendly herbicides and increased no–till farming, resulting in lower soil erosion, less runoff and less carbon dioxide release to the atmosphere.

- Decreased content of fungal toxins in food and feed, and correspondingly reduced incidence of illness in animals and humans.

- Recent research on the model plant *Arabidopsis* has identified the "master regulator" of the entire temperature transcriptome [8]. It was observed that the key mediator of plants' temperature–sensing ability is a specialized histone protein that wraps DNA into a tightly packed structure known as a nucleosome. This histone binds tightly to the plant's DNA at lower temperatures, inhibiting gene expression, but as the temperature rises, binding decreases and it detaches, permitting gene expression. These findings might help to explain how plants respond to higher ambient temperatures and facilitate the development of varieties that will grow at a greater range of temperatures – an excellent application of "omics" and genetic engineering to the challenges of global climate change.

In spite of these benefits and the absence of any unanticipated or unique negative effects, the technology has encountered various policy and public relations obstacles. A number of pseudo–crises – fomented by fear–mongering non–governmental organizations, irresponsible journalism and the expansionist tendencies of bureaucrats – have led to flawed public policy and over–regulation of genetic engineering techniques and their products.

THE SCIENTIFIC BASIS OF REGULATION

There exists a decades–old scientific consensus about the need for a more rational, risk– based approach to the regulation of both field trials and commercialization of genetically engineered plants. In 1987, the U.S. National Academy of Sciences (NAS) published a white paper on the planned introduction of genetically engineered organisms into the environment [9]. It noted that recombinant DNA techniques provide a powerful and safe means for modifying organisms and predicted that the technology would contribute substantially to improved health care, agricultural efficiency, and the amelioration of many pressing environmental problems. The paper had wide–ranging impacts in the United States and internationally. Its most significant conclusions and recommendations include:

- There is no evidence of the existence of unique hazards either in the use of recombinant DNA techniques or in the movement of genes between unrelated organisms.

- The risks associated with the introduction of recombinant DNA–modified organisms are the same in kind as those associated with the introduction of unmodified organisms and organisms modified by other methods.

- Assessment of the risks of introducing recombinant DNA–modified organisms into the environment should be based on the nature of the organism and of the environment into which the organism is to be introduced, and independent of the method of engineering per se.

In a 1989 follow–up to this white paper, the National Research Council (NRC), the research arm of the NAS, concluded that "no conceptual distinction exists between genetic modification of plants and microorganisms by classical methods or by molecular techniques that modify DNA and transfer genes," whether in the laboratory, in the field, or in large–scale environmental introductions [10]. The NRC report supported this statement with extensive discussions of experience with plant breeding and the cultivation of these pre–recombinant DNA genetically modified plants and microorganisms:

- "Crops modified by molecular and cellular methods should pose risks no different from those modified by classical genetic methods for similar traits. As the molecular methods are more specific, users of these methods will be more certain about the traits they introduce into the plants."

- "Recombinant DNA methodology makes it possible to introduce pieces of DNA, consisting of either single or multiple genes that can be defined in function and even in nucleotide sequence. With classical techniques of gene transfer, a variable number of genes can be transferred, the number depending on the mechanism of transfer; but predicting the precise number or the traits that have been transferred is difficult, and we cannot always predict the phenotypic expression that will result. With organisms modified by molecular methods, we are in a better, if not perfect position to predict the phenotypic expression."

- "Information about the process used to produce a genetically modified organism is important in understanding the characteristics of the product. However, the nature of the process is not a useful criterion for determining whether the product requires less or more oversight."

- As a consequence, "the *product* of genetic modification and selection should be the primary focus for making decisions about the environmental introduction of a plant or microorganism and not the *process* by which the products were obtained."

Thus, the NRC articulated some of the principles that should underlie the regulatory oversight field trials of plants, and subsequently these principles have been reiterated repeatedly by countless scientific bodies. The essence is that the mere fact that an organism has been modified by genetic engineering techniques should not determine how the organism is regulated. This was emphasized yet again in the comprehensive report from the U.S. National Biotechnology Policy Board (on which I served as a charter member), which was established by the U.S. Congress with representation from the public and private sectors. The report concluded: "The risks associated with biotechnology are not unique, and tend to be associated with particular products and their applications, not with the production process or the technology per se. In fact, biotechnology processes tend to reduce risks because they are more precise and predictable" [11]. The report went even further, concluding, "The health and environmental risks of not pursuing biotechnology–based solutions to the nation's problems are likely to be greater than the risks of going forward." This is true in general for this technology and its products, particularly for parts of the world where subsistence farming predominates.

Various other national and international groups, including the American Medical Association, the United Kingdom's Royal Society, and the UN's Food and Agriculture Organization and World Health Organization, have repeatedly echoed or extended these conclusions (summarized and referenced in [12]). For example, a joint statement from the International Council of Scientific Unions' (ICSU) Scientific Committee on Problems of the Environment (SCOPE) and the Committee on Genetic Experimentation (COGENE) concluded, "The properties of the introduced organisms and its target environment are the key features in the assessment of risk. Such factors as the demographic characterization of the introduced organisms; genetic stability, including the potential for horizontal transfer or outcrossing with weedy species; and the fit of the species to the physical and biological environment apply equally to both modified or unmodified organisms; and, in the case of modified organisms, they apply independently of the techniques used to achieve modification." That is, it is the organism itself, and not how it was constructed, is important.

Similarly, the report of a NATO Advanced Research Workshop concluded, "In principle, the outcomes associated with the introduction into the environment of organisms modified by recombinant DNA techniques are likely to be the same in kind as those associated with introduction of organisms modified by other methods. Therefore, identification and assessment of the risk of possible adverse outcomes should be based on the nature of the organism and of the environment into which it is introduced, and not on the method (if any) of genetic modification.

Other analyses have focused specifically on the food safety aspects of gene–spliced organisms and their derivatives. For example, in a 1993 report the Paris–based Organization for Economic Cooperation and Development (OECD) described several concepts related to food safety that are wholly consistent with the consensus discussed above.

- "Modern biotechnology broadens the scope of the genetic changes that can be made in food organisms and broadens the scope of possible sources of foods. This does not inherently lead to foods that are less safe than those developed by conventional techniques."

- "Evaluation of foods and food components obtained from organisms developed by the application of the newer techniques does not necessitate a fundamental change in established principles, nor does it require a different standard of safety."

Finally, a comprehensive analysis of food safety published in 2000 by the Institute of Food Technologists addressed both the scientific and regulatory implications of foods derived from genetically engineered organisms and specifically took current regulatory policies to task. The report concluded that the evaluation of genetically engineered organisms and the food derived from them "does not require a fundamental change in established principles of food safety; nor does it require a different standard of safety, even though, in fact, more information and a higher standard of safety are being required." The report went on to state unequivocally that theoretical considerations and empirical data do "not support more stringent safety standards than those that apply to conventional foods."

What could be clearer than this consensus about the appropriate basis for the oversight of genetically engineered plants in the field and in the food supply?

Principles of Regulation

In addition to the consensus described above specifically for the products of genetic engineering, there are certain general principles of regulation that should inform any regulatory scheme:

- The degree of regulatory scrutiny should be commensurate with the perceived level of risk.

- Similar things should be regulated in a similar way.

- If the scope of regulation – i.e. the regulatory net or the trigger that captures field trials or the finished product for review – is unscientific, it renders the entire approach unscientific. For example, if proposed field trials were reviewed only if the dossiers were received by regulators on certain days of the week – an absurd arrangement – the regulatory approach would fundamentally flawed and corrupt, no matter how enlightened and informed the actual reviews of the proposed field trials. This is a subtle but important point, because regulators often attempt to hide behind a veneer of scientific integrity, even when that veneer is breached by a scope that makes no sense.

CONSEQUENCES OF FLAWED REGULATION

The principles of regulation described above have been largely ignored. Current regulatory regimes are unscientific, process–based, and require case by case review for virtually all genetically engineered plants and microorganisms, no matter how obviously trivial the modification or benign the product might be. This flawed approach, which categorically ignores fundamental principles of regulation and the dictates of common sense, results in enormously inflated costs, lack of agricultural progress and needless human suffering.

Increased Research and Development Costs

The compliance costs of regulation for the development of an insect–resistant and a herbicide–resistant maize have been calculated to be between USD 6 and 15 million respectively, not including labeling [13]. This is several times more costly than for similar constructions made with conventional breeding, in spite of the latter being less precise and predictable.

Fewer Products in the Pipeline with Reduced Benefits for Farmers and Consumers

The costs and uncertainty created by the regulatory milieu have inhibited agricultural innovation and product development, decreased commercialization of already developed genetically engineered crops and decreased the potential for new, improved varieties of fruits and vegetables, tree fruits and nuts, and nursery and landscape crops. That is to say, development is economically viable primarily for commodity crops, which are grown on a vast scale.

In 2008, the global hectarage of biotech crops continued to show strong growth, reaching 125 million hectares, up from 114.3 million hectares the previous year [14]. Although it is difficult to calculate accurately the economic losses from the "opportunity costs" of flawed public policy, gratuitous regulation unquestionably imposes a huge punitive tax on a superior technology. With more rational, science–based regulation there would be an accelerated shift to genetically engineered crops, with cultivation on a far greater area and additional traits and species developed and commercialized. Put another way, the opportunity costs of flawed, unscientific public policy have been enormous, and as usual, have been imposed disproportionately on the poorer nations, which often lack the regulatory infrastructure to perform regulatory reviews.

Pseudo–crises and Litigation

Pseudo–crises–high–profile but inconsequential events publicized by anti–technology activists–have led to public relations debacles, flawed public policy, endless debate over non–issues such as the coexistence of genetically engineered and conventional crops, acceptable tolerances for "contamination," and labeling, as well as costly litigation. One well–known example is the StarLink case in which the US Environment Protection Agency granted approval of a maize variety for animal but not human consumption. When the variety, called StarLink, was later detected in human foodstuffs, the regulatory and civil penalties to the company that made the StarLink for this inconsequential "transgression" were substantial [15]. Other pseudo–crises include the (false) alarms over killing of Monarch butterflies and the contamination of land races in Mexico. All of these were based on inaccurate or fraudulent reports, or results taken out of proper context [16].

Vandalism and Intimidation of Academics

Field trials are constantly being vandalized, because the regulatory requirements, which are specific to recombinant DNA–modified products, dictate that the sites of trials become publically known. Researchers have been injured, and research destroyed. Two German universities responded to the threats of activists by banning the testing of recombinant DNA–modified plants, an appalling abdication of academic freedom [17].

Malnourishment, Illness and Deaths

Malnutrition claims thousands of lives per day, many of which could be saved if governments and international organizations would change their hostile attitudes and policies towards genetic engineering. The resulting greater availability of improved crop varieties would enhance food security for poor farmers.

THE "STANFORD MODEL" FOR RISK–BASED REGULATION

It is easy to complain about unscientific, non–risk–based regulatory regimes. But there are better alternatives, and science shows the way. One is the "Stanford Model" for risk–based regulation, which was developed in the 1990s [18]. The Stanford Model stratifies organisms according to their risk in field trials. This universe can be divided in two ways [19]:

- Horizontally, according to risk categories, with higher risk as one goes toward the top of the pyramid.

- By the oblique lines, dividing the universe of field trials according to technology: the green area is all field trials performed with organisms created by conventional breeding or tissue culture, for example, while the area to the far right corresponds to field trials with recombinant DNA modified organisms.

Conceptually, it should be clear that there is no particular increase in risk that is a function of technology–although as discussed above, the imprecision of older techniques of genetic modification does give rise to greater uncertainty about risk. There are high–risk organisms–for example foot and mouth disease virus, African killer bees, rusts that infect grains and highly invasive weeds such as kudzu–that require caution in field tests whether or not they have been genetically modified in any way. Although plants in general are of low or negligible risk, rarely they may be invasive (kudzu) or produce potent toxins (castor bean), Recombinant DNA technology affords no particular monopoly on safety, but on average, it is more precise and more predictable than the other techniques.

More than a decade ago, the Stanford University Project on Regulation of Agricultural Introductions developed a widely applicable regulatory model for the field–testing of any organism, regardless of the method or methods employed in its construction. The approach is patterned after quarantine systems such as the USDA's Plant Pest Act regulations, which are essentially binary; a plant that a researcher might wish to introduce into the field is either on the proscribed list of plants pests–and therefore requires a permit–or it is exempt. The more quantitative and nuanced "Stanford Model," which stratifies organisms into several risk categories, more closely resembles the approach that was taken in the National Institutes of Health/Centers for Disease Control (NIH/CDC) handbook, *Biosafety in Microbiological and Biomedical Laboratories* [20], which specifies the procedures and physical containment that are appropriate for research with microorganisms, including the most dangerous pathogens known. These microorganisms were stratified into risk categories by panels of scientists. Interestingly, unlike regulators' approach to gene–spliced organisms, the NIH/CDC approach – even for the most dangerous pathogens – is only to offer guidance to researchers but not to make adherence compulsory.

The Stanford Model–applied to plants in this first demonstration project–can be readily applied to accommodate different organisms, geographical regions, and preferences for more or less stringent regulation. In January 1997, the project assembled a group of approximately twenty agricultural scientists from five nations at a workshop held at the International Rice Research Institute (IRRI), in Los Baños, Philippines. The purpose of the workshop was to develop a broad, science–based approach that would evaluate all biological introductions, not just those that involve gene–spliced organisms. The need for such a broad approach was self–evident–there was already abundant evidence that severe ecological risks can be associated with plant pests and "exotics," or non–coevolved organisms.

As part of the pilot project, the IRRI conference participants evaluated and then stratified a variety of crops based on certain risk–related traits to be considered in order to estimate overall risk. Consensus was reached without serious difficulty–suggesting that it would be similarly easy to categorize other organisms as well.

The participants agreed at the outset that the following risk factors should be integral to a model algorithm for field–testing and commercial approval of all introductions:

1. Ability to colonize

2. Ecological relationships

3. Human effects

4. Potential for genetic change

5. Ease or difficulty of risk management

Each of the organisms evaluated during the conference was assessed for all five factors, which enabled the group to come to a global judgment about the organism's risk category. Most of the common crop plants addressed were found to belong to Category 1 (negligible risk), while a few were ranked in Category 2 (low but non–negligible risk). One plant (cotton) was judged to be in Category 1 if it was field–tested outside its center of origin, and Category 2 if tested in the vicinity of its center of origin.

It cannot be overemphasized that in the evolution of this Stanford Model, the factors taken into account were independent of either the nature of the genetic modification techniques employed, if any, or to the source(s) of the introduced genetic material. The participants agreed that risk was not systematically affected by whether conventional breeding techniques or recombinant DNA methods were being used to modify an organism. They also agreed that the source of the DNA, whether from phylogenetically close or distant organisms–that is, from organisms of different genera, families, orders, classes, phyla, or kingdoms–did not affect the risk posed by an organism.

In other words, the group's analysis supported the view that the risks associated with field–testing a genetically altered organism are independent of the process by which it was modified and of the phylogenetic distance between the donor and recipient organisms. The Stanford Model suggests the utility and practicality of an approach in which the degree of regulatory scrutiny over field trials is commensurate with the risks and independent of whether the organisms introduced are "natural," non–coevolved, or have been genetically improved by conventional methods or gene–splicing techniques. Variations and refinements of this approach are possible, of course; Professor Wayne Parrott has suggested that the risk category could be adjusted depending on the trait introduced– for example a gene that enhances weediness or that expresses a potent toxin or allergen (Parrott, personal communication).

What, then, are the practical implications of an organism being assigned to a "risk category?" The level of oversight faced by an investigator who intends to perform a field trial with an organism in one or another of the categories could include: complete exemption, a simple "postcard notification" to a regulatory authority (without affirmative, prior approval required), premarket review of only the first product in a given category, case–by–case review of all products in the category, or even prohibition (as is the case currently for experiments with foot–and–mouth disease virus in the United States).

A key feature of the Stanford Model is that it is sufficiently flexible to accommodate differences in regulatory authorities' preferences for greater or lesser regulatory stringency. Putting it another way, different national regulatory authorities could choose their preferred degree of risk aversion, some leaning more toward exemption and notification, others toward case–by–case review. However, as long as regulatory requirements are commensurate with the relative risk of each category and do not discriminate by treating organisms of equivalent risk differently, the regulatory methodology will remain within a scientifically defensible framework.

Under such a system, some currently unregulated introductions of traditionally bred cultivars and so–called "exotic," or non–coevolved, organisms considered to be of moderate or greater risk would likely become subject to regulatory review, whereas many gene–spliced organisms that now require case–by–case review would likely be regulated less stringently. The introduction of such a risk–based system would rationalize significantly the regulation of field trials, and it would reduce the regulatory and other disincentives to the use of molecular techniques for genetic modification.

By making possible accurate, scientific determinations of the risks posed by the introduction of an organism into the field, this regulatory model fosters enhanced agricultural productivity and innovation, while it protects valuable ecosystems. It offers regulatory bodies a highly adaptable, scientific paradigm for the oversight of plants, microorganisms and other organisms, whether they are "naturally occurring" or non–coevolved organisms, or have been genetically improved by either old or new techniques. The outlook for the new biotechnology applied to agriculture, especially environmental friendly innovations of particular benefit to the developing world, would be far better if governments and international organizations expended effort on perfecting such a model instead of on introducing and maintaining unscientific, palpably flawed regulatory regimes that inhibit research, stifle innovation and prevent needed advances in agriculture.

Advantages of the Stanford Model

- It stratifies all organisms according to risk and is indifferent to the technique (if any) of genetic alteration.

- It is flexible.

- It is scientifically defensible.

- It permits various degrees of risk–aversion depending on the need.

- It permits discretion–in a scientific context.

- It exempts field trials that should be exempt and captures field trials that should be reviewed.

One great advantage is that it is analogous to existing regulatory regimes, such as those for quarantine regulations for plant or animal pests, and also to the U.S. government's approach to handling dangerous pathogens or other microorganisms in the laboratory. In other words, the approach is not fundamentally new and has worked well in practice for decades.

SUMMARY

The attenuated growth of agricultural biotechnology worldwide stands as one of the great societal tragedies of the past quarter century. Irresponsible activism and unscientific, excessive, stultifying regulation, nationally and internationally, are the major reasons for the failure of agricultural biotechnology to achieve its potential. Scientists, regulators and politicians must find more rational and efficient ways to guarantee public health and environmental safety while encouraging new discoveries. Science shows the path, and society's leaders –secular and religious– must take us there.

REFERENCES

[1]   Biotechnology Industry Organization. Biotech Tools in Research and Development. http://www.bio.org/speeches/pubs/er/biotechtools.asp (accessed January 10, 2010).

[2]   Miller HI. Bertebos Conference in Falkenberg, Sweden, 7–9 September 2008, presentation available at http://www.ksla.se/sv/retrieve_file.asp?n=1669, (accessed January 16, 2010), slide 3.

[3]   Goodman RM *et al.* Gene transfer in crop improvement. Science 1987; 236: 48–54.

[4]   Fedoroff N. In: Tuteja N, Gill SS, Tuteja R, Eds. Stress Resistant Crops: Omics Approaches. Genetic Modification of Crops: Past, Present and Future 2010.

[5]   Miller HI. Bertebos Conference in Falkenberg, Sweden, 7–9 September 2008, presentation available at http://www.ksla.se/sv/retrieve_file.asp?n=1669, (accessed January 16, 2010), slide 11.

[6]   Miller HI. Bertebos Conference in Falkenberg, Sweden, 7–9 September 2008, presentation available at http://www.ksla.se/sv/retrieve_file.asp?n=1669, (accessed January 16, 2010), slide 13.

[7]   Huang J *et al.* http://www. agbioforum.org/v5n4/v5n4a04–huang.htm, (accessed January 15, 2010).

[8]   Kumar SV, Wigge PA. H2A.Z–containing nucleosomes mediate the thermosensory response in *Arabidopsis.* Cell 2010; 140: 136–147.

[9]   Anon. Introduction of Recombinant DNA–Engineered Organisms into the Environment: Key Issues, Washington DC: Council of the US Academy of Sciences/National Academy Press, 1987.

[10]   Anon. Field Testing Genetically Modified Organisms: Framework for Decisions. U.S. National Research Council. Washington, DC: U.S. National Academy of Sciences, 1989.

[11]   Anon. National Biotechnology Policy Board Outlines Priorities for Action. Biotechnology Law Report. March–April 1993; 12: 102–103. doi:10.1089/blr.1993.12.102.

[12]   Miller HI, Conko G. The Frankenfood Myth: How Protest and Politics Threaten the Biotech Revolution, Westport, CT (U.S.A.): Praeger Publishers, 2004.

[13]   Anon. Achieving risk–based regulation: Science shows the way, in "Golden Rice and other biofortified food crops for developing countries" 2009. Report from the Bertebos Conference in Falkenberg, Sweden, 7–9 September 2008, 69–74. The Royal Swedish Academy of Agriculture and Forestry in cooperation with the Bertebos Foundation.

[14]   Anon. ISAAA Brief 39–2008: Executive Summary Global Status of Commercialized Biotech/GM Crops: 2008 The First Thirteen Years, 1996 to 2008. International Service for the Acquisition of Agri–biotech Applications (ISAAA), http://www.isaaa.org/resources/publications/briefs/39/executivesummary/default.html (accessed January 23, 2010).

[15]   Kershen   DW.   The   Risks   of   Going   Non–GMO.   Available   from http://www.colby.edu/biology/BI402B/Kershen%202000.pdf, (accessed January 16, 2010).

[16]   Miller HI. Anti–social Bio–Activism. Genetic Engineering & Biotechnology News, December 1, 2007, available from http://www.genengnews.com/articles/chitem.aspx?aid=2296, (accessed January 16, 2010).

[17]   Miller HI. Thwarting Domestic Terrorism. Genetic Engineering & Biotechnology News, September 15, 2008, available from http://www.genengnews.com/articles/chitem.aspx?aid=2597, (accessed January 16, 2010).

[18]   Barton J *et al.* A model protocol to assess the risks of agricultural introductions. Nature Biotechnology 1997; 15: 845–848.

[19]   Miller HI. Bertebos Conference in Falkenberg, Sweden, 7–9 September 2008, presentation available at http://www.ksla.se/sv/retrieve_file.asp?n=1669, accessed January 16, 2010, slide 23.

[20]   Anon. Biosafety in Microbiological and Biomedical Laboratories. U.S. Department of Health and Human Services, Centers for Disease Control and Prevention and National Institutes of Health, Fourth Edition. Washington, DC: May 1999.

CHAPTER 2

Omics Approaches for Abiotic Stress Tolerance in Plants

Prasanta K. Subudhi*

School of Plant, Environmental, & Soil Sciences, 215 M.B. Sturgis Hall, Louisiana State University Agricultural Center, Baton Rouge, LA 70803, USA

Abstract: Plant growth and productivity are limited by various types of abiotic stresses. While significant advances have been made in understanding the plant adaptation in stress environments, there are still challenges to translate this acquired knowledge for improved crop performance and productivity under stress environments due to complex genetics involving multitude of genes and stress tolerance mechanisms and interaction with numerous environmental factors pose. Different omics technologies have evolved during the last few decades to systematically analyze and correlate the changes in the genome, transcriptome, proteome, and metabolome to the variability in plant's response to abiotic stresses. This chapter will provide an overview of various omics technologies and their application to increase the chances of developing abiotic stress tolerant plants. Although, we foresee significant potential of these innovative tools in making genetic improvement, challenges in integrating the vast amount of high throughput omics data should be overcome to understand the genetic network involved in abiotic stress tolerance.

INTRODUCTION

Plant productivity is seriously limited by various types of abiotic stresses all over the world [1]. Improvement in plant adaptation under major abiotic stresses such as drought, salinity, temperature extremes, submergence, nutrient deficiency, and toxicity is essential to ensure food security for the increasing world population. There is a serious concern over availability of more arable land due to soil erosion, degradation, and unsustainable farm practices. Global climatic pattern is becoming more unpredictable with increased occurrence of drought, flood, storm, heat waves, and sea water intrusion. Productivity of major food crops in semi-arid regions is expected to decline in next two decades [2]. Traditional breeding methods have been employed to exploit the natural genetic variation in germplasm but very few cultivars have demonstrated improved stress tolerance in field conditions [3]. While significant advances have been made in understanding the plant adaptation in stress environments, complex genetics involving multitude of genes and stress tolerance mechanisms and interaction with numerous environmental factors pose serious challenges to translate the acquired knowledge for improved crop productivity. Considering the importance and urgency to genetically improve abiotic stress tolerance in crop plants, novel strategies involving modern technologies will be needed for thorough understanding of plant's response to abiotic stresses. In particular, an array of new innovative omics technologies are evolving at rapid pace, which is empowering the scientists to systematically analyze the genome at various levels and their effect on phenotypic variability. These sophisticated technologies have brought about paradigm shift in conducting research to understand the complexity of biology of an organism and its response to environmental disturbances. Moreover, rapid progress in identification of abiotic stress tolerance genes, the underlying molecular mechanisms from diverse germplasm sources, and their efficiency in imparting stress tolerance in different genetic backgrounds and environments will be required to improve crop plant adaptation in stress environments using various omics technologies [4]. The aim of this chapter is to provide a critical overview of various available omics approaches that hold promise in developing stress tolerant plants in future through comprehensive understanding of the complexity of stress tolerance traits and mechanisms.

OMICS APPROACHES

Omics is a general term used for data acquisition, integration, and analysis in a high throughput manner to develop a global perspective of the biological processes in living organisms. The innovation in DNA sequencing is largely instrumental in ushering in the omics era. This has transcended the conventional paradigm of analyzing biological

*Address correspondence to: Dr. Prasanta K. Subudhi,** School of Plant, Environmental & Soil Sciences, 215 M.B. Sturgis Hall, Louisiana State University Agricultural Center, Baton Rouge, LA 70803, USA; Tel: 225-578-1303; Fax: 225-578-1403; E-mail: psubudhi@agctr.lsu.edu

Narendra Tuteja, Sarvajeet Singh Gill and Renu Tuteja [Eds.]

problems. Omics approaches enable to connect the molecular variation at global scale to trait expression in an integrated manner. A suite of omics sub-disciplines has evolved over the years due to the advances in instrumentation, techniques, and reagents to obtain a deeper understanding of the organization, evolution, and functions of the genome. The field of bioinformatics is also growing in parallel for rapid analysis, visualization and interpretation of the voluminous omics data. Depending on the sequence of translation of genetic information leading to a recognizable phenotype, different terms such as genome, transcriptome, proteome, and metabolome have been coined. 'Genome' was coined for the first time by Professor Hans Winkler in 1920 to describe an organism's complete set of chromosomes and the genes where as transcriptome, proteome and metabolome deals with the whole collection of all the transcripts or mRNA, proteins, and metabolites in a cell or tissue or organism, respectively.

GENOMICS

Genomics refers to the approaches describing the mapping, sequencing, and functional analysis of genes present in the genome of a given species. It focuses on the identification of all of the genes in an organism and the study of their functions. Genomics is divided into two broad categories; 'structural genomics' and 'functional genomics'. A number of reviews described the approaches and potential of this field for understanding the biology of plants [5-8].

Structural Genomics

Structural genomics is largely involved in genome sequencing, mapping, and cloning of the traits of interest for both biologists and agronomists. This phase established the foundation for genomics research and provided the necessary reagents and resources to transition into the functional genomics era.

Whole Genome Sequencing

Plant genomes are very diverse in terms of genome size, gene content, extent of repetitive sequences, and polyploidization events. Both repetitive sequences and polyploidization complicates the sequence assembly process. Exact sequence and location of all the genes in an organism is the first step towards understanding the functioning of the living organisms. Although EST sequencing a powerful approach for gene discovery, low abundance transcripts or transcripts under abiotic stresses are hardly represented in EST databases. The whole genome sequencing is the only way to obtain the complete repertoire of genes in an organism. It also provides information about the genome function, evolution, and organization including relative gene order and gene density which facilitates the application of positional cloning strategy.

There are two approaches for whole genome sequencing; Whole-genome shotgun (WGS) and clone-by-clone sequencing. Despite the difficulty in assembling the genome, the WGS strategy theoretically is expected to capture all the genes in the genome that are not possible in clone-based strategy. Genetic and physical maps and the associated BAC end sequences help in assembling and finishing the genome sequences in the clone based strategy [9]. Annotation of the genome sequences is cumbersome and difficult as it is often not easy to differentiate coding sequence from the noncoding intergenic sequences and introns. This can be facilitated by comparing the genome sequences with the cDNAs, ESTs, or the sequences of known coding genes.

The primary focus of plant genomics field at the beginning was to generate genome sequence data in model plant species such as *Arabidopsis* and rice or species of economic importance which have either already been completed or are underway (http: //www.ncbi.nlm.nih.gov/genomes/PLANTS/PlantList.html; http: //www.jgi.doe.gov/ sequencing/seqplans.html). *Arabidopsis* is the first plant species whose genome was sequenced [10]. This was a major mile stone in plant genomics field as it established the foundation for thorough comparison of plant biological processes and systematic identification of plant specific genes and their functions for genetic improvement of crops. This study sequenced 115.4 MB of the *Arabidopsis* genome and revealed 25,498 genes belonging to 11000 families. To date approximately 45% of all predicted *Arabidopsis* genes cannot be assigned a function [11]. In such cases, various functional genomics approaches are applied to assess the biological role of the genes by analyzing the gene products and expression patterns in any developmental stage or in response to environmental stress conditions.

Arabidopsis was followed by completion of the rice genome [12-14]. Whole-genome shotgun strategy was used to sequence and assemble 92-93% of the rice genome in *indica* subspecies [13,15] and japonica subspecies of rice [12]. Later, clone-by-clone strategy was adopted by the International Rice Genome Sequencing Project (IRGSP) to obtain the map-based sequences covering 95% of the 389 Mb rice genome including all of the euchromatin and two complete centromeres [14]. Both assemblies have predicted similar estimates of the number of genes. These estimates are 49,088 genes for *indica*, 37,794 genes for the *japonica* subspecies [15], and 37,544 for the clone-based assembly [14]. Ancient duplication was detected in all genome assemblies. Another important contribution of IRGSP study was the identification of single-nucleotide polymorphisms (SNPs) and simple sequence repeats (SSRs) and acceleration of application of the map-based sequence for the identification of genes underlying agronomic traits.

The draft sequence of the first tree species Poplar (*Populus trichocarpa*) was obtained by whole genome shotgun sequencing [16]. Recently, decoding of the genomes of two major food crops, corn and sorghum has been completed [17, 18]. Over 32,000 genes were predicted from the 2.3 GB of corn genome and 99.8% of sequences were assigned to reference chromosomes. It provided useful information about the composition and evolution of the corn genome. Several families of transposable elements constituted 85% of the corn genome and these elements played important role in capture and amplification of numerous gene fragments and affected the composition, sizes, and positions of centromeres [18]. The sequencing of the 730 Mb of the *Sorghum bicolor* (L.) genome by shotgun strategy placed 98% of the genes on chromosomes with the help of genetic, physical and syntenic information [17]. Although the gene order and density are similar to rice, its larger genome could be explained by accumulation of retrotransposons in heterochromatic regions. About 24% of genes are grass-specific and 7% are sorghum-specific. It is also suggested that sorghum's drought tolerance may be due to recent gene and microRNA duplications.

Mapping and Map-Based Cloning of Qtl For Abiotic Stress Tolerance

Abiotic stress tolerance in plants is inherently complex due to the cumulative effect of numerous genes that are modulated profoundly by the environmental factors. Due to the advances in molecular markers and sophisticated linkage mapping methods [19], naturally occurring variation for complex quantitative traits including abiotic stress tolerance in plant species is now amenable for thorough dissection and utilization for plant improvement. For analysis of abiotic stress tolerance traits, immortal mapping populations such as RILs are useful for reliable phenotyping in multiple environments with multiple replications. For the wild species, that are often reservoir of abiotic stress tolerance genes, introgression lines approach has been useful to transfer abiotic stress tolerance attributes to cultivated species [20]. Advanced backcross approach has been advocated to capture the abiotic stress tolerance genes from the wild species [21].

QTL mapping not only enables localization of factors for quantitative traits on linkage map, it also allows estimation of the effects of individual QTL and their joint effects. It contributes to the better understanding of the genetic basis of crop adaptation under stress environments by providing an improved strategy to dissect the genetic basis of traits through discovery of the target genomic regions for marker-assisted manipulation, cloning, and genetic engineering [22]. Positional cloning strategy, that involves fine mapping followed by physical mapping of the markers tightly linked to the mutant or natural allele, has been the leading approach to clone most of the plant QTLs for abiotic stress tolerance (Table 1). QTL cloning though a tedious and time consuming task, it has the advantage of providing much detailed insights about the molecular regulation of abiotic stress tolerance in addition to their utility for MAS, genetic engineering, and Ecotilling. It has been successful for those traits with large effect and high heritability. An alternative strategy for QTL cloning could be candidate genes approach, when information about the involvement of specific biochemical pathway or known genes controlling same or similar trait in other plant species is available. For example, the aluminum tolerance (*ALMT1*) or *HKT1* genes of *Arabidopsis* [23,24] are similar to those controlling these traits in cereals. Expression profile of genes localized on the QTL region could help in identification of candidate genes for abiotic stress tolerance [25]. QTL cloning has become more efficient and less time intensive with the development of improved PCR-based molecular marker systems, availability of saturated molecular maps in various crop species [reviewed in 26]. For the regions delimited by fine mapping, sequencing of the regions from both wild type and mutant could be employed to identify the target gene or mutation [27]. Various high throughput marker systems such as, SSR, InDel polymorphism, and SNPs are available in many plant species to accelerate fine mapping and map-based cloning [28-30]. High throughput array based genotyping procedure is being developed [31-33].

Table 1: Cloned quantitative trait loci for abiotic stress tolerance in crop plants.

Trait	Plant Species	Gene/QTL	Gene function	Reference
Submergence tolerance	Rice	*Sub1A*	ERF-related factor	[57]
Salt tolerance (Shoot K$^+$ concentration)	Rice	*Skc1*	HKT-type Na$^+$ transporter	[47]
Salt tolerance (low Na$^+$ in leaves)	Wheat (T. monococcum)	*Nax1*	HKT-type Na$^+$ transporter	[50]
Aluminum tolerance	Barley	*Alp*	Citrate efflux transporter	[73]
Aluminum tolerance	Wheat	*Alt$_{BH}$*	Malate efflux transporter	[72]
Aluminum tolerance	Sorghum	*Alt$_{SB}$*	Citrate efflux transporter	[74]
Boron toxicity tolerance	Barley	*Bot1*	Boron efflux transporter	[76]

Genetical genomics, which applies QTL mapping approach to the gene expression data, has been advocated to analyze the genetic regulation of natural variation in gene expression [34]. In plant species with fully sequenced genomes, mapping of QTLs controlling the transcript abundance of each gene (eQTLs) add a new dimension to the QTL approach enabling the analysis of the relationship between genome and transcriptome [35]. When map positions of QTL for the gene of interest and the expression QTL are compared, this approach provides clues about the genetic regulation of QTL expression. eQTLs can be grouped as Cis or Trans, where cis-eQTLs localizes near the gene itself and trans-eQTLs map at a location on the genome other than the actual physical position of the gene [36]. The eQTLs approach thus provides new opportunity to improve our understanding of transcriptional regulation and regulatory variation.

Association mapping based on linkage disequilibrium has been suggested as a complementary approach for genetic dissection of complex traits [37, 38]. It involves correlating the molecular variation with phenotypic variation in a population of diverse individuals using the historical recombination events. This approach has generated interest among the plant scientists due to ease of obtaining the haplotypes at many genetic loci using sequencing or high throughput SNP analysis. For the crop species where high density linkage maps, whole genome sequence, sequences of gene rich regions are available, candidate gene sequencing has been commonly used to determine the marker-trait association in crop species [39,40]. This approach has been used for mapping abiotic stress tolerance traits in forest tree species [41, 42].

Application of Structural Genomics In Abiotic Stress Tolerance Studies

With the release of whole genome sequence of sorghum [17], which is comparatively more abiotic stress tolerant than other cereals, genomics research in abiotic stress tolerance will be greatly accelerated. Several reviews have described the progress in application of QTL-based approach to abiotic stress tolerance in number of field crops [43-45].

Complexity of salt tolerance mechanisms is evident from a number of mapping studies in which several physiological components such as sodium transport and exclusion, tolerance to osmotic and oxidative stress, and tolerance to ion toxicity at tissue and cell level were targeted for mapping and cloning of both major and minor QTLs for better understanding of the genetic basis of salt tolerance in different plants [46]. One major QTL for shoot K$^+$ content, *SKC1*, codes for an HKT-type transporter and regulates K$^+$/Na$^+$ homeostasis under salt stress [47]. In wheat, QTL mapping successfully identified two salt tolerance loci controlling Na transport, *Nax1/TmHKT1;4* and *Nax2/TmHKT1;5* [48,49], which led to cloning of the former [50]. Fine mapping with microsatellite markers suggested that the high-affinity K$^+$ transporter *TmHKT1, 5 (HKT8)* was the candidate gene for *Nax2* and homologous to *Kna1/TaHKT1; 5* in bread wheat [51].

Drought tolerance in rice is addressed through analysis of several shoot and root related component traits such as osmotic adjustment, leaf water potential, cell-membrane stability, osmolytes, leaf rolling, leaf drying and relative water content, root thickness, weight, length, number (penetrated and total), and root penetration index in several mapping populations [reviewed in 43, 45, 52). Successful introgression of several QTLs for root traits improved crop performance under field condition [53, 54]. In the last decade, several genomic regions associated with pre-flowering and post-flowering drought tolerance or stay-green, lodging tolerance has been identified in sorghum [reviewed in 43]. The stay green QTLs showed consistent expression in different genetic backgrounds [55] and near-isogenic lines (NILs) of these QTL demonstrated improved the post flowering drought tolerance [56]. In pearl millet (*Pennisetum americanum*) and corn, QTL mapping and marker-assisted selection of the favorable QTL alleles could be useful for breeding programs to improve drought tolerance [reviewed in 45].

In rice, a major QTL for submergence tolerance on chromosome 9, cloned by map-based approach, consisted of a cluster of three genes *Sub1A, Sub1B,* and *Sub1C* belonging to ethylene response-factor (ERF) family and variation in *Sub1A* distinguished the tolerant cultivars from susceptible cultivars [57]. Marker assisted introgression of this QTL improved tolerance to submergence [57, 58]. A large collection of introgression lines developed in three elite rice genetic backgrounds for identification of candidate genes and the cloning of QTLs involved in the stress response [59].

Frost tolerance in diploid wheat (*Triticum monococcum* L.) and barley is controlled by multiple QTLs. The fact that several C-repeat Binding Factor (CBF) genes colocalized with the QTLs in wheat [60-62] and barley [63, 64] indicated that these CBF genes may be responsible for frost tolerance. Chilling tolerance was analyzed by QTL approach at seedling stage in corn [65-67], sorghum [68], and rice in both seedling and booting stage [69-71].

Major QTLs for aluminum tolerance have been mapped in a number of crop species [reviewed in 44]. The cloned aluminum tolerance genes encode either malate transporters in wheat and *Arabidopsis* [23,72] or citrate transporters belonging to the multidrug and toxic compound extrusion (MATE) family as in barley and sorghum [73,74]. In rye, a cluster of *ALMT1* genes control aluminum tolerance [75].

Functional Genomics

Functional genomics refers to the development and application of global (genome-wide) experimental approaches to assess the gene function by making use of the information and reagents provided by structural genomics [75]. Greatly accelerated by the genome sequencing project, it is well positioned to add meaning to the massive amount of sequence data resulting from genome sequencing project. Although computerized tools are commonly used to analyze the genome sequence to predict the gene function, still experimental confirmation is needed to unambiguously assign gene function. Although expression profiling is a functional genomics tools; it will be described under the transcriptomics approach. In this section we will discuss the approaches to determine gene function through phenotypic analysis of mutants. These methods include targeted gene deletions, insertional mutagenesis, and RNA interference.

Targeted Gene Replacement by Homologous Recombination

Gene replacement by homologous recombination is a powerful tool to study gene function. Although some success has been reported in plants [76,77], use of this method on a large scale is largely inefficient due to low frequency of recombination events, too many random integration events, non-homologous end joining, and lack of gene specific selection system [78]. Further understanding of the molecular mechanism involved in homologous recombination and improved selection strategies are needed for wide spread use of this technology [79-81].

Gene Knockouts and Mutagenesis

Random mutational approach is a promising approach in plant functional genomics. Two different methods are commonly used to generate the mutants; Classical mutagenesis employs physical and chemical mutagens whereas heterologous transposons (Ac/Ds, En/Spm, or Mu) [82] or the T-DNA of *Agrobacterium tumefaciens* are used in insertional mutagenesis [83]. Although chemical mutagens are highly efficient in generating a broad range of changes in the genome in an unbiased manner, discovery of such DNA changes is cumbersome in a large genome. However, development of sensitive technologies such as Targeting Induced Local regions in Genomes (TILLING) allows genome-wide screening for point mutations [84]. It is suitable for any organism that can be heavily mutagenized. A typical high throughput TILLING procedure involves pooling of individuals in the M2 population, amplification of pools using gene specific primers that are end-labeled with fluorescent molecules, denaturation, renaturation followed by digestion with a single-strand specific nuclease, and electrophoresis of cleaved products using denaturing polyacrylamide gel electrophoresis or a LICOR gel analyzer system [85]. Differential double end labeling of amplification products allows rapid visual confirmation as mutations are detected on complementary strands and so can be easily distinguished from amplification artifacts. The migration of cleaved products indicates the approximate location of nucleotide polymorphisms. The TILLING procedure has also been useful for the discovery and cataloguing of natural polymorphisms, a method called Ecotilling [86].

Compared to classical mutagenesis, insertional mutagenesis is a rapid method of gene cloning. This approach involves generation of loss-of-function mutations in plants followed by recovery of the affected genes. Various cloning or PCR-based strategies using primers designed from the insertional element and the gene of interest help in identification of the mutated genes. Some limitations of the insertional mutagenesis approach are predominance of loss-of function alleles, insertion bias, inability to characterize lethal mutations, and difficulty in getting full saturation of the genome. The size and organization of the genome influences the level of saturation in an organism. Compared to transposon mutagenesis, however, fewer and stable insertions, easier maintenance, and less insertional bias are advantages of the T-DNA approach. The approach is modified further through the use of enhancer or gene trap to facilitate the screening of the mutagenized populations [87, 88]. Additionally, the flanking regions of the insertions can be systematically sequenced from pooled or individual lines and then mapped by comparison with the genomic sequences. Database of this information will allow quick identification of lines containing the insertion for the target gene [89, 90]. Since gene knockouts don't always display visible phenotype, gene function in such cases can be inferred by measuring gene expression and activity [91].

Other gene silencing methods such as co-suppression [92], virus-induced gene silencing [93] double-stranded RNA-mediated interference [94-96], and chimeric oligonucleotides [97] have been used to study gene function used in plants.

Application of Functional Genomics in Abiotic Stress Tolerance Studies

With the availability of whole genome sequences, systematic analysis of the annotated genes is now possible using reverse genetics approach. Among the crop plants, rice is in the forefront in development of large collections of indexed T-DNA or transposon insertion mutants to study gene function [reviewed in 98] and identification of mutants for abiotic stress responses from these genetic resources [99,100]. A number of databases have been established to facilitate the identification of insertion mutations in genes using available flanking sequence tag information: The rice reverse genetics database OryGenes DB (http: //orygenesdb.cirad.fr/index.html) [101,102], RiceGE/SIGnAL (http: //signal.salk.edu/cgi-bin/RiceGE), Gramene (http: //www. gramene.org). The Oryza Tag Line mutant database (http: //urgi.versailles.inra.fr/Oryza TagLine/) contains phenotypic data of 30 000 T-DNA enhancer trap lines, complemented by GUS/GFP expression data [103]. Since the databases are extensive containing information about large number of mutant lines generated from many sources, it should be useful to extrapolate this information to analyze gene function in other grasses.

TILLING projects are underway to identify both mutations and natural genetic variation in crop plants such as corn, rice, wheat, barley, soybean, *Medicago truncatula*, mung bean (*Vigna radiata* L.), and melon (*Cucumis melo* L.) [reviewed in 26]. In *Medicago truncatula*, a leucine-rich repeat RLK gene (*Srlk*) that was rapidly induced in roots by salt stress was identified using RNAi and TILLING approach [104]. The function of stress-responsive rice genes, such as *OsMT2B*, a metallothionein gene with role in the scavenging of reactive oxygen species (ROS) [105], and *OsTPC1*, a putative voltage-gated Ca(2+)-permeable channel protein involved in the regulation of elicitor-induced hypersensitive cell death [106] was elucidated using *Tos17* insertion lines. Using T-DNA-tagged transgenic rice lines, two cold responsive genes, namely *OsDMKT1* (putative demethylmenaquinone methyl transferase) and *OsRLK1* (putative LRR-type receptor-like protein kinase) have been identified [107]. Knockout mutants of the *OsGSK1* gene, a negative regulator of brassinosteroid signaling, showed cold, salt, drought and heat tolerance, while overexpression of the full-length cDNA lead to stunted growth [108].

TRANSCRIPTOMICS

The transcriptomics approach deals with comprehensive analysis of gene expression in a cell. Unlike the genome of an organism, which is highly stable, transcriptome is highly dynamic because mRNA population changes are dependent on the developmental stage and environmental conditions. Northern hybridization is the oldest technique of transcript profiling, in which a labeled probe is hybridized to an RNA target. This technique is sensitive, time consuming, and still considered the gold standard for confirmation and quantification of a small number of genes. Quantitative real-time PCR (QRT-PCR) has been developed as a robust, fast, and reproducible technique for quantification of expression of a single gene. The above low throughput techniques are still used as validating the results obtained from global approaches. Advances in genomics technologies allow measurement of transcript levels of thousands of genes at the same time. These high throughput transcriptomics approaches can be grouped into two categories. First approach makes the direct analysis of transcript levels inferred from nucleotide sequencing and fragment sizing. Examples of the category using nucleotide sequencing are EST, SAGE, and MPSS where as Differential Display, cDNA-AFLP involves fragment sizing. In the second category, gene expression is indirectly assessed in DNA microarray or DNA chips using the principle of nucleic acid hybridization of mRNA or cDNA fragments. The advantages and disadvantages of various transcriptomic tools have been discussed in several reviews [109,110]

EST Sequencing

Large scale sequencing of cDNA fragments or ESTs is a rapid way of discovering new genes and assessing their expression levels in a particular tissue. In this approach, single-pass sequences of 300-500 bp from one or both ends are determined from randomly chosen cDNA clones from diverse tissues. EST information is available for a number of organisms in public databases (NCBI and TIGR gene Indices). EST resources are valuable resources for gene discovery, development of molecular markers, and for gene expression analysis in those species for which complete genome sequences are not available. Full length cDNAs are useful for genome annotation, identifying exon-intron junctions and splice variants, and antisense RNA genes that might play a role in gene regulation and imprinting. Comparison of frequency of occurrence of ESTs in different kinds of tissues under different environmental condition or developmental stages can give a preliminary assessment of gene expression pattern. In order to represent the rare transcripts, large scale ESTs sequencing will be required.

SAGE

Serial analysis of gene expression (SAGE) is a similar sequence based method like EST sequencing [111]. It is a high throughput and cost saving technique for quantitative assessment and comparison of expressed genes. In this technique, 10-14 bp tags from a unique position within each species of mRNA are extracted, concatenated, and cloned before sequencing. These tags are used to identify corresponding genes in the database. Relative abundance of individual tags is used to determine the expression pattern of different genes. This approach can be followed for those organisms with an extensive cDNA sequence database or genome sequences.

MPSS

Massively Parallel Signature Sequencing (MPSS), developed by Lynx Therapeutics Inc. California, is a powerful technique for transcription profiling on a genome wide scale [112]. It involves cloning of individual cDNA molecules on microbeads and sequencing, in parallel, short tags from these cDNAs. Improved speed and depth of coverage in this technique compared to SAGE is due to generation of massive amount of data coupled with incorporation of good quality automation and data management.

Differential Display

Differential display method [113] discriminates the mRNA population by differential separation of representative cDNA fragments. It involves four steps: (1) reverse transcription of mRNA with anchored oligo dT primers, (2) amplification of cDNAs with arbitrary primers, (3) separation of amplified cDNAs in polyacrylamide gels, and (4) isolation of differentially expressed fragments from the gel followed by sequencing. Although this method provided a cost effective alternative for gene discovery and was used extensively by plant scientists, it has limited utility as a transcription profiling tool. Lack of reproducibility, lack of sensitivity, and occurrence false positives are major criticisms of this technology but several modifications have been made to improve its sensitivity and reproducibility to some degree [109].

cDNA-AFLP

cDNA-AFLP (amplified fragment length polymorphism) is an improvement over Differential Display due to ligation of adapters to restriction fragments and use of specific primers sets, and amplification under stringent PCR conditions [114]. Variation of this technology using single restriction enzyme has been developed [115]. cDNA AFLP has been very popular due to improved reproducibility, sensitivity, and good correlation with northern analysis. Further enhancement in throughput and automation has been possible through use of fluorescent labeling, multiplexing, and capillary based electrophoresis [116]. It has been modified to monitor one restriction fragment for each cDNA [117]. It allows systematic survey of the transcriptome using selective fragment amplification. This method has been automated for gene expression analysis [READS, 118; Gene Calling, 119; TOGA, 120].

DNA Microarrays

The principle of northern hybridization was exploited to develop DNA microarray technology to monitor expression of genes on a global scale. Basically there are two types of microarray formats: cDNA arrays and Oligoarrays. In cDNA microarrays, cDNA fragments were robotically spotted and immobilized on microscope slides [121] whereas oligonucleotides are directly synthesized in high density on to a solid matrix using photolithographic mask to determine the correct sequence [122]. cDNA arrays require maintenance and handling of microtitre plates, large scale PCR reactions, and clone validation and were widely used at the early stage, but oligoarrays are becoming the most preferred platform in plant science. Direct synthesis of oligos on solid surfaces makes it easier to print exact quantity of material than the cDNA clones. Oligoarrays can be used for detecting SNPs, slice variants, and members of gene families. For those species with whole genome sequences, maximum coverage of the predicted genes can be attained in oligoarrays. cDNA array may be advantageous in experiments involving heterologous species [123] and for annotation of genome sequences.

Despite its power and usefulness, microarray technology is both expensive and time intensive. Besides several technical problems such as contamination of DNA in spots on arrays, uneven hybridizations, and spurious hybridizations, it requires multiple biological and technical replications for generating reliable data. Genes of multigene families can result in cross hybridizations but hybridization to unique gene sequences can be achieved by using either 3' or 5' untranslated sequences as probes or oligonucleotide probes. Problem associated with the complexity of tissue types in multicellular organisms can be eliminated by adoption of in situ technologies, such as laser capture to isolate and amplify mRNA from specific cell types. Methods have been developed to amplify target RNA sample from a small quantity of total RNA [124]. Due to massive amount of data points generated in microarray experiments, downstream statistical analysis is particularly intimidating for meaningful interpretation and conclusion.

There has been extensive discussion in several papers regarding the factors associated with the production, hybridization, analysis of microarrays, and their application in plant science research [125-127]. It involves the isolation of mRNA from the tissue samples, preparation of the fluorescent dye labeled target cDNA or cRNA, and hybridization to the probes on the microarray. Individual cDNA or cRNAs hybridize with the corresponding probe on the microarray proportionate to their representation in the sample and finally specialized scanner quantifies the hybridization signal of each individual spot. The use of two different fluorescently labeled targets in single hybridization allows detection of gene expression on a global level of one sample in relation to the other. This technique can reliably detect mRNA abundance down to 1: 100,000 which is equivalent to approximately 0.15 mRNA molecules per cell.

DNA microarrays are an extremely powerful platform for elucidation of gene function, biochemical pathways, and regulatory network in different tissues at different developmental stages or environmental conditions [128,129]. Several approaches can be employed to determine genes function using global expression data. Two important approaches are: (a) association of temporal and spatial expression pattern with a specific phenotype or response, and (b) association of unknown genes with 'known' genes based on co-expression and co-regulation [130]. Thus genes with similar pattern of expression over time under certain environmental condition can be prospective candidates for involved in the same pathway. Similarly, those up-regulated or down-regulated genes prior to expression of genes of a particular pathway can be possible regulatory gene candidates. Another significant development in this field is the advent of whole genome tiling arrays which will allow a comprehensive understanding of various biological processes by obtaining unbiased and accurate information about the transcriptome and its regulation at chromosome level [131]. Another application of gene expression data is its use as a signature profile, in which expression pattern of thousands of genes can be used to classify genetic mutants [132].

To facilitate integration and exchange of microarray expression data among the researchers, several databases have been established in public domain. The main databases are the Gene Expression Omnibus (GEO) (http: //www.ncbi.nlm.nih.gov/ projects/geo) and ArrayExpress (http: //www.ebi.ac.uk/arrayexpress). Databases have been also established for specific plant species [130].

Application of Transcriptomics to Characterize Abiotic Stress Response

Several databases in the public domain store the information regarding plant ESTs and characterized genes [National Center for Biotechnology Information (NCBI) Unigenes, http: //www.ncbi.nlm.nih.gov/; The Institute for Genomic Research (TIGR) Gene Indices, www.tigr.org; Sputnik, http: //mips.gsf.de/proj/sputnik], which have been the basis of gene discovery and global gene expression studies related to abiotic stresses.

EST sequencing has been applied to analyze response to salinity, drought, and cold stress in rice (see review 133). Many of these stress-responsive ESTs overlapped with the abiotic stress induced sequences obtained from expression profiling studies in *Arabidopsis*, barley, maize, and rice [134]. Several large scale EST sequencing efforts in wheat and its close relatives provided useful information about the transcriptome under abiotic stresses at different developmental stages [135-138]. Structural and functional analysis of wheat genome based on ESTs resulted in identification of 278 ESTs related to abiotic stress (cold, heat, drought, salinity, and aluminum) from 7671 ESTs previously mapped to wheat chromosomes [136]. The highest number of abiotic stress related loci were found in homologous chromosome group 2 (142 loci) and the lowest number were found in group 6 (94 loci). When the genome-specific ESTs were considered, the B genome showed the highest number of unique ESTs (7 loci), while none were found in the D genome. Digital expression analyses of wheat ESTs datasets from different sources provided an overview of metabolic changes and specific pathways that are regulated under stress conditions in wheat and other cereals [135]. Comparison of ESTs from drought tolerant and one drought sensitive wild emmer genotype along with a modern wheat variety indicated both similar as well as differential expression patterns among the genotypes [137]. EST approach has been used to identify abiotic stress responsive genes in other crop species [soybean, 139; pearl millet, 140]. EST in halophytes such as *Spartina alterniflora* [141], *Thellungiella halophila* [142, 143], mangroves [144], and resurrection plant *Selaginella lepidophylla* [145], may serve as a rich genetic resource for the identification of novel genes associated with environmental stress tolerance. For those plant species where whole genome sequencing is not expected in near future, the EST libraries and the sequence information are valuable resource for comprehensive analysis of the stress-response transcriptome. The abiotic stress responsive ESTs from plants adapted to extreme abiotic stress could be potential targets for genetic engineering.

Matsumura *et al.* (1999) [146] used SAGE for the first time in rice and comparison of anerobically and aerobically treated seedlings revealed 24 differentially expressed genes of which 18 genes were anerobically induced and six genes were repressed. Most highly induced genes were prolamin, expansin, and glycine-rich cell wall protein in addition to several with no match to any genes in database. A comprehensive analysis of the drought-response transcriptome of chickpea using SuperSAGE and microarray revealed that the transcriptional program for signal transduction, transcription regulation, osmolyte accumulation, and ROS scavenging wassignificantly changed in chickpea roots under drought stress [147]. Comparable results were also obtained from SAGE and a cDNA microarray and this combinatorial approach was demonstrated to be more efficient and accurate in analyzing gene-expression patterns [148]. MPSS databases have been created in three plant species (*Arabidopsis*, rice and grape) [http: //mpss.udel.edu; 149]. But these resources have been rarely used for global transcription profiling under various abiotic stresses.

Microarrays studies have been widely used to analyze the global transcriptional changes under various abiotic stresses (Table **2**). The first microarray study [150] in salt resistant Pokkali and salt sensitive IR 29 revealed clear difference in expression pattern of transcripts at the initial phase of salt stress but with passage of time it became less pronounced. The delayed response in terms of kinetics of gene expression in IR 29 could be responsible for its sensitivity to salt stress. Similarity in response to different abiotic stresses was evident particularly for the drought and salt stress response in other microarray studies [151,152]. Since barley has efficient regulatory mechanisms for genes for maintaining ion homeostasis compared with rice, genetic modification of transcriptional regulators of early salt-responsive genes was suggested to improve salt tolerance in rice [153]. Function was assigned to a network of >200 genes that were responsive to both environmental stress and developmental stimuli [154]. Rice oligoarray was used to compare the salt stressed transcriptome of salt-tolerant (FL478) and salt-sensitive (IR29) rice varieties at vegetative stage [155]. Response of IR 29 was strikingly different from FL478 with induction of a large number genes induced in the former. Salt stress activated a number of genes in flavonoid pathway in IR 29 but not in FL 478 during vegetative growth stage. Another study by the same group [156] revealed induction of large number of genes at panicle initiation stage in salt sensitive *indica* and *japonica* cultivars compared to salt tolerant varieties. Many of these salt responsive genes are ion homeostasis related genes.

Table 2: Microarray studies to analyze global transcriptional changes under abiotic stress conditions in plants.

Plant	Stress	Platform used	Reference
Rice *(Oryza sativa)*	Salt stress	cDNA array	[150]
	Multiple abiotic stresses	Oligoarray	[154]
	Cold, drought, salinity and ABA treatment	cDNA array	[151]
	ABA treatment	cDNA array	[175]
	Chilling stress	cDNA array	[176]
	Salt stress	cDNA array	[177]
	Drought stress	Oligoarray	[157]
	Drought stress	cDNA array	[159]
	Salt stress	Oligoarray	[155, 156]
	Salt stress	cDNA array	[153]
	Low nitrogen stress	cDNA array	[160]
	Drought stress	Oligoarray	[158]
Wheat *(Triticum aestivum)*	Low-temperature stress	cDNA array	[178]
	Salt stress	Oligoarray	[163]
	Drought stress	Oligoarray	[162]
	Wheat drought stress	Oligoarray	[161]
Maize *(Zea mays)*	Water stress	cDNA array	[179]
	Low temperature stress	oligoarray	[164]
Barley *(Hordeum vulgare)*	Drought and salinity	cDNA array	[165]
	Iron deficiency	cDNA array	[180]
Sorghum *(Sorghum bicolor)*	Dehydration, salt and ABA treatment	cDNA array	[166]
Medicago truncatula	Descication tolerance	Oligo array	[171]
	Aluminum toxicity, resistance and tolerance	Oligo array	[170]
Potato *(Solanum tuberosum)*	Cold, heat and salt stress	cDNA array	[181]
	Salinity	cDNA array	[167]
Soybean *(Glycine soja)*	Iron deficiency chlorosis	cDNA array	[169]
Brassica rapa	Cold, salt, and drought	Oligoarray	[168]
Chickpea **(**Cicer arietinum)	High-salinity, cold and drought	cDNA array	[172]
Hot pepper *(Capsicum annum)*	Cold stress	cDNA array	[182]
Populus euphratica	Multiple abiotic stresses	cDNA array	[183]
Thellungiella halophylla	Abiotic stresses and hormonal responses	cDNA array	[174]
Arabidopsis and T. halophylla	Salt stress	cDNA array	[173]

In another study [157], microarray was used to identify genes associated with QTLs for osmotic adjustment (OA). Several candidate genes such as a LEA protein, a protein phosphatase 2C, a *Sar1* homolog, and SnRNP auxiliary factor were identified at each of the five OA QTLs. Transcription profiling of drought-tolerant (IR57311 and LC-93-4) and drought-sensitive (Nipponbare and Taipei 309) rice cultivars under drought stress using 20K NSF oligoarray indicated that more genes were significantly drought regulated in the sensitive than in the tolerant cultivars but more photosynthesis related genes were down-regulated in the tolerant than in the sensitive cultivars [158]. The global gene expression analysis during pollination/fertilization under drought and wounding stress in rice revealed that dehydration stress influenced the regulation of more than one-half of the pollination/fertilization-related genes [159].

The expression profiles of an *indica* rice cultivar Minghui 63 was analyzed at seedling stage under both low N stress and normal N (control) to understand plant's response under low N stress [160]. This study concluded that the genes involved in photosynthesis and energy metabolism were down-regulated rapidly, many of the genes involved in early responses to biotic and abiotic stresses were up-regulated while other stress responsive genes were down-regulated. Regulatory genes including transcription factors and ones involved in signal transduction were both up- and down-regulated, and the genes known to be involved in N uptake and assimilation showed little response to the low N stress.

A recent microarray study concluded that bread and durum wheat genotypes differ in their transcriptional response to drought stress which could be influenced by large number of genes located on the D Genome [161]. In two other studies [162,163], several novel candidate genes in wheat for drought tolerance and salt tolerance were identified. Analysis of maize leaf transcriptome under cold stress revealed that many transcripts with significant alteration in expression pattern were of unknown function [164]. Few photosynthesis-related genes were repressed by chilling, but chloroplast-related genes not directly engaged in photosynthesis were induced at later stages of the response. Genome-wide transcript abundance in response to drought and salinity stress was also monitored in barley [165] and sorghum [166]. Responses to ABA, high salinity, and osmotic stress in sorghum was both overlapping and distinct as revealed from the changes in expression pattern of genes involved in signal transduction, chromatin structure, transcription, translation, and RNA metabolism. In barley study, jasmonate-responsive, metallothionein-like, LEA and ABA responsive genes were significantly upregulated under drought stress in contrast to drastic down regulation for the photosynthesis related genes.

In a microarray study in potato, induction of various kinds of transcription factors in osmotic stress response and plant defense pathways indicated crosstalk between abiotic and biotic stress responses during salt exposure, which activated several adaptation mechanisms including heat shock proteins, LEA proteins, dehydrins and PR proteins [167]. In *Brassica rapa,* various transcriptional regulatory mechanisms and common signaling pathway were suggested to be jointly involved be under the abiotic stresses [168].

In soybean, 43 differentially expressed genes could be identified by comparing the transcription profile between two near isogenic lines differing in response to iron deficiency and 57% of these genes have close similarity to known stress induced genes [169]. In *Medicago truncatula*, novel genes associated with aluminum toxicity, resistance, and tolerance were identified [170] and another study used transcription profiling to analyze the metabolic and regulatory processes occurring during the transition from desiccation-sensitive to desiccation-tolerant stages in *Medicago truncatula* seeds [171]. In the former study, genes involved in cell-wall modification and abiotic and biotic stress responses were up-regulated while genes associated with primary metabolism, secondary metabolism, and protein synthesis and processing, and the cell cycle were down regulated. The later study suggested that there is probably partial overlap of ABA-dependent and -independent regulatory pathways involved in both drought and desiccation tolerance and there is massive repression of genes belonging to numerous classes including cell cycle, biogenesis, primary and energy metabolism during late seed maturation stage.

Using microarray, genes coding for various functional and regulatory proteins with differential expression between tolerant and susceptible chickpea genotypes under cold, drought and salinity stresses were identified [172]. Comparison of model plant *Arabidopsis thaliana* with its closely related halophyte *Thellungiella halophila* has shown that constitutive expression of stress related genes that are stress inducible in *Arabidopsis*, might be responsible for its higher degree of stress tolerance [173,174].

PROTEOMICS

The term 'proteome' was first coined to describe the entire complement of proteins present in a cell, tissue, or subcellular compartment of an organism under defined conditions [184]. Proteomics not only involves large-scale identification of proteins but also deals with analysis of all protein isoforms and post-translational modifications, protein-protein interactions, enzymatic assays for the functional determination, localisation studies of gene products and promoter activity and structural information of protein complexes [185, 186]. Proteomics research began with the development of the two-dimensional (2D) gel electrophoresis techniques to separate the crude protein mixtures [187]. In the 2D-gel electrophoresis, polypeptides are separated in the first dimension by a nondenaturing technique, usually isoelectric focusing followed by further resolution according to molecular weight using denaturing SDS-PAGE in the second dimension. Later with the development of mass spectrometry, determination of the identity of proteins at a large scale with increased sensitivity was only possible [188]. A standard proteomics procedure involves separation, quantitation, and identification of proteins in a complex mixture using 2-D gel electrophoresis in conjunction with mass spectrometry. The advancement in MS techniques [189] coupled with database searching have played a crucial role in proteomics for protein identification. Databases have been constructed containing all expressed proteins from plant organs and cell organelles from various species (Table **3**). A number of reviews on plant proteomics give a thorough insight of the technology and its potential [190-195].

Table 3: Websites of interest for omics research in plants

Genomics-related websites	
National Center for Biotechnology Information	http: //www.ncbi.nlm.nih.gov
EMBL nucleotide sequence database	http: //www.ebi.ac.uk/embl
Gramene	http: //www.gramene.org
Graingenes	http: //wheat.pw.usda.gov/
Plant Genomes Central in NCBI	http: //www.ncbi.nlm.nih.gov/genomes/PLANTS/PlantList.html
The Arabidopsis Information Resource (TAIR)	http: //www.arabidopsis.org/index.jsp
Arabidopsis Transposon Insertion Database	http: //atidb.cshl.org/
Rice Genome Project (RGP)	http: //rgp.dna.affrc.go.jp/
Rice reverse genetics database OryGenes DB	http: //orygenesdb.cirad.fr/index.html
RiceGE	http: //signal.salk.edu/cgi-bin/RiceGE
The Oryza Tag Line mutant database	http: //urgi.versailles.inra.fr/OryzaTagLine/
Rice GAAS	http: //ricegaas.dna.affrc.go.jp
TIGR rice genome annotation	http: //blast.jcvi.org/euk-blast/index.cgi?project=osa1
Oryzabase	http: //www.shigen.nig.ac.jp/rice/oryzabase/top/top.jsp
Rice Annotation Database (RAD)	http: //rad.dna.affrc.go.jp
Maize genome resources	http: //www.maizegenome.org/,
Maize sequence	http: //www.maizesequence.org/index.html
Maize Genetics and Genomics Database	http: //www.maizegdb.org/genome/
Phytozome	http: //www.phytozome.net/sorghum
Sorghum Genomics	http: //sorgblast3.tamu.edu
Gene Ontology	www.geneontology.org
Transcriptomics-related websites	
Gene expression omnibus	http: //www.ncbi.nlm.nih.gov/projects/geo
ArrayExpress	http: //www.ebi.ac.uk/arrayexpress
PLEXdb	http: //www.barleybase.org/plexdb/html/index.php
TIGR Arabidopsis arrays	http: //www.jcvi.org/arabidopsis/qpcr/
Stanford Microarray Database	http: //smd.stanford.edu/index.shtml
Rice transcriptional database	http: //microarray.rice.dna.affrc.go.jp
Rice Expression Database (RED)	http: //red.dna.affrc.go.jp/RED/
NSF Rice Oligonucleotide Array Project	www.ricearray.org/

Table 3: cont....

Virtual center for cellular expression profiling in rice	http: //bioinformatics.med.yale.edu/riceatlas/
Barleybase	http: //www.barleybase.org
Zeamage	www.maizearray.org
TIGR Solanaceae Genomics Resource	http: //www.jcvi.org/potato/
Soybean Genomics and Microarray Database	http: //psi081.ba.ars.usda.gov/SGMD/default.htm
Tomato Expression Database	http: //ted.bti.cornell.edu
Genevestigator	https: //www.genevestigator.com/gv/index.jsp
Proteomics-related websites	
Swissprot	http: //us.expasy.org/sprot/
Proteome Analysis at EBI	http: //www.ebi.ac.uk/proteome/
Arabidopsis Membrane Protein Library	http: //www.cbs.umn.edu/arabidopsis/
Database of *A. thaliana* Annotation	http: //luggagefast.Stanford.EDU/group/arabprotein/
ExPASy *A. thaliana* 2D-proteome database	http: //www.expasy.ch/cgi-bin/map2/def?ARABIDOPSIS
PlantsP: Functional Genomics of Plant Phosphorylation	http: //PlantsP.sdsc.edu/
Metabolomics-related websites	
KEGG	http: //www.genome.jp/kegg/kegg2.html
BRENDA	http: //www.brenda-enzymes.org/
AraCyc Arabidopsis metabolic pathway annotations	http: //www.arabidopsis.org/biocyc/index.jsp
Platform Plant Metabolomics	http: //www.metabolomics.nl/
Metabolic modeling	http: //www.hort.purdue.edu/cfpesp/models/models.htm
MetAlign tool for GC- or LC-MS data analysis	http: //www.rikilt.wur.nl/UK/services/MetAlign+download
The Platform for RIKEN Metabolomics	http: //prime.psc.riken.jp/?action=metabolites_index
The Golm Metabolome Database	http: //csbdb.mpimp-golm.mpg.de/csbdb/gmd/gmd.html
Iowa Gene Expression Toolkit	http: //metnet.vrac.iastate.edu
Solcyc Solanaceae metabolic pathway annotations	http: //www.sgn.cornell.edu/tools/solcyc

Analysis of large scale variation in proteome has been useful in plant genomics, genetics, and physiological studies [196]. Common applications of proteomics in plants are to obtain a snapshot of proteins present in a biological material or any subcellular compartment for comparative proteomics studies. Comparative proteomics studies aim to compare the protein profiles between different samples such as stressed versus unstressed, tissues of different developmental stages, or wild vs. mutants. In studies involving responses to environmental changes, affected proteins can be identified and metabolic pathway can be deduced from their sequences and function. Proteins have been useful as genetic markers for genome mapping studies. Taking the advantage of protein polymorphism, factors controlling quantity of the encoded proteins (protein quantity loci or PQL) regulated at a certain development or physiological stage could be mapped, and colocalization of PQLs and QTLs may lead to identification of candidate proteins for the target traits [197]. Proteomics is rapidly becoming an essential complement to other functional genomics approaches in biological research. Proteomic analysis is particularly vital because the observed phenotype is a direct result of the action of the proteins rather than the genome sequence.

A major limitation of the current technology is the reduced coverage and inability to detect low abundance proteins. High resolution 2-D gels can resolve about 1,000 proteins that are highly abundant in a crude mixture. Even under optimal condition, approximately 25% of the proteome may be observed [198]. However, development in direct mass spectrometric analysis is increasing sensitivity, robustness, and data handling [186]. Certain groups of proteins are extremely resistant to extraction, solubilization or subsequent separation. Many low abundance proteins such as transcription factors, signal transduction proteins, receptors, and extremely acidic and basic proteins are not detectable in standard proteomics techniques. Other difficulties include variability in samples, degradation of samples, various post-translational modifications, vast dynamic range, tissue types, developmental and temporal specificity which make the comprehensive proteomics studies challenging [186]. Another significant challenge of proteomics is that due to highly dynamic nature of proteome, it is not yet possible to assess the true proteome dynamics of cellular responses due to unavailability of high throughput and highly sensitive techniques. With the improvement in protein extraction and separation technology, it is now possible to resolve large number of proteins in 2D gels [199] and fluorescent 2D differential in-gel electrophoresis allows analysis and comparison of three proteomes in the same gel [200]. Gel free approaches such as multidimensional protein identification technology (MudPIT) [201] has been developed as an alternative to identify many proteins despite its problem in reproducibility of quantitation.

A number of proteome wide platforms have been developed to complement mass spectrometric platform. Yeast two-hybrid system [202] can detect weak interactions between low abundance proteins. Analogous to DNA microarrays, protein microarrays [203] allow rapid interrogation of protein activity. These arrays involve printing of antibodies or proteins which are then probed with a complex protein mixture to study of protein-protein interactions. The intensity or identity of the resulting protein-protein interactions may be determined by fluorescence imaging or mass spectrometry. *In vivo* read outs are also possible using protein arrays through use of green fluorescent protein (GFP) signals or protein association through fluorescent resonance energy transfer (FRET) between protein fusions to different wavelength variants of GFP [186].

Application of Proteomics to Characterize Abiotic Stress Response

New insights have been obtained on plant adaptation to abiotic stresses through application of proteomics approach to organelles and tissues in several plant species. Proteome of mitochondria of *Pisum sativum* [204], poplar leaves [205], rice anthers and leaves [206,207] have been analyzed to study plants response to cold stress. The effect of salinity stress, especially in crop plants, was investigated by comparative proteome studies in various tissue types in rice [leaves, 208; leaf lamina, 209; roots, 210,211,212], wheat [roots, 213; leaves, 214], and barley [root, 215]. These studies revealed species- as well as tissue-specific stress responses and identification of novel stress specific protein candidates, under and over-expressed proteins in response to particular stress or combination of stresses, alterations in protein phosphorylation patterns, and salinity stress-responsive protein in root apoplast with a putative function in stress signaling. In wheat, salt stress induced changes in proteome of tolerant variety was different from that of susceptible variety [216,217]. A proteomic study to analyze the effect of salt stress in leaf sheath, root, and leaf blade of three *indica* rice cultivars Nipponbare,IR36, and Pokkali using 2-DE and N-terminal and internal amino acid sequence analysis indicated a coordinated response to salt stress due to expression of specific proteins in specific organs [218]. Proteomics provided excellent opportunities to study the response of plants to stresses caused by drought, salinity, heat, ozone, UV light, heavy metals, nutrient deficiencies, elevated CO_2 conditions, low P and high Al [reviewed in 219]. Several stress responsive proteins identified in these studies open up new targets for genetic engineering to improve stress tolerance in plants.

Although proteomics has been exploited in abiotic stress tolerance studies in plants, large-scale proteomic studies are still limited. Application of proteomic approach particularly the comparative proteomics studies provided essential information about stress induced alterations in protein quantity and quality, and specific modifications of proteome [220].

METABOLOMICS

Metabolome is defined as the full complement of low molecular weight molecules or metabolites in an organism [221] and the associated field 'metabolomics' aims to identify the gene function through analysis of metabolite diversity and quantity by employing high throughput techniques [222]. Since metabolites are the ultimate gene products, simultaneous profiling of all metabolites gives a correct assessment of the developmental and physiological state of an organism under defined conditions at all levels of complexity such as cell or cell compartments, tissues or organisms. The metabolite diversity and quantity, that ultimately determines the response of an organism to environmental stresses, are controlled by both genetic and environmental factors.

The metabolome is highly dynamic because metabolites represent the catabolic and anabolic activities inside a cell at a given time. The diversity in metabolite as intermediates or the final product of the biochemical pathways, controlled by many structural and regulatory genes as well as produced in response to environmental stimuli, complicates the metabolomic studies. Metabolites are often grouped into two categories: primary metabolites or secondary metabolites. Primary metabolites are those essential in central carbohydrate metabolism [223,224] and the primary metabolism pathway is highly conserved between different species. But large variation is observed in both amount and type of secondary metabolites, which are believed to play important role in plant adaptation due to their involvement in response to environmental factors [225,226]. But our knowledge about the role played by individual metabolite in plant adaptation is limited [227].

Due to complexity involved in plant metabolism, four different metabolomic strategies are commonly followed to answer specific question: Metabolite target analysis, metabolite or metabolic profiling, metabolomics, and metabolite fingerprinting [222]. In target analysis strategy, a small set of known metabolites are determined and quantified using an analytical technique. This is employed for screening certain phytohormones or to study the primary effect of genetic or environmental influence. Metabolic profiling, on the other hand, gives a snapshot of the chemical composition of the tissues. It often involves analysis of a group of both known and unknown metabolites with similar chemical properties. For example, it may be lipids, carbohydrates or isoprenoids or it could be products of a particular biosynthetic pathway. This strategy can help in elucidating the function of a whole pathway or intersecting pathways. When comprehensive analysis of metabolites in terms of both identification and quantification in a biological system is needed, metabolomics strategy is chosen to determine and quantify all possible identified or unknown metabolites using a range of techniques such as Liquid Chromatography-Mass Spectrometry (LC-MS), Gas Chromatography-Mass Spectrometry (GC-MS), and/or Nuclear Magnetic Resonance (NMR). The last strategy, metabolic fingerprinting, aims to generate metabolic signature or mass profile of samples for screening purpose. A large population of samples can be rapidly classified according to the origin and biological relevance using this strategy.

Metabolomic analysis can have a range of applications: analysis of response to environmental stresses, plant-pathogen interactions, effect of genetic modification, elucidating gene function, and assessment of substantial equivalence of genetically modified organisms [228]. Another obvious application is the use of metabolomic data to identify new targets in the biochemical pathways for metabolic engineering to enhance stress tolerance, nutritional value of foods, and decreasing the need for pesticide or fertilizer application.

A range of analytical technologies are available to analyze the metabolites in different organisms, tissues, or fluids [229]. Extraction method and analytical tool must be carefully selected to analyze the targeted metabolites. The currently used analytical techniques in plant metabolic profiling are NMR and Mass Spectropmetry often combined with chromatography [230]. GC-MS is frequently used for detection of primary metabolites such as amino acids and carbohydrates [231,232]. But LC-MS allows detection of wide range of diverse secondary metabolites such as (poly) phenols, flavonoids, alkaloids, and phenylpropanoids [233]. NMR technique helps in both identification and quantification of metabolites present in both liquid and solid state. This technique is used to study the structure of detected compounds [234] and is non destructive, allowing subsequent analysis by other analytical techniques.

In plant kingdom, 100,000-200,000 secondary metabolites are synthesized [235], which are chemically diverse, often species specific, and many are unknown for its physical details. Although metabolomic analysis is comparatively fast, cheap, reliable, and precise, unambiguous and simultaneous identification of all metabolites in a biological system poses a significant challenge. Major limitations of plant metabolomics are the lack of knowledge regarding a vast number of metabolites synthesized in the plant kingdom [236] and nonavailability of reference standards for many metabolites [237], which prohibit definite quantification of large number of metabolites and utility as a predictive tool [238]. Since current analytical technologies require tissue extract for generating metabolic profile, it is not possible to perform metabolomic analysis at the cellular or subcellular levels where the metabolites are involved in pathway activity, metabolic flux etc. Since the metabolome is very diverse and the constituents have wide dynamic range, which is severely limited by the sample matrix or the presence of interfering and competing compounds, there is no method to extract and measure all metabolites. Different analytic techniques have been developed to overcome these limitations [239]. However, improvement in analytical techniques is needed to increase the sampling of the metabolome in general but also the cellular or subcellular compartments.

Theoretically, it may be possible to link metabolomic changes in biochemical pathways to the enzymes and the genetic factors, but explanation of such metabolomic variation in physiological context or establishing linkage between specific metabolites and phenotypes is often difficult [240]. Similar to other functional genomics tools, a typical metabolomics experiment generate massive amount of data which requires bioinformatic tools to store, handle, and analyze the data for meaningful interpretation. Metabolomics is emerging as a useful tool in functional annotation of genes and analysis of plants response to environmental perturbation [241]. It has the potential to complement the other omics technologies in improving the plant response to environmental stress through enhanced understanding of the biochemical pathways.

Application of Metabolomics to Characterize Abiotic Stress Response

Using data of the metabolite levels in leaves of unstressed plants in four different species: the moss *Physcomitrella patens*, *Arabidopsis*, barley, and wheat [228], it was demonstrated that each species has a distinct metabolite profile with barley and wheat being the most similar. Moss and *Arabidopsis* recorded lower level of most metabolites. Higher amount of some compounds such as urea, glycerol, tyramine, allantoin, tocopherol, xylitol, fucose, and inositol in moss compared to other three species, provided some indication that these metabolites may be responsible for its high abiotic stress tolerance. Comparison of the metabolic profiles of transgenic rice and barley plants constitutively expressing a sodium-pumping ATPase (*PpENA1*) of *Physcomitrella patens* revealed distinct pattern of metabolites level in both species compared to control [242].

The relationship between metabolite content and stress tolerance has been investigated in a number of wild species of tomato and introgression lines [243, 244]. High degree of variation in the level of metabolites in selected wild species and introgression lines was observed and 900 QTL influencing the amount many metabolites were identified. The study was further extended to study the inheritance of metabolite QTL [245] and their results suggest that introgression breeding might be a useful approach for engineering enhanced metabolic traits in crop species

An investigation of the metabolic response to drought stress in tomato indicated modest change in metabolic profile of hybrid of the cross between M82 and a wild species *Solanum penelli* compared to M82 under drought stress [246]. The level of aminoacids, TCA cycle intermediates, sugars, polyols, proline, and its biosynthetic precursor glutamate, glycine betaine is increased as expected because of their stress protective role as osmoptotectants. Thus the metabolomic analysis can be utilized to develop metabolite biomarkers to make genetic progress to improve abiotic stress tolerance. Based on the carbohydrate profiling results combined with the measurements of oxidative products and antioxidative enzymes, it was suggested that a more effective ROS scavenging system may be responsible for chilling-tolerance in rice genotypes [247].

Metabolic profiling indicated conserved and divergent metabolic responses among *Arabidopsis thaliana, Lotus japonicas,* and *Oryza sativa* [248]. A change in the balance between amino acids and organic acids may be a conserved metabolic response of plants to salt stress. A number of metabolomic studies have been conducted to analyze plants response to temperature stress, water and salinity, sulfur, phosphorus, oxidative, and heavy metal stress as well as a combination of multiple stresses [reviewed in 249].

CHALLENGES IN INTEGRATING OMICS APPROACHES

Reductionist approach, the mainstay of the molecular biology in the past, has been proved to be extremely successful, leading to the discovery of different genetic components involved in abiotic stress tolerance. But the fundamental biological question of discovering the connections between different genes and gene products involved in functioning of a plant system still lingers. Particularly in case of abiotic stress tolerance, this endeavour is highly challenging due to the complexity of multiple mechanisms and interplay of environmental factors and developmental stages. The obviously attractive tools for high throughput profiling of genome, proteome, transcriptome, and metabolome on a global scale is increasingly used by scientists, but interpretation of the whole range of omics data to create a holistic view of a biological process has been limited to date.

System biology is emerging as a valuable approach to realize the full potential of post-genomic revolution that strives for understanding the complexity of biological processes in living organisms at genomic, transcriptomic, proteomic, and metabolite level [250]. This approach is getting wide attention as different omics technologies are allowing collection of large quantity of data at every level of translation of genetic information to trait expression. To meet the challenges involved in integrating the omics information, progress in bioinformatics will be essential [251] to obtain a more complete picture of the regulation of the abiotic stress tolerance mechanisms in crop plants. Many websites (Table **3**) provide useful information about the tools and the resources for omics-related research in plants. A web-based system, Plant MetGenMAP (http: //bioinfo.bti.cornell.edu/tool/MetGenMAP), has been developed to comprehensively integrate and analyze large-scale gene expression and metabolite profile data sets along with diverse biological information to provide understanding of a biological process [252].

While focus was on DNA in the genomics era, it is now apparent that translation of genotype to phenotype is not always straight forward. As we move from genotype toward phenotype, complexity arises due to increased level of epistasis, which makes prediction more difficult [240]. High-throughput omics in combination with genetic analyses would allow analyzing the functional consequences of natural genetic variation at transcript, protein, and metabolite level [253], but the collection of proteins and metabolites data on omics scale is not yet possible. In higher organisms, adoption of this approach to simultaneously profile and analyze transcriptome, proteome, and metabolome might be difficult as metabolites are not produced in the same cell types. Due to the diversity in properties, limited resolving power of instrumentation, it is not yet possible to catalog all the proteins and metabolites even for a fully sequenced genome. Post-transcriptional and post-translational modifications further complicate the analysis. Due to the dynamic nature of transcript, proteins, and metabolites, it is quite difficult to get a complete picture of transcriptome, proteome, and metabolome.

Despite the challenges in interpreting the omics data collected in parallel, combining genome wide expression data with metabolic data could result in generating biochemical network model to explain the natural genetic variation [254,255]. Successful combination of metabolomics combined with genetic approach revealed useful information about the genetic control of metabolite production [244,256]. A number of genes involved in metabolism have been functionally identified in *Arabidopsis* by integration of transcriptomics and metabolomics [257]. A system based approach coupled with multivariate data analysis is suggested to unravel the mechanisms responsible for plant desiccation tolerance [258]. A systematic study involving transcriptomics, proteomics, and metabolomics tools to investigate the molecular responses of O_3 in rice leaves showed that O_3 triggers a chain reaction of altered gene, protein, and metabolite expressions involved in multiple cellular processes [259]. In another study, both metabolites and proteins profiling revealed that alteration of metabolites and proteins may contribute to multiple stress tolerances in transgenic rice [260].

CONCLUSIONS AND FUTURE PERSPECTIVES

Technological breakthroughs in omics technologies are providing necessary impetus to transform the biological research. These omics technologies have demonstrated their potential in accelerating basic science research as well as formulating applied genomics strategies to improve abiotic stress tolerance in crop plants. Decoding of the whole genome in case of several plant species has been completed and many are under way. Based on these genome sequences, the complete inventory of genes have been predicted in these plant species, but assigning function to these gene sequences will continue to be the most difficult task because DNA sequence is not always translated to proteins and mRNA levels may or may not correlate with the protein or metabolite levels. Despite extensive investigation conducted in *Arabidopsis* by researchers worldwide, only half of the genes have been functionally annotated based on homology to known genes, and 11% of these genes have been functionally validated [261]. Therefore, the major goal of plant biologists is to decipher the function of all the genes including those responsible for abiotic stress tolerance to improve world food security. For plant species with whole genome sequences and large EST resources and high density molecular maps, other functional genomics resources have been developed or under development to experimentally determine the function of all the genes. Although overexpression or knock-out strategy have been commonly used for validating gene function, several complementary omics technologies are enabling in depth dissection of the flow of biological information in plants at all levels between genome and phenome. The power of genetic analysis is now greatly improved due to application of high throughput omics analyses in experimental populations to make scientific advances in the field of abiotic stress tolerance. Since a single omics approach in isolation will not be productive to make a comprehensive understanding of abiotic stress tolerance mechanisms, systematic analyses of the genome, transcriptome, proteome, and metabolome in parallel would aid in assigning the function of genes unambiguously even if these is no difference in phenotype between wild type and the mutant or transgenic plants.

As discussed in earlier sections, the omics technologies has opened up numerous possibilities to apply these tools to unravel the novel molecular components regulating the complex abiotic stress tolerance attributes. It enables the plant geneticists to obtain information about the steps involved in translation of genomic information to biological function [262]. There is now a better understanding of the plants response to abiotic stresses at cellular level, but the wealth of data gathered so far using various omics tools has not yet been able to elucidate the functioning of the plant as a whole. The vast amount of data from various omics approaches need to be interconnected and organized into central databases in order to allow easy extraction and comparison of data for meaningful analysis [263]. In case of abiotic stress tolerance, which is much more complex than any other complex quantitative traits, the system-based approach will be essential to discover the molecular genetics networks through collection and analysis of complex omics data at all levels. With the recent advances in bioinformatics to integrate the high throughput omics data, system-wide analysis from genome to phenome [264] is now possible to provide clues to improve abiotic stress tolerance in plants.

The ultimate goal of plant scientists is to identify the pathways linking the genes to the phenotype of any target trait so that the traits can be manipulated for crop improvement. However, it is now well recognized that identification of the genetic network and their components to improve plants adaptation in stress environments will be challenging. Genomic strategy to improve abiotic stress tolerance in plants mostly relied on the identification, introgression, and cloning of stress tolerant genes or QTLs. QTL studies enabled systematic analysis of the naturally occurring variation in crop germplasm to localize various genomic regions controlling abiotic stress tolerance traits or mechanisms for introgression in to elite cultivars to improve their performance under abiotic stress environments [45,265,266]. Cloning and introgression of QTLs for abiotic stress tolerance component traits [53,54,57] will continue to be a major strategy for crop improvement in future and would be accelerated further through integration of other omics technologies for genetic dissection of these complex traits. For example, transcript, protein or metabolite profiling of the mapping population or introgression lines differing abiotic stress tolerance will provide better candidates for stress tolerance [22]. With the accumulation of genomic resources (Table **3**) and genomic sequences in every important crop species, cloning of abiotic stress tolerance QTL would be achieved more efficiently in future. Introgression of a major QTL or a combination of beneficial QTLs by pyramiding is an efficient way of improving crop plants [4]. New strategies such as advanced backcross QTL analysis [21] or Introgression library approach [20] are being employed to discover beneficial alleles for abiotic stress tolerance from the wild relatives and landraces. However, the real challenge to their utilization will depend on identification of the stress tolerant alleles which will not compromise the growth and yield of crop plants. Most marker-assisted breeding strategies to date use DNA markers. With the demonstration of linkage between natural variation at the metabolite level and growth-related phenotypes [244,267,268], development of metabolomics-based breeding tools may be possible for application breeding programs in future [269].

Identification and introgression of stress tolerant genes from model plants, extremophiles, and crop germplasm through transgenesis has been a popular strategy of many researchers to improve abiotic stress tolerance [45]. While these studies have been useful to study the cellular response at molecular level and demonstrated enhanced abiotic stress tolerance in laboratory conditions, none of these results has been translated into commercial product in any field crop [270]. Since single gene transformants using constitutive promoters has not demonstrated the level of abiotic stress tolerance required for retaining crop productivity in field conditions, pyramiding of multiple transgenes using organ-specific, developmental stage specific or stress responsive promoters should be explored to provide protection against abiotic stresses. *Arabidopsis* and other model plants have contributed significantly to the understanding of abiotic stress tolerance through discovery of abiotic stress tolerance pathways and genes and are expected to continue in future [271], but the knowledge from these models can hardly be applied to crop plants [272]. The successful crop improvement strategy will depend on identifying the superior crop orthologs in wild and cultivated germplasm and their introgression.

Plants are often exposed to multiple abiotic stresses. Although overlapping of plants responses to different stresses is well known, there has been little effort to breed plants with improved tolerance to multiple abiotic stresses. Future research effort should be directed using the omics approaches to elucidate plant's response to abiotic stresses at different developmental stages, root architecture and plasticity, genotype x environment interactions, and temporal and spatial regulation of abiotic stress tolerance genes. High throughput omics technologies coupled with easily accessible integrated databases should now facilitate the elucidation of the complex stress regulatory network and their components to understand the mechanism of stress tolerance. The real benefits of these technologies, however, will only be realized when the knowledge and the tools resulting from the advances in omics field are translated into a products with improved abiotic stress tolerance in field environment.

ACKNOWLEDGEMENTS

The financial support for this study from the USDA-CSREES is gratefully acknowledged. This manuscript is approved for publication by the Director of Louisiana Agricultural Experiment Station as manuscript number 2010-XXX-XXXX.

REFERENCES

[1]   Boyer JS. Plant productivity and environment. Science 1982; 218: 443-448.

[2]   Lobell DB, Burke MB, Tebaldi C, Mastrandrea MD, Falcon WP, Naylor RL. Prioritizing climate change adaptation needs for food security in 2030. Science 2008; 319: 607-610.

[3]   Flowers TJ, Yeo AR. Breeding for salinity resistance in crop plants: where next? Aust J Plant Physiol 1995; 22: 875-884.

[4]   Takeda S, Matsuoka M. Genetic approaches to crop improvement: responding to environmental and population changes. Nat Rev Genet 2008; 9: 444-457.

[5]   Bouchez D, Hofte H. Functional genomics in plants. Plant Physiol 1998; 118: 725-732.

[6]   Terryn N, Rouze P, Montagu MV. Plant genomics. FEBS Lett 1999; 452: 3-6.

[7]   Borevitz JO, Ecker JR. Plant genomics: The third wave. Annu Rev Genomics Hum Genet 2004; 5: 443-477.

[8]   Steinmetz LM, Davis RW. Maximizing the potential of functional genomics. Nat Rev Genet 2004; 5: 190-201.

[9]   Green ED. Strategies for the systematic sequencing of complex genomes. Nat Rev Genet 2001; 2: 573-583.

[10]   Arabidopsis Genome Initiative. Analysis of the genome sequence of the flowering plant *Arabidopsis thaliana*. Nature 2000; 408: 796-815.

[11]   Somerville C, Dangl J. Genomics: plant biology in 2010. Science 2000; 290: 2077-2078.

[12]   Goff SA *et al.* A draft sequence of the rice genome (*Oryza sativa* L. ssp. *japonica*). Science 2002; 296: 92-100.

[13]   Yu J *et al.* A draft sequence of the rice genome (*Oryza sativa* L. ssp. *indica*). Science 2002; 296: 79-92.

[14]   International Rice Genome Sequencing Project. The map-based sequence of the rice genome. Nature 2005; 436: 793-800.

[15]   Yu J *et al.* The Genomes of *Oryza sativa*: a history of duplications. PLos Biol 2005; 3: 266-281.

[16]   Tuskan GA *et al.* The genome of black cottonwood, *Populus trichocarpa* (Torr & Gray). Science 2006; 313: 1596-604.

[17]   Paterson AH *et al.* The *Sorghum bicolor* genome and the diversification of grasses. Nature 2009; 457: 551-556.

[18]   Schnable PS *et al.* The B73 Maize genome: complexity, diversity, and dynamics Science 2009; 326: 1112-1115.

[19]   Doerge RW. Mapping and analysis of quantitative trait loci in experimental populations. Nat Rev Genet 2002; 3: 43-52.

[20]   Zamir D. Improving plant breeding with exotic genetic libraries. Nat Rev Genet 2001; 2: 983-989.

[21]   Tanksley SD, Nelson JC. Advanced backcross QTL analysis: a method for the simultaneous discovery and transfer of valuable QTLs from unadapted germplasm into elite breeding lines. Theor Appl Genet 1996; 92: 191–203.

[22]   Salvi S, Tuberosa R. To clone or not to clone plant QTLs: present and future challenges. Trends Plant Sci 2005; 10: 297-304.

[23]   Hoekenga OA *et al.* AtALMT1, which encodes a malate transporter, is identified as one of several genes critical for aluminum tolerance in *Arabidopsis*. Proc Natl Acad Sci USA 2006; 103: 9738-9743.

[24]   Rus A, Baxter I, Muthukumar B, Gustin J, Lahner B, Yakubova E, Salt DE. Natural variants of *AtHKT1* enhance Na accumulation in two wild populations of *Arabidopsis*. PLoS Genet 2006; 2: 1964-1973

[25]   Gorantla M, Babu PR, Lachagari VBR, Feltus FA, Paterson AH, Reddy AR. Functional genomics of drought stress response in rice: Transcript mapping of annotated unigenes of an *indica* rice (*Oryza sativa* L. cv. Nagina 22). Curr Sci 2005; 89: 496-514.

[26]   Papdi C, Joseph MP, Salamo IP, Vidal S, Szabados L. Genetic technologies for the identification of plant genes controlling environmental stress responses. Funct Plant Biol 2009; 36: 696-720.

[27]   Jander G. Gene identification and cloning by molecular marker mapping. Methods Mol Biol (Clifton, NJ) 2006; 323: 115-126.

[28]   Cho RJ *et al.* Genome-wide mapping with biallelic markers in *Arabidopsis thaliana*. Nat Genet 1999; 23: 203-207.

[29]   Peters JL, Cnudde F, Gerats T. Forward genetics and map-based cloning approaches. Trends Plant Sci 2003; 8: 484-491.

[30]   Feltus FA, Singh HP, Lohithaswa HC, Schulze SR, Silva TD, Paterson AH. A Comparative genomics strategy for targeted discovery of single-nucleotide polymorphisms and conserved-noncoding sequences in orphan crops. Plant Physiol 2006; 140: 1183-1191.

[31]   Borevitz JO, Liang D, Plouffe D, Chang HS, Zhu T, Weigel D, Berry CC, Winzeler E, Chory J. Large-scale identification of single-feature polymorphisms in complex genomes. Genome Res 2003; 13: 513-523.

[32]   Salathia N, Lee HN, Sangster TA, Morneau K, Landry CR, Schellenberg K, Behere AS, Gunderson KL, Cavalieri D, Jander G, Queitsch C. Indel arrays: an affordable alternative for genotyping. Plant J 2007; 51: 727-737.

[33]   Ríos G, Naranjo MA, Iglesias DJ, Ruiz-Rivero O, Geraud M, Usach A, Talon M (2008) Characterization of hemizygous deletions in citrus using array-comparative genomic hybridization and microsynteny comparisons with the poplar genome. BMC Genomics 2008; 9: 381.

[34]   Jansen RC, Nap JP: Genetical genomics: the added value from segregation. Trends Genet 2001; 17: 388-391.

[35]   Hansen BG, Halkier BA, Kliebenstein DJ. Identifying the molecular basis of QTLs: eQTLs add a new dimension. Trends Plant Sci 2007; 13: 72-77.

[36]   Rockman MV, Kruglyak L. Genetics of global gene expression. Nat Rev Genet 2006; 7: 862-872.

[37]   Flint-Garcia SA, Thornsberry JM, Buckler IV ES. Structure of linkage disequilibrium in plants. Ann Rev Plant Biol 2003; 54: 357-374.

[38]   Gupta PK, Rustgi S, Kulwal PL. Linkage disequilibrium and association studies in higher plants: present status and future prospects 2005; Plant Mol Biol 57: 461-485.

[39]   Thornsberry JM, Goodman MM, Doebley J, Kresovich S, Nielsen D, Buckler ES. Dwarf8 polymorphisms associate with variation in flowering time. Nat Genet 2001; 28: 286-289.

[40]   Aranzana MJ *et al.* Genome-wide association mapping in *Arabidopsis* identifies previously known flowering time and pathogen resistance genes. PLos Genet 2005; 1: e60.

[41]   Gonzalez-Martinez SC, Huber D, Ersoz E, Davis JM, Neale DB. Association genetics in *Pinus taeda* L. II. Carbon isotope discrimination. Heredity 2008; 101: 19-26.

[42]   Eckert AJ, Bower AD, Wegrzyn JL, Pande B, Jermstad KD, Krutovsky KV, Clair JBS, Neale DB. Asssociation genetics of coastal Douglas Fir (*Pseudotsuga menziesu* var. menziesii, Pinaceae). I. Cold-hardiness related traits. Genetics 2009; 182: 1289-1302.

[43]   Pathan MS, Subudhi PK, Courtois B, Nguyen HT. Molecular dissection of abiotic stress tolerance in sorghum and rice: A case study. In: Nguyen HT, Blum A, Eds. Physiology and Biotechnology Integration for Plant Breeding, Marcel Dekker, Inc., New York. 2004; pp 525-569.

[44]   Collins NC, Tardieu F, Tuberosa R. Quantitative trait loci and crop performance under abiotic stress: where do we stand? Plant Physiol 2008a; 147: 469-486.

[45]   Dwivedi S, Upadhyaya H, Subudhi P, Gehring C, Bajic V, Ortiz R. Enhancing abiotic stress tolerance.in cereals through breeding and transgenic interventions. In: Janick J. Ed, Plant Breeding Reviews 2010; 33: 31-114.

[46]   Munns R, Tester M. Mechanisms of salinity tolerance. Annual Rev Plant Biol 2008; 59: 651–681.

[47]   Ren ZH, Gao JP, Li LG, Cai XL, Huang W, Chao DY, Zhu MZ, Wang ZY,Luan S, Lin HX. A rice quantitative trait locus for salt tolerance encodes a sodium transporter. Nat Genet 2005; 37: 1141-1146.

[48]   Lindsay MP, Lagudah ES, Hare RA, Munns R. A locus for sodium exclusion (Nax1), a trait for salt tolerance, mapped in durum wheat. Funct Plant Biol 2004; 31: 1105-1114.

[49]   James RA, Davenport RJ, Munns R. Physiological characterization of two genes for Na$^+$ exclusion in durum wheat, *Nax1* and *Nax2*. Plant Physiol 2006; 142: 1537-1547.

[50]   Huang S, Spielmeyer W, Lagudah ES, James RA, Platten JD, Dennis ES, Munns R A sodium transporter (*HKT7*) is a candidate for *Nax1*, a gene for salt tolerance in durum wheat. Plant Physiol 2006; 142: 1718-1727.

[51]   Byrt CS, Platten JD, Spielmeyer W, James RA, Lagudah ES, Dennis ES, Tester M, Munns R. HKT1;5-like cation transporters linked to Na$^+$ exclusion loci in wheat, *Nax2* and *Kna1*. Plant Physiol 2007; 143: 1918-1928.

[52]   Subudhi PK, Sasaki T, Khush GS. Rice. In: Kole CR, Ed. Genome Mapping and Molecular Breeding in Plants, Springer-Verlag GMBH, Tiergartenstr. 17, 69121 Heidelberg, Germany, 2006; pp 1-78.

[53]   Steele KA, Price AH, Shashidhar HE, Witcombe JR. Marker-assisted selection to introgress rice QTLs controlling root traits into an Indian upland rice variety. Theor Appl Genet 2006; 112: 208-221.

[54]   Steele KA, Virk DS, Kumar R, Prasad SC, Witcombe JR. Field evaluation of upland rice lines selected for QTLs controlling root traits. Field Crops Res 2007; 101: 180-186.

[55]   Subudhi PK, Rosenow DT and Nguyen HT. Quantitative trait loci for the stay-green trait in sorghum (*Sorghum bicolor* L. Moench): consistency across genetic backgrounds and environments. Theor Appl Genet 2000; 101: 733-741.

[56]   Harris K, Subudhi PK, Borrell A, Jordan D, Rosenow D, Nguyen H, Klein P, Klein R, Mullet J. Sorghum stay-green QTL individually reduce post-flowering drought-induced leaf senescence. J Exp Bot 2007; 58: 327-338

[57]   Xu K, Xu X, Fukao T, Canlas P, Maghirang-Rodriguez R, Heuer S, Ismail AM, Bailey-Serres J, Ronald PC, Mackill DJ. *Sub1A* is an ethylene-response-factor-like gene that confers submergence tolerance to rice. Nature 2006; 442: 705-708.

[58]   Neeraja CN, Maghirang-Rodriguez R, Pamplona A, Heuer S, Collard BCY, Septiningsih EM, Vergara G, Sanchez D, Xu K, Ismail AM, Mackill DJ. A marker-assisted backcross approach for developing submergence-tolerant rice cultivars. Theor Appl Genet 2007; 115: 767-776.

[59]   Li ZK, Fu BY, Gao YM, Xu JL, Ali J, Lafitte HR, Jiang YZ, Rey JD, Vijayakumar CH, Maghirang R, Zheng TQ, Zhu LH. Genome-wide introgression lines and their use in genetic and molecular dissection of complex phenotypes in rice (*Oryza sativa* L.). Plant Mol Biol 2005; 59: 33-52.

[60]   Miller AK, Galiba G, Dubcovsky J. A cluster of 11 CBF transcription factors is located at the frost tolerance locus Fr-Am2 in *Triticum monococcum*. Mol Genet Genomics 2006; 275: 193-203

[61]   Baga M, Chodaparambil SV, Limin AE, Pecar M, Fowler DB, Chibbar RN. Identification of quantitative trait loci and associated candidate genes for low-temperature tolerance in cold-hardy winter wheat. Funct Integr Genomics 2007; 7: 53-68.

[62]   Knox AK, Li C, Vagujfalvi A, Galiba G, Stockinger EJ, Dubcovsky J. Identification of candidate CBF genes for the frost tolerance locus *Fr-Am2* in *Triticum monococcum*. Plant Mol Biol 2008; 67: 257-270.

[63]   Tondelli A, Francia E, Barabaschi D, Aprile A, Skinner JS, Stockinger EJ, Stanca AM, Pecchioni N. Mapping regulatory genes as candidates for cold and drought stress tolerance in barley. Theor Appl Genet 2006; 112: 445-454.

[64]   Francia E, Barabaschi D, Tondelli A, Laido` G, Rizza F, Stanca AM, Busconi M, Fogher C, Stockinger EJ, Pecchioni N. Fine mapping of a HvCBF gene cluster at the frost resistance locus Fr-H2 in barley. Theor Appl Genet 2007; 115: 1083-1091.

[65]   Hund A, Frascaroli E, Leipner J, Jompuk C, Stamp P, Fracheboud Y. Cold tolerance of the photosynthetic apparatus: pleiotropic relationship between photosynthetic performance and specific leaf area of maize seedlings. Mol Breed 2005; 16: 321-331.

[66]   Jompuk C, Fracheboud Y, Stamp P, Leipner J. Mapping of quantitative trait loci associated with chilling tolerance in maize (*Zea mays* L.) seedlings grown under field conditions. J Exp Bot 2005; 56: 1153–1163.

[67]   Presterl T, Ouzunova M, Schmidt W, Moller EM, Rober FK, Knaak C, Ernst K, Westhoff P, Geiger HH. Quantitative trait loci for early plant vigor of maize grown in chilly environments. Theor Appl Genet 2007; 114: 1059-1070.

[68]   Knoll J, Ejeta G. Marker-assisted selection for early-season cold tolerance in sorghum: QTL validation across populations and environments. Theor Appl Genet 2008; 116: 541-553.

[69]   Andaya VC, Tai TH. Fine mapping of the qCTS12 locus, a major QTL for seedling cold tolerance in rice. Theor Appl Genet 2006; 113: 467-475.

[70]   Kuroki M, Saito K, Matsuba S, Yokogami N, Shimizu H, Ando I, Sato Y. A quantitative trait locus for cold tolerance at the booting stage on rice chromosome 8. Theor Appl Genet 2007; 115: 593-600.

[71]   Lou QJ, Chen L, Sun ZX, Xing YZ, Li J, Xu XY, Mei HW, Luo LJ. A major QTL associated with cold tolerance at seedling stage in rice (*Oryza sativa* L.). Euphytica 2007; 158: 87-94.

[72]   Sasaki T, Yamamoto Y, Ezaki B, Katsuhara M, Ahn SJ, Ryan PR, Delhaize E, Matsumoto H. A wheat gene encoding an aluminum-activated malate transporter. Plant J 2004; 37: 645-653.

[73]   Furukawa J, Yamaji N, Wang H, Mitani N, Murata Y, Sato K, Katsuhara M, Takeda K, Ma JF. An aluminum-activated citrate transporter in barley. Plant Cell Physiol 2007; 48: 1081-1091.

[74]   Magalhaes JV, Liu J, Guimaraes CT, Lana UGP, Alves VMC, Wang YH, Schaffert RE, Hoekenga OA, Pineros MA, Shaff JE, Klein PE, Carneiro NP, Coelho CM, Trick HN, Kochian LV. A gene in the multidrug and toxic compound extrusion (MATE) family confers aluminum tolerance in sorghum. Nat Genet 2007; 39: 1156-1161.

[75]   Collins NC, Shirley NJ, Saeed M, Pallotta M, Gustafson JP. An ALMT1 Gene cluster controlling aluminum tolerance at the *Alt4* locus of rye (*Secale cereale* L.) Genetics 2008b; 179: 669-682.

[76]   Sutton T, Baumann U, Hayes J, Collins NC, Shi BJ, Schnurbusch T, Hay A, Mayo G, Pallotta M, Tester M, Langridge P. Boron-toxicity tolerance in barley arising from efflux transporter amplification. Science 2007; 318: 1446-1449.

[77]   Hieter P, Bogushi M. Functional genomics: it's all how you read it. Science 1997; 278: 601-602.

[78]   Kempin SA, Liljegren SJ, Block LM, Rounsley SD, Yanofsky MF, Lam E. Targeted disruption in *Arabidopsis*. Nature 1997; 389: 802-803.

[79]   Beetham PR, Kipp PB, Sawycky XL, Arntzen CJ, May GD. A tool for functional plant genomics: Chimeric RNA/DNA oligonucleotides cause *in vivo* gene-specific mutations Proc Natl Acad Sci USA 1999; 96: 8774-8778.

[80]   Recombination in the plant genome and its application in biotechnology. Crit Rev Plant Sci 1999; 18: 1-31.

[81]   Terada R, Johzuka-Hisatomi Y, Saitoh M, Asao H, Iida S. Gene targeting by homologous recombination as a biotechnological tool for rice functional genomics. Plant Physiol 2007; 144: 846-856

[82]   Hanin M, Paszkowski J. Plant genome modification by homologous recombination. Curr Opin Plant Biol 2003; 6: 157-162.

[83]   Shaked H, Melamed-Bessudo C, Levy AA. High-frequency gene targeting in *Arabidopsis* plants expressing the yeast *RAD54* gene. Proc Natl Acad Sci USA 2005; 102: 12265-12269.

[84]   Martienssen RA. Functional genomics: probing plant gene function and expression with transposons. Proc Natl Acad Sci USA 1998; 95: 2021-2026.

[85]   Azpiroz-Leehan R, Feldman KA. T-DNA insertion mutagenesis in Arabidopsis: going back and forth. Trends Genet 1997; 13: 152-156.

[86]   McCallum CM, Comai L, Greene E, Henikoff S. Targeting induced local lesions in genomes (TILLING) for plant functional genomics. Plant Physiol 2000; 123: 439-442.

[87]   Till BJ, Zerr T, Comai L, Henikoff S. A protocol for TILLING and Ecotilling in plants and animals. Nat Prot 2006; 1: 2465-2477.

[88]   Henikoff S, Till BJ, Comai L. TILLING: Traditional mutagenesis meets functional genomics. Plant Physiol 2004; 135: 630-636.

[89]   Springer PS. Gene traps: tools for plant development and genomics. Plant Cell 2000; 12: 1007-1020.

[90]   Weigel D, Ahn JH, Blázquez MA, Borevitz JO, Christensen SK, Fankhauser C, Ferrándiz C, Kardailsky I, Malancharuvil EJ, Neff MM, Nguyen JT, Sato S, Wang ZY, Xia Y, Dixon RA, Harrison MJ, Lamb CJ, Yanofsky MF, Chory J. Activation tagging in *Arabidopsis*. Plant Physiol 2000; 122: 1003-1014.

[91]   Parinov S, Sevugan M, Ye D, Yang WC, Kumaran M, Sundaresan V. Analysis of flanking sequences from dissociation insertion lines: a database for reverse genetics in Arabidopsis. Plant Cell 1999; 11: 2263-2270.

[92]   Tissier AF, Marillonnet S, Klimyuk V, Patel K, Torres MA, Murphy G, Jones JD. Multiple independent defective suppressor-mutator transposon insertions in Arabidopsis: a tool for functional genomics. Plant Cell 1999; 11: 1841-1852.

[93]   Bouché N, Bouchez D. Arabidopsis gene knockout: phenotypes wanted. Curr Opin Plant Biol 2001; 4: 111-117

[94]   Meins F. RNA degradation and models for post-transcriptional gene-silencing. Plant Mol Biol 2000; 43: 261-273

[95]   Baulcombe DC. Fast forward genetics based on virus induced gene silencing. Curr Opin Plant Biol 1999; 2: 109-113

[96]   Waterhouse PM, Graham MW, Wang MB. Virus resistance and gene silencing in plants can be induced by simultaneous expression of sense and antisense RNA. Proc Natl Acad Sci USA 1998 95: 13959-13964.

[97]   Chuang CF, Meyerowitz EM. Specific and heritable genetic interference by double-stranded RNA in *Arabidopsis thaliana*. Proc Natl Acad Sci USA 2000; 97: 4985-4990

[98]   Ogita S, Uefuji H, Yamaguchi Y, Koizumi N, Sano H. Producing decaffeinated coffee plants. Nature 2003; 423: 823.

[99]   Zhu T, Mettenburg K, Peterson DJ, Tagliani L, Baszczynski CL. Engineering herbicide-resistant maize using chimeric RNA/DNA oligonucleotides. Nat Biotech 2000; 18: 555-558.

[100]   Krishnan A, Guiderdoni E, An G, Hsing YC, Han C, Lee MC, Yu S, Upadhyaya N, Ramachandran S, Zhang Q, Sundaresan V, Hirochika H, Leung H, Pereira A. Mutant resources in rice for functional genomics of the grasses[W] Plant Physiol 2009; 149: 165-170.

[101]   Alonso JM, Ecker JR. Moving forward in reverse: genetic technologies to enable genome-wide phenomic screens in *Arabidopsis*. Nat Rev Genet 2006; 7: 524–536.

[102]   Ülker B, Peiter E, Dixon DP, Moffat C, Capper R, Bouché N, Edwards R, Sanders D, Knight H, Knight MR. Getting the most out of publicly available T-DNA insertion lines. Plant J 2008; 56: 665–677.

[103]   Droc G, Ruiz M, Larmande P, Pereira A, Piffanelli P, Morel JB, Dievart A,Courtois B, Guiderdoni E, Périn C. OryGenes DB: a database for rice reverse genetics. Nucl Acids Res 2006; 34: D736-D740.

[104]   Droc G, Perin C, Fromentin S, Larmande P. OryGenesDB 2008 update: database interoperability for functional genomics of rice. Nucl Acids Res 2009; 37: D992-D995.

[105]   Larmande P, Gay C, Lorieux M, Périn C, Bouniol M, Droc G, Sallaud C, Perez P, Barnola I, Biderre-Petit C, Martin J, Morel JB, Johnson AA, Bourgis F, Ghesquière A, Ruiz M, Courtois B, Guiderdoni E. Oryza Tag Line, a phenotypic mutant database for the genoplante rice insertion line library. Nucl Acids Res 2008; 36: D1022-D1027.

[106]   de Lorenzo L, Merchan F, Laporte P, Thompson R, Clarke J, Sousa C, Crespi M. A novel plant leucine-rich repeat receptor kinase regulates the response of *Medicago truncatula* roots to salt stress. Plant Cell 2009; 21: 668-680.

[107]   Wong HL, Sakamoto T, Kawasaki T, Umemura K, Shimamoto K. Down-regulation of metallothionein, a reactive oxygen scavenger, by the small GTPase *OsRac1* in rice. Plant Physiol 2004; 135: 1447-1456.

[108]   Kurusu T, Yagala T, Miyao A, Hirochika H, Kuchitsu K. Identification of a putative voltage-gated Ca^{2+} channel as a key regulator of elicitor-induced hypersensitive cell death and mitogen-activated protein kinase activation in rice. Plant J 2005; 42: 798-809.

[109]   Lee S, Kim SH, Kim SJ, Lee K, Han SK. Trapping and characterization of cold-responsive genes from T-DNA tagging lines in rice. Plant Sci 2004; 166: 69-79.

[110] Koh S, Lee SC, Kim MK, Koh JH, Lee S, An G, Choe S, Kim SR. T-DNA tagged knockout mutation of rice *OsGSK1*, an orthologue of *Arabidopsis BIN2*, with enhanced tolerance to various abiotic stresses. Plant Mol Biol 2007; 65: 453-466.

[111] Donson J, Fang Y, Espiritu-Santo G, Xing W, Salazar A, Miyamoto S, Armendarez V, Volkmuth W. Comprehensive gene expression analysis by transcript profiling. Plant Mol Biol 2002; 48: 75-97.

[112] Meyers BC, Vu TH, Tej SS, Ghazal H, Matvienko M, Agrawal V, Ning JC, Haudenschild CD. Analysis of the transcriptional complexity of *Arabidopsis thaliana* by massively parallel signature sequencing. Nat Biotechnol 2004; 22: 1006-1011.

[113] Velculescu VE, Zhang L, Vogelstein B, Kinzler KW. Serial analysis of gene expression. Science 1995; 270: 448-487.

[114] Brenner S *et al.* Gene expression analysis by massively parallel signature sequencing (MPSS) on microbead arrays. Nat Biotechnol 2000; 18: 630-634

[115] Liang P, Pardee AB. Differential display of eukaryotic messenger RNA by means of the polymerase chain reaction. Science 1992; 257: 967-971

[116] Bachem CW, Hoeven RS van der, Bruijn SM de, Vreugdenhil D, Zabeau M, Visser RG. Visualization of differential gene expression using a novel method of RNA fingerprinting based on AFLP: analysis of gene expression during potato tuber development. Plant J 1996; 9: 745-753.

[117] Habu Y, Fukada-Tanaka S, Hisatomi Y, Iida S. Amplified restriction fragment length polymorphism-based mRNA fingerprinting using a single restriction enzyme that recognizes a 4-bp sequence. Biochem Biophy Res Commun 1997; 234: 516-521.

[118] Cho Y, Meade JD, Walden JC, Chen X, Guo Z, Liang P. Multicolor fluorescent differential display. Biotechniques 2001; 30: 562-573.

[119] Breyne P, Zabeau M. Genome-wide expression analysis of plant cell cycle modulated genes. Curr Opin Plant Biol 2001; 4: 136-142.

[120] Prashar Y, Weissman SM. Analysis of differential gene expression by display of 3' end restriction fragments of cDNAs. Proc Natl Acad Sci USA 1996; 93: 659-663.

[121] Shimkets RA, Lowe DG, Tai JT, Sehl P, Jin H, Yang R, Predki PF, Rothberg BE, Murtha MT, Roth ME, Shenoy SG, Windemuth A, Simpson JW, Simons JF, Daley MP, Gold SA, McKenna MP, Hillan K, Went GT, Rothberg JM. Gene expression analysis by transcript profiling coupled to a gene database query. Nat Biotechnol 1999; 17: 798-803.

[122] Sutcliffe JG, Foye PE, Erlander MG, Hilbush BS, Bodzin LJ, Durham JT, Hasel KW. TOGA: an automated parsing technology for analyzing expression of nearly all genes. Proc Natl Acad Sci USA 2000; 97: 1976-1981.

[123] Schena M, Shalon D, Davis RW, Brown PO. Quantitative monitoring of gene expression patterns with a complementary DNA microarray. Science 1995; 270: 467-470.

[124] Lockhart DJ, Dong H, Byrne MC, Follettie MT, Gallo MV, Chee MS, Mittmann M, Wang C, Kobayashi M, Horton H, Brown EL. Expression monitoring by hybridization to high-density oligonucleotide arrays. Nat Biotech 1996; 14: 1675-1680.

[125] Girke T, Todd J, Ruuska S, White J, Benning C, Ohlrogge J. Microarray analysis of developing Arabidopsis seeds. Plant Physiol 2000; 124: 1570-1581

[126] Hertzberg M, Sievertzon M, Aspeborg H, Nilsson P, Sandberg G, Lundeberg J. cDNA microarray analysis of small plant tissue samples using a cDNA tag target amplification protocol. Plant J 2001; 25: 585-591.

[127] Brazma A, Hingamp P, Quackenbush J, Sherlock G, Spellman P, Stoeckert C, Aach J, Ansorge W, Ball CA, Causton HC, Gaasterland T, Glenisson P, Holstege FCP, Kim IF, Markowitz V, Matese JC, Parkinson H, Robinson A, Sarkans U, Schulze-Kremer S, Stewart J, Taylor R, Vilo J, Vingron M. Minimum information about a microarray experiment (MIAME)-toward standards for microarray data. Nat Genet 2001; 29: 365-371.

[128] van de Peppel J, Kemmeren P, van Bakel H, Radonjic M, van Leenen D, Holstege FCP. Monitoring global messenger RNA changes in externally controlled microarray experiments. EMBO Rep 2003; 4: 387-393.

[129] Robinson PN, Wollstein A, Bohme U, Beattie B. Ontologizing gene-expression microarray data: characterizing clusters with Gene Ontology. Bioinformatics 2004; 20: 979-981.

[130] Schena M, Heller RA, Theriault TP, Konrad K, Lachenmeier E, Davis RW. Microarrays: biotechnology's discovery platform for functional genomics. Trends Biotechnol 1998; 16: 301-306.

[131] Devaux F, Marc P, Jacq C. Transcriptomes, transcription activators and microarrays. FEBS Lett 2001; 498: 140-144.

[132] Rensink WA, Buell CR. Microarray expression profiling resources for plant genomics. Trends Plant Sci 2005; 10: 603-609.

[133] Gregory BD, Yazaki J, Ecker JR. Utilizing tiling microarrays for whole-genome analysis in plants. Plant J 2008; 53: 636-644.

[134] Hughes TR, Marton MJ, Jones AR, Roberts CJ, Stoughton R, Armour CD, Bennett HA, Coffey E, Dai HY, He YDD, Kidd MJ, King AM, Meyer MR, Slade D, Lum PY, Stepaniants SB, Shoemaker DD, Gachotte D, Chakraburtty K, Simon J, Bard M, Friend SH. Functional discovery via a compendium of expression profiles. Cell 2000; 102: 109-126.

[135] Vij S, Tyagi AK. Emerging trends in the functional genomics of the abiotic stress response in crop plants. Plant Biotech J 2007; 5: 361-380.

[136] Gorantla M, Babu PR., Lachagari VBR, Reddy AMM, Wusirika R, Bennetzen, JL, Reddy AR. Identification of stress-responsive genes in an *indica* rice (*Oryza sativa* L.) using ESTs generated from drought-stressed seedlings. J Exp Bot 2007; 58: 253-265.

[137] Houde M, Belcaid M, Ouellet F, Danyluk J, Monroy AF, Dryanova A, Gulick P, Bergeron A, Laroche A, Links MG, MacCarthy L, Crosby WL, Sarhan F. Wheat EST resources for functional genomics of abiotic stress. BMC Genomics 2006; 7: 149.

[138] Ramalingam J, Pathan MS, Miftahudin Feril O, Ross K, Ma XF, Mahmoud AA, Layton J, Rodriguez-Milia MA, Chikmawati T, Valliyodan B, Skinner R, Matthews DE, Gustafson JP, Nguyen HT. Structural and functional analyses of the wheat genomes based on expressed sequence tags (ESTs) related to abiotic stresses. Genome 2006; 49: 1324-1340.

[139] Ergen NZ, Budak H. Sequencing over 13 000 expressed sequence tags from six subtractive cDNA libraries of wild and modern wheats following slow drought stress. Plant Cell Environ 2009; 32: 220-236.

[140] Chao S, Lazo GR, You F, Crossman CC, Hummel DD, Lui N, Laudencia-Chingcuanco D, Anderson JA, Close TJ, Dubcovsky J, Gill BS, Gill KS, Gustafson JP, Kianian SF, Lapitan NLV, Nguyen HT, Sorrells ME, McGuire PE, Qualset CO, Anderson OD. Use of a large-scale Triticeae expressed sequence tag resource to reveal gene expression profiles in hexaploid wheat (*Triticum aestivum* L.). Genome 2006; 49: 531-544.

[141] Ji W, Li Y, Li J, Dai CH, Wang X, Bai X, Cai H, Yang L, Zhu YM. Generation and analysis of expressed sequence tags from NaCl-treated *Glycine soja*. BMC Plant Biol 2006; 6: 4.

[142] Mishra RN, Reddy PS, Nair S, Markandeya G, Reddy AR, Sopory SK, Reddy MK. Isolation and characterization of expressed sequence tags (ESTs) from subtracted cDNA libraries of *Pennisetum glaucum* seedlings. Plant Mol Biol 2007; 64: 713-732.

[143] Baisakh N, Subudhi PK, Varadwaj P. Primary responses to salt stress in a halophyte, smooth cordgrass (*Spartina alterniflora* Loisel.). Funct Integr Genom 2008; 8: 287-300.

[144] Inan G *et al.* Salt Cress: a halophyte and cryophyte Arabidopsis relative model system and its applicability to molecular genetic analyses of growth and development of extremophiles. Plant Physiol 2004; 135: 1718-1737.

[145] Wong CE, Li Y, Whitty BR, Diaz-Camino C, Akhter SR, Brandle JE, Golding GB, Weretilnyk EA, Moffatt BA, Griffith M. Expressed sequence tags from the Yukon ecotype of *Thellungiella* reveal that gene expression in response to cold, drought and salinity shows little overlap. Plant Mol Biol 2005; 58: 561-574.

[146] Mehta PA, Sivaprakash K, Parani M, Venkataraman G, Parida AK. Generation and analysis of expressed sequence tags from the salt-tolerant mangrove species *Avicennia marina* (Forsk) Vierh. Theor Appl Genet 2005; 110: 416-424.

[147] Iturriaga G, Cushman MAF, Cushman JC. An EST catalogue from the resurrection plant *Selaginella lepidophylla* reveals abiotic stress-adaptive genes. Plant Sci 2006; 170: 1173-1184.

[148] Matsumura H, Nirasawa S, Terauchi R. Transcript profiling in rice (*Oryza sativa* L.) seedlings using serial analysis of gene expression (SAGE). Plant J 1999; 20: 719-726.

[149] Molina C, Rotter B, Horres R, Udupa SM, Besser B, Bellarmino L, Baum M, Matsumura H, Terauchi R, Kahl G, Winter P. SuperSAGE: the drought stress-responsive transcriptome of chickpea roots. BMC Genomics 2008; 9: Art. No. 553.

[150] Eom EM, Lee JY, Park HS, Byun YJ, Ha-Lee YM, Lee DH. Comparison between SAGE and cDNA microarray for quantitative accuracy in transcript profiling analyses. J Plant Biol 2006; 49: 498-506.

[151] Nakano M, Nobuta K, Vemaraju K, Tej SS, Skogen JW, Meyers BC. Plant MPSS databases: signature-based transcriptional resources for analyses of mRNA and small RNA. Nucl Acids Res 2006; 34: D731-D735.

[152] Kawasaki S, Borchert C, Deyholos M, Wang H, Brazille S, Kawai K, Galbraith D, Bohnert HJ. Gene expression profiles during the initial phase of salt stress in rice. Plant Cell 2001; 13: 889-905.

[153] Rabbani MA, Maruyama K, Abe H, Khan MA, Katsura K, Ito Y, Yoshiwara K, Seki M, Shinozaki K, Yamaguchi-Shinozaki K. Monitoring expression profiles of rice genes under cold, drought, and high-salinity stresses and abscisic acid application using cDNA microarray and RNA gel-blot analyses. Plant Physiol 2003; 133: 1755-1767.

[154] Chao DY, Luo YH, Shi M, Luo D, Lin HX. Salt-responsive genes in rice revealed by cDNA microarray analysis. Cell Res 2005; 15: 796-810.

[155] Ueda A, Kathiresan A, Bennett J, Takabe T. Comparative transcriptome analyses of barley and rice under salt stress. Theor Appl Genet 2006; 112: 1286-1294.

[156] Cooper B, Clarke JD, Budworth P, Kreps J, Hutchison D, Park S, Guimil S, Dunn M, Luginbuhl P, Ellero C, Goff SA, Glazebrook J. A network of rice genes associated with stress response and seed development. Proc Natl Acad Sci USA 2003; 100: 4945-4950.

[157] Walia H, Wilson C, Condamine P, Liu X, Ismail AM, Zeng L, Wanamaker SI, Mandal J, Xu J, Cui X, Close TJ. Comparative transcriptional profiling of two contrasting rice genotypes under salinity stress during the vegetative growth stage. Plant Physiol 2005; 139: 822-835.

[158] Walia H, Wilson C, Zeng LH, Ismail AM, Condamine P, Close TJ. Genome-wide transcriptional analysis of salinity stressed japonica and indica rice genotypes during panicle initiation stage. Plant Mol Biol 2007; 63: 609-623.

[159] Hazen SP, Pathan MS, Sanchez A, Baxter I, Dunn M, Estes B, Chang HS, Zhu T, Kreps JA, Nguyen HT. Expression profiling of rice segregating for drought tolerance QTLs using a rice genome array. Funct Integr Genomics 2005; 5: 104-116.

[160] Degenkolbe T, Do PT, Zuther E, Repsilber D, Walther D, Hincha DK, Kohl KI. Expression profiling of rice cultivars differing in their tolerance to long-term drought stress. Plant Mol Biol 2009; 69: 133-153.

[161] Lan L, Li M, Lai Y, Xu W, Kong Z, Ying K, Han B, Xue Y. Microarray analysis reveals similarities and variations in genetic programs controlling pollination/fertilization and stress responses in rice (*Oryza sativa* L.). Plant Mol Biol 2005; 59: 151-164.

[162] Lian XM, Wang SP, Zhang JW, Feng Q, Zhang LD, Fan DL, Li XH, Yuan DJ, Han B, Zhang QF. Expression profiles of 10,422 genes at early stage of low nitrogen stress in rice assayed using a cDNA microarray. Plant Mol Biol 2006; 60: 617-631.

[163] Aprile A, Mastrangelo AM, De Leonardis AM, Galiba G, Roncaglia E, Ferrari F, De Bellis L, Turchi L,Giuliano G, Cattivelli L. Transcriptional profiling in response to terminal drought stress reveals differential responses along the wheat genome. BMC Genomics 2009; 10: Art. No. 279.

[164] Mohammadi M, Kav NNV, Deyholos MK. Transcriptional profiling of hexaploid wheat (*Triticum aestivum* L.) roots identifies novel, dehydration-responsive genes. Plant Cell Environ 2007; 30: 630-645.

[165] Kawaura K, Mochida K, Ogihara Y. Genome-wide analysis for identification of salt-responsive genes in common wheat. Funct Integr Genomics 2008; 8: 277-286.

[166] Trzcinska-Danielewicz J, Bilska A, Fronk J, Zielenkiewicz P, Jarochowska E, Roszczyk M, Jonczyk M, Axentowicz E, Skoneczny M, Sowinski P. Global analysis of gene expression in maize leaves treated with low temperature I. Moderate chilling (14 degrees C). Plant Sci 2009; 177: 648-658.

[167] Ozturk ZN, Talame V, Deyholos M, Michalowski CB, Galbraith DW, Gozukirmizi N, Tuberosa R, Bohnert HJ. Monitoring large-scale changes in transcript abundance in drought- and salt-stressed barley. Plant Mol Biol 2002; 48: 551-573.

[168] Buchanan CD, Lim S, Salzman RA, Kagiampakis I, Morishige DT. *Sorghum bicolor*'s transcriptome response to dehydration, high salinity and ABA. Plant Mol Biol 2005; 58: 699-720.

[169] Legay S, Lamoureux D, Hausman JF, Hoffmann L, Evers D. Monitoring gene expression of potato under salinity using cDNA microarrays. Plant Cell Rep 2009; 28: 1799-1816.

[170] Lee SC, Lim MH, Kim JA, Lee SI, Kim JS, Jin M, Kwon SJ, Mun JH, Kim YK, Kim HU, Hur Y, Park BS. Transcriptome analysis in *Brassica rapa* under the abiotic stresses using *Brassica* 24k oligo microarray. Mol Cells 2008; 26: 595-605.

[171] O'Rourke JA, Charlson DV, Gonzalez DO, Vodkin LO, Graham MA, Cianzio SR, Grusak MA, Shoemaker RC. Microarray analysis of iron deficiency chlorosis in near-isogenic soybean lines. BMC Genomics 2007; 8: Art. No. 476.

[172] Chandran D, Sharopova N, Ivashuta S, Gantt JS, VandenBosch KA, Samac DA. Transcriptome profiling identified novel genes associated with aluminum toxicity, resistance and tolerance in *Medicago truncatula*. Planta 2008; 228: 151-166.

[173] Buitink J, Leger JJ, Guisle I, Vu BL, Wuilleme S, Lamirault G, Le Bars A, Le Meur N, Becker A, Kuester H, Leprince O. Transcriptome profiling uncovers metabolic and regulatory processes occurring during the transition from desiccation-sensitive to desiccation-tolerant stages in *Medicago truncatula* seeds. Plant J 2006; 47: 735-750.

[174] Mantri NL, Ford R, Coram TE, Pang ECK. Transcriptional profiling of chickpea genes differentially regulated in response to high-salinity, cold and drought. BMC Genomics 2007; 8: Art. No. 303.

[175] Taji T, Seki M, Satou M, Sakurai T, Kobayashi M, Ishiyama K, Narusaka Y, Narusaka M, Zhu JK, Shinozaki K. Comparative genomics in salt tolerance between *Arabidopsis* and *Arabidopsis*-related halophyte salt cress using *Arabidopsis* microarray. Plant Physiol 2004; 135: 1697-1709.

[176] Wong CE, Li Y, Labbe A, Guevara D, Nuin P, Whitty B, Diaz C, Golding GB, Gray GR, Weretilnyk EA, Griffith M, Moffatt BA. Transcriptional profiling implicates novel interactions between abiotic stress and hormonal responses in *Thellungiella*, a close relative of *Arabidopsis*. Plant Physiol 2006; 140: 1437-1450.

[177] Yazaki J, Kishimoto N, Nagata Y, Ishikawa M, Fujii F, Hashimoto A, Shimbo K, Shimatani Z, Kojima K, Suzuki K, Yamamoto M, Honda S, Endo A, Yoshida Y, Sato Y, Takeuchi K, Toyoshima K, Miyamoto C, Wu J, Sasaki T, Sakata K, Yamamoto K, Iba K, Oda T, Otomo Y, Murakami K, Matsubara K, Kawai J, Carninci P, Hayashizaki Y, Kikuchi S. Genomics approach to abscisic acid-and gibberellins responsive genes in rice. DNA Res 2003; 10: 249-261.

[178] Yamaguchi T, Nakayama K, Hayashi T, Yazaki J, Kishimoto N, Kikuchi S, Koike S. cDNA microarray analysis of rice anther genes under chilling stress at the microsporogenesis stage revealed two genes with DNA transposon Castaway in the 5'-flanking region. Biosci Biotech Biochem 2004; 68: 1315-1323.

[179] Chao DY, Luo YH, Shi M, Luo D, Lin HX. Salt-responsive genes in rice revealed by cDNA microarray analysis. Cell Res 2005; 15: 796-810.

[180] Gulick PJ, Drouin S, Yu Z, Danyluk J, Poisson G, Monroy AF, Sarhan F. Transcriptome comparison of winter and spring wheat responding to low temperature. Genome 2005; 48: 913-923.

[181] Yu LX, Setter TL. Comparative transcriptional profiling of placenta and endosperm in developing maize kernels in response to water deficit. Plant Physiol 2003; 131: 568-582.

[182] Negishi T, Nakanishi H, Yazaki J, Kishimoto N, Fujii F, Shimbo K, Yamamoto K, Sakata K, Sasaki T, Kikuchi S, Mori S. Nishizawa NK. cDNA microarray analysis of gene expression during Fe-deficiency stress in barley suggests that polar transport of vesicles is implicated in phytosiderophore secretion in Fe deficient barley roots. Plant J 2002; 30: 83-94.

[183] Rensink WA, Iobst S, Hart A, Stegalkina S, Liu J, Buell CR. Gene expression profiling of potato responses to cold, heat, and salt stress. Funct Integr Genomics 2005; 5: 201-207.

[184] Hwang EW, Kim KA, Park SC, Jeong MJ, Byun MO. Kwon HB. Expression profiles of hot pepper (*Capsicum annum*) genes under cold stress conditions. J Biosci 2005; 30: 657-667.

[185] Brosche M, Vinocur B, Alatalo ER, Lamminmaki A, Teichmann T, Ottow EA, Djilianov D, Afif D, Bogeat-Triboulot MB, Altman A, Polle A, Dreyer E, Rudd S, Paulin L, Auvinen P, Kangasjarvi J. Gene expression and metabolite profiling of *Populus euphratica* growing in the Negev desert. Genome Biol 2005; 6: R101.

[186] Wilkins MR, Pasquali C, Appel RD, Ou K, Golaz O, Sanchez JC, Yan JX, Gooley AA, Hughes G, Humphery-Smith I, Williams KL, Hochstrasser DF. From proteins to proteomes: large scale protein identification by two-dimensional electrophoresis and amino acid analysis. Biotechnology 1996; 14: 61-65.

[187] Pandey A, Mann M. Proteomics to study genes and genomes. Nature 2000; 405: 837-846.

[188] Tyers M, Mann M. From genomics to proteomics. Nature 2003; 422: 193-197.

[189] O'Farrell PH. High-resolution 2-dimensional electrophoresis of proteins. J Biol Chem 1975; 250: 4007-4021.

[190] Yates JR III. Mass spectrometry and the age of the proteome. J Mass Spectrom 1998; 33: 1-19.

[191] Aebersold R, Mann M. Mass spectrometry-based proteomics. Nature 2003; 422: 198-207.

[192] Thiellement H, Bahrman N, Damerval C, Plomion C, Rossignol M, Santoni V, deVienne D, Zivy M. Proteomics for genetic and physiological studies in plants. Electrophoresis 1999; 20: 2013-2026.

[193] van Wijk KJ. Challenges and prospects of plant proteomics. Plant Physiol 2001; 126: 501-508.

[194] Roberts JK. Proteomics and a future generation of plant molecular biologists. Plant Mol Biol 2002; 48: 143-154.

[195] Heazlewood JL, Miller AH. Integrated plant proteomics-Putting the green genomes to work. Funct Plant Biol 2003; 30: 471-482.

[196] Canovas FM, Dumas-Gaudot E, Recorbet G, Jorrin J, Mock HP, Rossignol M. Plant proteome analysis. Proteomics 2004; 4: 285-298.

[197] Khan PSSV, Hoffmann L, Renaut J, Hausman JF. Current initiatives in proteomics for the analysis of plant salt tolerance. Curr Sci 2007; 93: 807-817.

[198] Zivy M, deVienne D. Proteomics: a link between genomics, genetics and physiology. Plant Mol Biol 2000; 44: 575-580.

[199] de Vienne D, Leonardi A, Damerval C, Zivy M. Genetics of proteome variation for QTL characterization: application to drought-stress responses in maize. J Exp Bot 1999; 50: 303-309.

[200] Patterson SD. How much of the proteome do we see with discovery-based proteomic methods and how much do we need to see? Curr Proteomics 2004; 1: 3-12.

[201] Herbert BR, Harry JL, Packer NH, Godey AA, Pederson SK, Williams KL. What place for polyacrylamide in proteomics? Trends Biotechnol 2001; 19: S3-S9.

[202] Unlu M, Morgan ME, Minden JS. Difference gel electrophoresis: A single gel method for detecting changes in protein extracts. Electrophoresis 1997; 18: 2071-2077.

[203] Bayer E, Bottrill AR, Walshaw J, Vigouroux M, Naldrett MJ, Thomas CL, Mule AJ. *Arabidopsis* cell wall proteome defined using multidimensional protein identification technology. Proteomics 2005; 6: 301-311.

[204] Fields S, Song O. A novel genetic system to detect protein–protein interactions. Nature 1989; 340: 245-246.

[205] MacBeath G. Protein microarrays and proteomics. Nat Genet 2002; 32(Suppl.): 526-532.

[206] Taylor NL, Heazlewood JL, Day DA, Millar AH Differential impact of environmental stresses on the pea mitochondrial proteome. Mol Cell Proteomics 2005; 4: 1122–1133.

[207] Renaut J, Hausman JF, Wisniewski ME. Proteomics and low temperature studies: Bridging the gap between gene expression and metabolism. Physiol Plant 2006; 126: 97-109.

[208] Imin N, Kerim T, Rolfe BG, Weinman JJ. Effect of early cold stress on the maturation of rice anthers. Proteomics 2004; 4: 1873-1882.

[209] Cui S, Huang F, Wang J, Ma X, Cheng Y, Liu J. A proteomic analysis of cold stress responses in rice seedlings. Proteomics 2005; 5: 3162-3172.

[210] Salekdeh GH, Siopongco J, Wade LJ, Ghareyazie B, Bennett J. Proteomic analysis of rice leaves during drought stress and recovery. Proteomics 2002; 2: 1131-1145.

[211] Parker R, Flowers TJ, Moore AL, Harpham NVJ. An accurate and reproducible method for proteome profiling of the effects of salt stress in the rice leaf lamina. J Exp Bot 2006; 57: 1109-1118.

[212] Chitteti BR, Peng ZH. Proteome and phosphoproteome differential expression under salinity stress in rice (*Oryza sativa*) roots. J Proteome Res 2007; 6: 1718-1727.

[213] Yan S, Tang Z, Su W, Sun W. Proteomic analysis of salt stress-responsive proteins in rice roots. Proteomics 2005; 5: 235-244.

[214] Zhang L, Tian LH, Zhao JF, Song Y, Zhang CJ, Guo Y. Identification of an apoplastic protein involved in the initial phase of salt stress response in rice root by two-dimensional electrophoresis. Plant Physiol 2009; 149: 916-928.

[215] Wang MC, Peng ZY, Li CL, Li F, Liu C, Xia GM. Proteomic analysis on a high salt tolerance introgression strain of *Triticum aestivum/Thinopyrum ponticum*. Proteomics 2008; 8: 1470-1489.

[216] Caruso G, Cavaliere C, Guarino C, Gubbiotti R, Foglia P, Lagana A. Identification of changes in *Triticum durum* L. leaf proteome in response to salt stress by two-dimensional electrophoresis and MALDI-TOF mass spectrometry. Anal Bioanal Chem 2008; 391: 381-390.

[217] Witzel K, Weidner A, Surabhi GK, Borner A, Mock HP. Salt stress-induced alterations in the root proteome of barley genotypes with contrasting response towards salinity. J Exp Bot 2009; 60: 3545-3557.

[218] Ouerghi Z, Remy R, Ouelhazi L, Ayadi A, Brulfert J. Two-dimensional electrophoresis of soluble leaf proteins isolated from two wheat species (*Triticum durum and Triticum aestivum*) differing in sensitivity towards NaCl. Electrophoresis 2000; 21: 2487-2491.

[219] Majoul T, Chahed K, Zamiti E, Ouelhazi L, Ghrir R. Analysis by two-dimensional electrophoresis of the effect of salt stress on the polypeptide patterns in roots of a salt-tolerant and salt-sensitive cultivar of wheat. Electrophoresis 2000; 21: 2562-2565.

[220] Abbasi F, Komatsu S. A proteomic approach to analyze salt responsive proteins in rice leaf sheath. Proteomics 2004; 4: 2072-2081.

[221] Qureshi MI, Qadir S, Zolla L. Proteomics-based dissection of stress-responsive pathways in plants. J Plant Physiol 2007; 164: 1239-1260.

[222] Peck SC. Update on proteomics in *Arabidopsis*, where do we go from here? Plant Physiol 2005; 138: 591-599.

[223] Oliver SG, Winson MK, Kell DB, Baganz F. 1998. Systematic functional analysis of the yeast genome. Trends Biotech 1998; 16: 373-378.

[224] Fiehn O. Metabolomics-the link between genotype and phenotype. Plant Mol Biol 2002; 48: 155-171.

[225] Koch K. Sucrose metabolism: regulatory mechanisms and pivotal roles in sugar sensing and plant development. Curr Opin Plant Biol 2004; 7: 235-246.

[226] Rontein D, Dieuaide-Noubhani M, Dufourc EJ, Raymond P, Rolin D. The metabolic architecture of plant cells: stability of central metabolism and flexibility of anabolic pathways during the growth cycle of tomato cells. J Biol Chem 2002; 277: 43948-43960.

[227] Wink M. Plant breeding: importance of plant secondary metabolites for protection against pathogens and herbivores. Theor Appl Genet 1988; 75: 225-233.

[228] Mitchell-Olds T, Pedersen D. The molecular basis of quantitative genetic variation in central and secondary metabolism in *Arabidopsis*. Genetics 1998; 149: 739-747.

[229] Mitchell-Olds T, Schmitt J. Genetic mechanisms and evolutionary significance of natural variation in *Arabidopsis*. Nature 2006; 441: 947-952.

[230] Roessner U, Bowne J. What is metabolomics all about? BioTechniques 2009; 46: 363-365.

[231] Roessner U, Beckles DM. Metabolite measurements. In: Schwender J, Ed. Plant Metabolic Networks. Springer, NY. 2009; pp39-69.

[232] Fiehn O, Kopka J, Trethewaey RN, Willmitzer L. Identification of uncommon plant metabolites based on calculation of elemental compositions using gas chromatography and quadrupole mass spectrometry. Anal Chem 2000; 72: 3573-3580.

[233] Schauer N, Zamier D, Fernie AR. Metabolic profiling of leaves and fruit of wild species tomato: A survey of the *Solanum lycopersicum* complex. J Exp Bot 2005; 56: 297-307.

[234] Tikunov Y, Lommen A, deVos CH, Verhoeven HA, Bino RJ, Hall RD, Bovy AG. A novel approach for nontargeted data analysis for metabolomics. Large-scale profiling of tomato fruit volatiles. Plant Physiol 2005; 139: 1125-1137.

[235] Moco S, Bino RJ, Vorst O, Verhoeven HA, de Groot J, van Beek TA, Vervoort J, de Vos CH. A liquid chromatography-mass spectrometry-based metabolome database for tomato. Plant Physiol 2006; 141: 1205-1218.

[236] Ward JL, Baker JM, Beale MH. Recent applications of NMR spectroscopy in plant metabolomics. FEBS J 2007; 274: 1126-1131.

[237] Oksman-Caldentey KM, Inzé D. Plant cell factories in the post-genomic era: new ways to produce designer secondary metabolites. Trends Plant Sci 2004; 9: 433-440.

[238] Fernie AR, Trethewey RN, Krotzky AJ, Willmitzer L. Metabolite profiling: from diagnostics to systems biology. Nat Rev Mol Cell Biol 2004; 5: 763-769.

[239] Hall RD. Plant metabolomics: from holistic hope, to hype, to hot topic. New Phytol 2006; 169: 453-468.

[240] Last RL, Jones AD, Shachar-Hill Y. Towards the plant metabolome and beyond. Nature Rev Mol Cell Biol 2007; 8: 167-174.

[241] Sumner LW, Mendes P, Dixon RA. Plant metabolomics: large-scale phytochemistry in the functional genomics era. Phytochem 2003; 62: 817-836.

[242] Keurentjes JJB. Genetical metabolomics: closing in on phenotypes. Curr Opin Plant Biol 2009; 12: 223-230.

[243] Schauer N, Fernie AR. Plant metabolomics: towards biological function and mechanism. Trends Plant Sci 2006; 11: 508-516.

[244] Jacobs A, Lunde C, Bacic A, Tester M, Roessner U. The impact of constitutive expression of a moss Na^+ transporter on the metabolomes of rice and barley. Metabolomics 2007; 3: 307-317.

[245] Schauer N, Steinhauser D, Strelkov S, Schomburg D, Allison G, Moritz T, Lundgren K, Roessner-Tunali U, Forbes MG, Willmitzer L, Fernie AR, Kopka J. GC-MS libraries for the rapid identification of metabolites in complex biological samples. FEBS Lett 2005; 579: 1332-1337.

[246] Schauer N, Semel Y, Roessner U, Gur A, Balbo I, Carrari F, Pleban T, Perez-Melis A, Bruedigam C, Kopka J, Willmitzer L, Zamir D, Fernie AR. Comprehensive metabolic profiling and phenotyping of interspecific introgression lines for tomato improvement. Nature Biotechnol 2006; 24: 447-454.

[247] Schauer N, Semel Y, Balbo I, Steinfath M, Repsilber D, Selbig J, Pleban T, Zamir D, Fernie AR. Mode of inheritance of primary metabolic traits in tomato. Plant Cell 2008; 20: 509-523.

[248] Semel Y, Schauer N, Roessner U, Zamier D, Fernie AR. Metabolite analysis for the comparison of irrigated and non-irrigated field grown tomato of varying genotype. Metabolomics 2007; 3: 289-295.

[249] Morsy MR, Jouve L, Hausman JF, Hoffmann L, Stewart JM. Alteration of oxidative and carbohydrate metabolism under abiotic stress in two rice (*Oryza sativa* L.) genotypes contrasting in chilling tolerance. J Plant Physiol 2007; 164: 157-167

[250] Sanchez DH, Siahpoosh MR, Roessner U, Udvardi M, Kopka J. Plant metabolomics reveals conserved and divergent metabolic responses to salinity. Physiol Planta 2008; 132: 209-219.

[251] Shulaeva V, Cortesa D, Miller G, Mittler R. Metabolomics for plant stress response. Physiol Planta 2008; 132: 199–208.

[252] Kitano H. Systems biology: a brief overview. Science 2002; 295: 1662-1664.

[253] Faccioli P, Stanca AM, Morcia C, Terzi V. From DNA sequence to plant phenotype: bioinformatics meets crop science. Current Bioinform 2009; 4: 173-176.

[254] Joung JG, Corbett AM, Fellman SM, Tieman DM, Klee HJ, Giovannoni JJ, Fei ZJ. Plant MetGenMAP: An integrative analysis system for plant systems biology. Plant Physiol 2009; 151: 1758-1768.

[255] Jansen RC. Studying complex biological systems using multifactorial perturbation. Nat Rev Genet 2003; 4: 145-151.

[256] Rowe HC, Hansen BG, Halkier BA, Kliebenstein DJ. Biochemical networks and epistasis shape the *Arabidopsis thaliana* metabolome. Plant Cell 2008; 20: 1199-1216.

[257] Keurentjes JJB, Fu J, Terpstra IR, Garcia JM, Ackerveken G, Snoek LB, Peeters AJM, Vreugdenhil D, Koornneef M, Jansen RC. Regulatory network construction in *Arabidopsis* by using genome-wide gene expression quantitative trait loci. Proc Natl Acad Sci USA 2007; 104: 1708-1713.

[258] Keurentjes JJB, Fu J, de Vos CH, Lommen A, Hall RD, Bino RJ, vander Plas LHW, Jansen RC, Vreugdenhil D, Koornneef M. The genetics of plant metabolism. Nat Genet 2006; 38: 842-849.

[259] Saito K. Plant functional genomics based on integration of metabolomics and transcriptomics: toward plant systems biology. In: Nakanishi S, Kageyama R, Watanabe D. Eds, Systems Biology, Springer Japan 2009; pp 135-142.

[260] Moore JP, Le NT, Brandt WF, Driouich A, Farrant JM. Towards a systems-based understanding of plant desiccation tolerance. Trends Plant Sci 2009; 14: 110-117.

[261] Cho K, Shibato J, Agrawal GK, Jung YH, Kubo A, Jwa NS, Tamogami S, Satoh K, Kikuchi S, Higashi T, Kimura S, Saji H, Tanaka Y, Iwahashi H, Masuo Y, Rakwal R. Integrated transcriptomics, proteomics, and metabolomics analyses to survey ozone responses in the leaves of rice seedling. J Proteome Res 2008; 7: 2980-2998.

[262] Takahashi H, Hotta Y, Hayashi M, Kawai-Yamada M, Komatsu S, Uchimiya H. High throughput metabolome and proteome analysis of transgenic rice plants (*Oryza sativa* L.). Plant Biotechnol 2005; 22: 47-50.

[263] The Multinational Arabidopsis Steering Committee. The multinational coordinated *Arabidopsis thaliana* functional genomics project. Annual Report 2007.

[264] Keurentjes JJB, Koornneef M, Vreugdenhil D. Quantitative genetics in the age of omics. Curr Opin Plant Biol 2008; 11: 123-128.

[265] Holtorf H, Guitton MC, Reski R. Plant functional genomics. Naturwissenschaften 2002; 89: 235-249.

[266] Edwards D, Batley J. Plant bioinformatics: from genome to phenome. Trends Biotechnol 2004; 22: 232-237.

[267] Baum M, Von Korff M, Guo P, Lakew B, Udupa SM, Sayed H, Choumane W, Grando S, Ceccarelli S. Molecular approaches and breeding strategies for drought tolerance in barley. In: Varshney RK, Tuberosa R, Eds. Genomics-assisted crop improvement, Springer, Dordrecht, The Netherlands, 2007; pp 51-80.

[268] Xu Y, Crouch JH. Marker-assisted selection in plant breeding: from publications to practice. Crop Sci 2008; 48: 391-407

[269] Meyer RC, Steinfath M, Lisec J, Becher M, Witucka-Wall H, Torjek O, Fiehn O, Eckardt A, Willmitzer L, Selbig J, Altmann T. The metabolic signature related to high plant growth rate in *Arabidopsis thaliana*. Proc Natl Acad Sci USA 2007; 104: 4759-4764.

[270] Sergeeva LI, Keurentjes JJ, Bentsink L, Vonk J, van der Plas LH, Koornneef M, Vreugdenhil D. Vacuolar invertase regulates elongation of *Arabidopsis thaliana* roots as revealed by QTL and mutant analysis. Proc Natl Acad Sci USA 2006; 103: 2994-2999.

[271] Fernie AR, Schauer N. Metabolomics-assisted breeding: a viable option for crop improvement? Trends Genet 2009; 25: 39-48.

[272] Flowers TJ, Flowers SA. Why does salinity pose such a difficult problem for plant breeders? Agril Water Manage 2005; 78: 15-24.

CHAPTER 3

Plant Responses to Abiotic Stresses: Shedding Light on Salt, Drought, Cold and Heavy Metal Stress

Narendra Tuteja*, Sarvajeet Singh Gill and Renu Tuteja

International Centre for Genetic Engineering and Biotechnology (ICGEB), Aruna Asaf Ali Marg, New Delhi - 110 067, India

Abstract: Agriculture using genetically modified crops is emerging as an effective measure to counteract the negative impact of abiotic stresses on crop production. Abiotic stresses mainly salt, drought, cold and heavy metals are the major cause of crop failure which restrict crops to reach their full genetic potential. Salt, drought and heavy metals exert their negative impact essentially by disrupting the ionic and osmotic equilibrium of the cell, whereas, cold causes mechanical constraint to the membrane. Plants respond to abiotic stresses through multifaceted molecular signaling pathways. Therefore, understanding of molecular signaling pathways and identification of key molecules and their specific roles is important for crop improvement. Several genes responsible for abiotic stress tolerance have been identified which code for antioxidants, enzymes that modify lipids in the cell membrane, stress-response transcription factors, proteins that maintain ion homeostasis, heat shock proteins, or enzymes that synthesize important stress-response compounds. Transgenic plants having some of these genes have been produced and found to be abiotic stress tolerant. Present chapter reviews the plant responses to abiotic stresses such as salinity, drought, cold and heavy metal stresses and tolerance mechanisms through omics approaches.

INTRODUCTION

Increasing crop production is now the highest agricultural priority worldwide because of huge population. United Nations Food and Agriculture Organization (FAO) reported that more than one billion people now suffer malnutrition [1]. According to the United Nation's World Population Prospects report, the world population is currently increasing at an alarming rate of approximately seventy four million people per year and expected to reach more than 9 billion near 2050 (http://www.un.org/esa/population/unpop.htm), whereas, global food productivity is declining due to the negative effect of various environmental stress factors. It is estimated that global warming decreased the yield of major crops like maize, wheat and barley by ~40 million metric tons per year between 1981 and 2002 (http://environmentalresearch web.org/cws/article/news/27343). Therefore, minimizing the losses in crop productivity is a matter of concern for all nations to feed the several billion people living on this planet [2]. It is well documented that among abiotic stresses extreme temperatures (freezing, cold, heat), water availability (drought, flooding), and ion toxicity (salinity, heavy metals) are the major causes which adversely affect the plant growth and productivity worldwide [3,4,5,6]. It has been estimated that the relative decreases in potential maximum yields associated with abiotic stress factors vary between 54-82% [7]. Therefore, it is clear that there is an urgent need to increase abiotic stress tolerance in plants. Recently, in 2008, Nina Fedoroff, Science and Technology adviser to the US Secretary of State, emphasized the acute need for a "Second Green Revolution". Climate change and the decreased availability of fertile land will create a problem for future crop production. In fact, these stresses threaten the sustainability of agricultural industry. The challenge now is to produce additional food under stress conditions and in less soil. Therefore, it is now necessary to obtain stress-tolerant varieties to cope with this upcoming problem of food security.

First of all it is important to understand the notion of stress. In agriculture, the stress factors are a menace for plants and prevent them from reaching their full genetic potential and thus limit the crop productivity worldwide [3]. It is estimated from the comparison of record yields and average yields for various crop plants that crops mainly attain only 20% of their genetic potential for yield due to various biotic and abiotic stress factors. Plants respond to various

---

*Address correspondence to: **Dr. Narendra Tuteja,** International Centre for Genetic Engineering and Biotechnology (ICGEB), Aruna Asaf Ali Marg, New Delhi - 110 067, India; E-mail: narendra@icgeb.res.in

stress factors such as salinity, heat, cold, drought, excess water, heavy metals toxicity, wounding, excess light, nutrient loss, anaerobic conditions and radiations through multifaceted molecular signaling pathways. Plants have adapted to respond to these stresses at the molecular, cellular, physiological and biochemical level, enabling them to survive [8]. In general, the stress signal is first perceived by the receptors present on the membrane of the plant cells. Following this the signal information is transduced downstream resulting in the activation of various stress responsive genes. The products of these stress genes ultimately lead to stress tolerance response or plant adaptation and help the plant to survive and surpass unfavorable conditions [3,9]. The response could also result in growth inhibition or cell death, which will depend upon how many and which kinds of genes are up- or down-regulated in response to the stress. The various stress responsive genes can be broadly categorized as early and late induced genes. Early genes are induced within minutes of stress signal perception, which include various transcription factors. Late genes include the major stress responsive genes such as RD (responsive to dehydration)/KIN (cold induced)/COR (cold responsive), which encodes and modulate the proteins needed for synthesis, for example LEA-like proteins, antioxidants, membrane stabilizing proteins and osmolytes [2]. Overall, the stress response is a coordinated action of many genes encoding signaling proteins/factors, including protein modifiers (methylation, ubiquitination, glycosylation, etc.), adaptors and scaffolds [3,4]. Understanding of molecular signaling pathways and identification of key molecules and their specific roles may provide a treasure trove of opportunity for molecular breeding approaches to increase the efficiency of crop plants under stressful conditions without yield penalty. Therefore, efforts should be taken to reduce hunger and promote growth and development of crop plants.

In the present review we tried to focus on plant response to abiotic stress like cold, salt drought and heavy metal stresses and their toxic effects on plants. In this regard, the response of various genes involved in acclimation to various stresses has been discussed. Following salt, drought, cold and heavy metal stresses, the role of calcium, SOS pathway, osmolytes and antioxidants has also been taken into consideration.

GENERIC PATHWAY FOR PLANT RESPONSE TO ABIOTIC STRESSES

Plants are sessile organisms therefore, cannot avoid adverse environmental conditions such as salinity, drought, cold and heavy metal contamination. To sense these environmental signals, higher plants have evolved a complex signaling network, which may also cross talk. A generic pathway in response to salt, drought and cold stresses is depicted in Fig. **1**. Stress signal transduction pathways starts with signal perception by receptors (phytochromes, histidine kinase, receptor-like kinase, G-protein coupled receptors [GPCR], hormone receptors, etc). The heterotrimeric G-proteins mediate the coupling of signal transduction from activated GPCR to appropriate downstream effectors and thereby play an important role in signaling [10]. After this the signals enter the side of the cell and activate the mechanism, which helps to generate secondary signaling molecules (inositol phosphatase, reactive oxygen species and abscisic acid, etc). These secondary molecules can modulate the intracellular Ca^{2+} level by receptor mediated Ca^{2+} release or it can bypass Ca^{2+} in early signaling steps and initiate a protein phosphorylation cascade (protein phosphatase, MAPK, CDPK, SOS3/PKS etc.), which activates specific stress-responsive genes for cellular protection through transcription control (MYC/MTB, CBF/DREB) [3,4,11,12]. Salinity and drought exert their influence on a cell mainly by disrupting the ionic and osmotic equilibrium resulting in a stress condition [3]. Thus, excess of Na^+ ions and osmotic changes in the form of turgor pressure are the initial triggers of this pathway. This leads to a cascade of events, which can be grouped under ionic and osmotic signaling pathways, the outcome of which is ionic and osmotic homeostasis, leading to stress tolerance. These stresses are marked by symptoms of stress injury including chlorosis and necrosis and may also exert negative influence on cell division resulting in growth retardation of plant [3]. Reduction in shoot growth, especially leaves, is beneficial for plant as it reduces the surface area exposed for transpiration hence minimizing water loss. Plants may also sacrifice or shed their older leaves, which is another adaptation in response to drought. Stress injury may occur through denaturation of cellular proteins/enzymes or through the production of ROS, Na^+ toxicity and disruption of membrane integrity. In response to injury stress plants trigger a detoxification process, which may include change in the expression of LEA/dehydrin type gene synthesis of molecular chaperones, proteinases, enzymes for scavenging ROS and other detoxification proteins. This process functions in the control and repair of stress induced damage and results in stress tolerance. Cold stress mainly results in disruption of membrane integrity leading to severe cellular dehydration and osmotic imbalance. Cold acclimation results in the triggering of various genes, which result in a restructuring of the cellular membranes by change in the lipid composition and generation of osmolytes, which prevent cellular dehydration and therefore leading to stress tolerance (Fig. **1**).

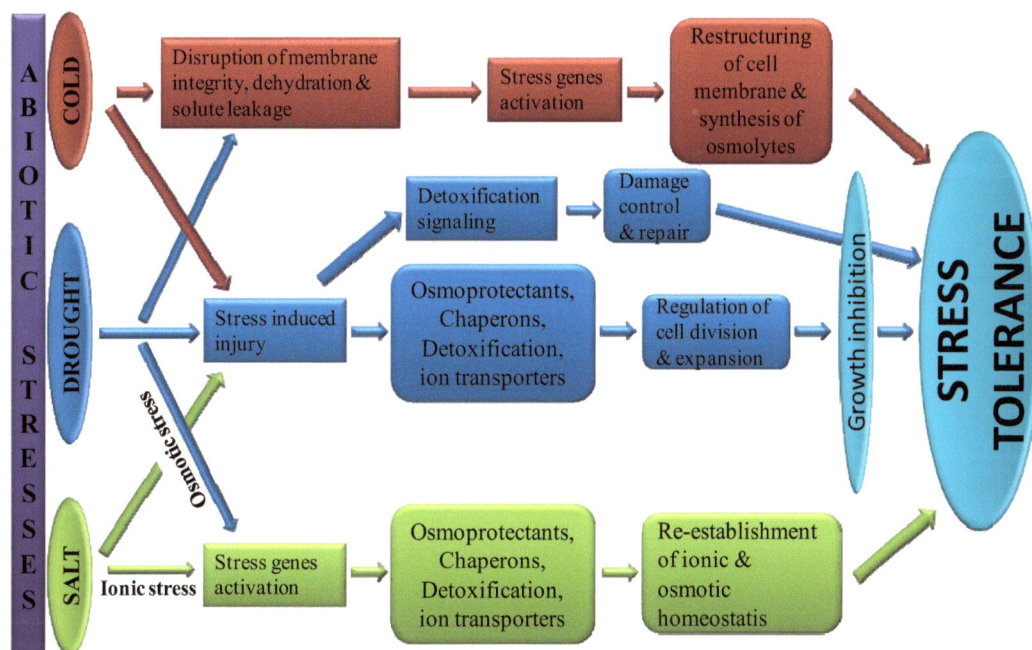

Figure 1: Generic pathway under salt, drought and cold stress.

Plants suffer from dehydration or osmotic stress under drought, salinity and also under low-temperature conditions which cause reduced availability of water for cellular function and maintenance of cellular turgor pressure. Stomata closure, reduced supply of CO_2 and eventual slowing down of the rate of biochemical reactions [3]. Prolonged periods of dehydration lead to high production of reactive oxygen species (ROS) in the chloroplasts causing irreversible cellular damage and photo inhibition. Overall, in response to all these stresses several stress responsive genes are upregulated whose products can directly or indirectly help the plant through stress tolerance. Understanding the molecular mechanism for abiotic stress tolerance is still a major challenge in biology. Many chemicals are also critical for plant growth and development and play an important role in integrating various stress signals and controlling downstream stress responses by modulating gene expression machinery and regulating various transporters/pumps and biochemical reactions. Some of the chemicals include calcium (Ca^{2+}), cyclic nucleotides, polyphosphoinositides, nitric oxide (NO), sugars, abscisic acid (ABA), jasmonates (JA), salicylic acid (SA) and polyamines [7]. Microarray technology employing cDNAs or oligonucleotides is a powerful tool for analysing gene expression profiles of plants exposed to abiotic stresses. A 7000 full-length cDNA microarray was utilized to identify 299 drought-inducible genes, 54 cold-inducible genes, 213 high salinity-inducible genes, and 245 ABA-inducible genes in Arabidopsis [13,14]. More than half of these drought-inducible genes were also induced by high salinity and/or ABA treatments, implicating significant cross-talk between the drought, high salinity, and ABA response pathways. Recently, Shinozaki and Yamaguchi-Shinozaki [15] have beautifully summarized recent progress in the gene networks involved in drought stress response and tolerance. By using transgenic technology, Bhatnagar-Mathur *et al.* [16] have also described the recent progress in the improvement of abiotic stress tolerance in plants, which includes discussion on the evaluation of abiotic stress response and the protocols for testing the transgenic plants for their tolerance under close-to-field conditions. Ca^{2+}-dependent protein kinases (CDPKs) sense the Ca^{2+} concentration changes in plant cells and play important roles in signaling pathways for disease resistance and various stress responses as indicated by emerging evidence. Among the 20 wheat CDPK genes studied, 10 were found to respond to drought, salinity and ABA treatments [17].

SALT STRESS AND ITS ADVERSE EFFECTS ON PLANTS

It is important, first of all, to understand the notion of salt stress. Soils having high concentration of soluble salts (EC_e is ≥ 4 dS m^{-1}, which is equivalent to ~40 mM NaCl and generates an osmotic pressure of approximately 0.2 MPa) are referred as saline soils and their effect results in yield loss of crop plants. It has been reported that over 800 million hectares of land throughout the world are salt-affected, either by salinity (397 million ha) or by sodicity (434 million ha) [18]. Saline soils significantly limit crop production and consequently have negative effects on food security. Sodic soils or alkaline soils contain high concentrations of free carbonate and bicarbonate and excess of sodium. The pH of this soil is greater than 8.5 and sometimes up to 10.7. Saline soils are dominated by sodium cations and usually soluble chloride and sulphate anions and the pH values of these soils are much lower than in sodic soils. The reasons of soil salinity are as follows:

- Improper irrigation which includes insufficient water application and drainage, less efficient irrigation, over irrigation or previous exposure of the land to sea water (sea water contains approximately 3% of NaCl and in terms of molarity of different ions, Na^+ is about 460 mM, Mg^{2+} is 50 mM and Cl^- around 540 mM along with smaller quantities of other ions)

- High evaporation

- Soil compaction and poor drainage

- Use of improper cropping patterns and rotations and poor land levelling

- Chemical contamination and

- Excessive leaching during reclamation techniques on land with insufficient drainage.

According to FAO [18], world is losing at least 3 ha of arable land every minute because of soil salinity. Many crop species are very sensitive to soil salinity and they are known as glycophytes whereas the salt-tolerant plants are known as halophytes. In general glycophytes cannot grow at 100 mM NaCl, whereas the halophytes can grow at salt concentration over 250 mM NaCl. These plants require salt concentrations (250 mM NaCl) for optimal growth and show retarded growth without such salt concentration but this adaptation is still unknown. It has been reported that crop plants exhibit different sensitivity to salt stress. Among cereals, rice is the most sensitive crop to salinity stress, whereas, barley referred as most tolerant. Furthermore, bread wheat showed moderate tolerance, whereas, durum wheat is less so. Interestingly, tall wheatgrass, a halophytic relative of wheat, is one of the most tolerant of the monocotyledonous species and its growth proceeds at concentrations of salt as high as in seawater. The salinity sensitive plants restrict the uptake of salt and strive to maintain an osmotic equilibrium by the synthesis of compatible solutes such as proline, glycine betaine and sugars. The salinity tolerant plants have the capacity to sequester and accumulate salt into the cell vacuoles thus preventing the build up of salt in the cytosol and maintaining a high cytosolic K^+/Na^+ ratio in their cells. Generally, the salinity tolerance is inversely related to the extent of Na^+ accumulation in the shoot [4]. The basic physiology of high salinity stress and drought stress plants overlaps with each other. High salinity also leads to increased cytosolic Ca^{2+}, which initiates the salt stress signal transduction pathways for stress tolerance.

In Arabidopsis the transcript profile of various genes under salt and other stresses have been made openly available through several databases, such as TAIR, NASC, and Genevestigator [19]. The various genes have been reported to be up-regulated in response to salinity stress, which are listed in Ma *et al.* (2006) [19]. Earlier, we have shown a novel role of a DNA helicase and G-proteins in salinity stress tolerance [20,21]. Recently, it has been shown that overexpression of the trehalose-6-phosphate phosphatase gene *OsTPP1* confers salt, osmotic and abscisic acid tolerance in rice and results in the activation of stress responsive genes [22]. Recently, conservation in mechanism of salt response and stress tolerance has been observed between bryophytes and higher plants [23]. It is well documented that salinity stress cause adverse effects and thus reduce the crop production to a considerable extent. The deleterious effects observed in response to high salinity stress on plants growth are described below:

- Salinity stress interferes with plant growth and development as it can also lead to physiological drought condition and ion toxicity and therefore causes both hyperionic and hyperosmotic stress which lead to plant demise [24, 25].

- High salt deposition in soil leads to deposition of low water potential zone in the soil. This makes it increasingly difficult for the plant to acquire water as well as nutrients.

- High salt also decreases the soil porosity and thereby reduces soil aeration.

- Salinity causes ion-specific stresses resulting in altered K^+/Na^+ ratio. The external Na^+ can negatively impact intracellular K^+ influx.

- The K^+ ions are one of the essential elements required for growth. Alteration in K^+ ions (due to impact of high salinity stress) can disturb the osmotic balance, function of stomata and function of some enzymes.

- Salinity leads to build up of Na^+ and Cl^- concentrations in the cytosol, which can be ultimately detrimental for the cell. The Na^+ can dissipate the membrane potential and therefore facilitates the uptake of Cl^- down the gradient.

- Higher concentrations of sodium ions (above 100 mM) are toxic to cell metabolism and can inhibit the activity of many essential enzymes, cell division and expansion, membrane disorganization and osmotic imbalance [24, 26].

- Higher concentrations of sodium ions can also lead to reduction in photosynthesis and increase in the production of reactive oxygen species and polyamines [27].

- High salinity can also injure cells in transpiring leaves, which lead to growth inhibition. This is the salt-specific or ion-excess effect of salinity, which causes toxic effect of salt inside the plant. The salt can concentrate in the old leaves and the leaves die, which is crucial for the survival of a plant [28].

- High salinity affects the cortical microtubule organization and helical growth in Arabidopsis [29].

Salt Stress and Oxidative Stress

It is evidence that salt stress can induce conditions of oxidative stress such as generation and/or accumulation of reactive oxygen species (ROS), including hydrogen peroxide (H_2O_2), superoxide anion ($O_2{\cdot}^-$) and hydroxyl radicals ($OH\cdot$) [30] (Fig. **3**). ROS typically result from the excitation of O_2 to form singlet oxygen ($^1O_2$) or transfer of one, two or three electrons to O_2 to form $O_2{\cdot}^-$, H_2O_2 or $OH\cdot$ respectively. The enhanced production of ROS during stresses can pose a threat to plants. The unquenched ROS spontaneously react with organic molecules and causes membrane lipid peroxidation, protein oxidation, enzyme inhibition and DNA, RNA damage, etc. [4,31]. Plants possess very efficient enzymatic (superoxide dismutase, SOD; catalase, CAT; ascorbate peroxidase, APX; glutathione reductase, GR; monodehydroascorbate reductase, MDHAR; dehydroascorbate reductase, DHAR; glutathione peroxidase, GPX; guaicol peroxidase, GOPX and glutathione-*S*- transferase, GST) and non-enzymatic (ascorbic acid, ASH; glutathione, GSH; phenolic compounds, alkaloids, non-protein amino acids and α-tocopherols) antioxidant defense systems which work in concert to control the cascades of uncontrolled oxidation and protect plant cells from oxidative damage by scavenging of ROS. The ROS scavengers can increase the plant resistance to salinity stress. ROS also influence the expression of a number of genes and therefore control the many processes like growth, cell cycle, programmed cell death (PCD), abiotic stress responses, pathogen defense, systemic signaling and development. Overexpression of the aldehyde dehydrogenase gene in Arabidopsis has been reported to confer salinity tolerance. The aldehyde dehydrogense catalyzes the oxidation of toxic aldehydes, which accumulate as a result of side reactions of ROS with lipids and proteins. The enhancement of stress tolerance in transgenic tobacco plants has been shown by overexpressing Chlamydomonas glutathion peroxidase in chlorop last or cytosol [31]. Significant increase in SOD activity under salt stress has been observed in various plants *viz.* mulberry [32], *Cicer arietinum* [33] and *Lycopersicon esculentum* [34]. Eyidogan and Oz [35] noted three SOD activity bands in *Cicer arietinum* under salt stress and recognized the higher band as MnSOD and the others as FeSOD and Cu/ZnSOD. Furthermore, significant increase in the activities of Cu/ZnSOD and MnSOD isozymes under salt stress was observed. Pan *et al.* [36] studied the effect of salt and drought stress on *Glycyrrhiza uralensis* Fisch and found significantly increased SOD activity but an additional MnSOD isoenzyme was detected under only salt stress. Transgenic rice plants overexpressing *OsMT1a* demonstrated enhanced tolerance to drought with significantly increased SOD activity [37]. It was found that transgenic wheat protoplasts with Mn-SOD and GR overexpression in chloroplasts showed less oxidative damage, higher H_2O_2 content and significant increase in SOD and GR activities under photooxidative stress [38]. Overexpression of a Mn-SOD in transgenic *Arabidopsis* plants also showed increased salt tolerance [39]. Furthermore, they showed that Mn-SOD activity as well as the activities of Cu/Zn-SOD, Fe-SOD, CAT and POD were significantly higher in transgenic *Arabidopsis* plants than control [39]. Overexpression of Mn-SOD in transformed *Lycopersicon esculentum* plants also showed enhanced tolerance against salt stress [40]. Transgenic *Arabidopsis* plants over-expressing *OsAPXa* or *OsAPXb* exhibited increased salt tolerance. It was found that the overproduction of *OsAPXb* enhanced and maintained APX activity to a much higher extent than *OsAPXa* in transgenic plants under different NaCl concentrations. It was suggested that the rice cytosolic *OsAPXb* gene has a more functional role than *OsAPXa* in the improvement of salt tolerance in transgenic plants

[41]. Overexpression of MDAR in transgenic tobacco increased the tolerance against salt and osmotic stresses [42]. There have been many reports of the production of salt stress tolerant transgenic plants having different antioxidant enzymes (Table **1**).

Table 1: Transgenic plants overexpressing antioxidant enzymes for abiotic stress tolerance

Gene	Source	Target transgenic	Response in transgenic plants	Reference
MnSOD (Superoxide dismutase)	Tamarix androssowii	Populus davidiana x P. bolleana	Salt stress due to increased SOD activity	[171]
Cu/Zn SOD	Avicennia marina	Oryza sativa Pusa Basmati-1	MV, salinity and drought stress tolerance	[172]
MnSOD + FeSOD	Nicotiana Plumbaginifolia and Arabidopsis thaliana	Medicago sativa L.	Moderate drought stress tolerance due to maintained photosynthesis	[173]
Mn SOD + APX	Nicotiana tabacum	Festuca arundinacea Schreb. cv. Kentucky-31	MV, H_2O_2, Cu, Cd and As tolerance due to increased SOD and APX activity	[174]
CAT3 (Catalase)	Brassica juncea	N. tabacum	Cd stress tolerance	[175]
katE	E. coli	N. tabacum 'Xanthi	Salt stress tolerance	[176]
APX1 (Ascorbate peroxidase)	Hordeum vulgare L.	A. thaliana	Salt tolerance due to higher APX, SOD, CAT and GR and less H_2O_2 and MDA content	[177]
swpa4	Ipomoea batatas	N. tabacum	MV, H_2O_2, Salt and drought stress tolerance with many times higher POD specific activity	[178]
cAPX	Pisum sativum	Lycopersicon esculentum cv. Zhongshu No. 5	Heat, drought and chilling stress tolerance due to increased APX activity	[179]
GR (Glutathione reductase)	E. coli	Triticum aestivum cv. Oasis protoplast	Salt stress tolerance with higher GSH content	[180]
GR	A. thaliana ecotype Columbia	Gossypium hirsutum L. cv. Coker 312	Chilling stress tolerance	[181]
DHAR (Dehydro ascorbate reductase)	A. thaliana	N. tabacum	Drought and salt tolerance with higher DHAR activity	[182]
DHAR	O. sativa	A. thaliana L. ecotype Wassilewskija	Salt tolerance due to slight increase in DHAR activity and total ascorbate	[183]
GST (Glutathione-S-transferase)	Suaeda salsa	O. sativa cv. Zhonghua No.11	Salt and paraquat stress tolerance due to GST, CAT and SOD activity	[184]
GST + GPX	N. tabacum	N. tabacum L. cv. Xanthi NN	Increased thermal or salt-stress tolerance due to glutathione and ascorbate content	[185]
GPX (Glutathione peroxidase)	Chlamydomonas	N. tabacum cv. Xanthi	Tolerant to MV under moderate light intensity, chilling stress under high light intensity or salt stress due to low MDA and high photosynthesis and antioxidative system	[186]
GPX-2	Synechocystis PCC 6803	A. thaliana	Tolerance H_2O_2, Fe ions, MV, chilling, high salinity or drought stresses	[187]

Salinity Stress and Signal Transduction

Salinity in soil or water is one of the major abiotic stresses that reduce plant growth and crop productivity worldwide therefore; it is the need of the time to develop the stress tolerant varieties to counteract the inhibitory effects of salt stress and to meet the upcoming problem of food security. Salt tolerance or susceptibility in plants is a coordinated action of various genes including those encoding calcium-binding proteins and other components of stress-signaling pathways. These components may cross talk with each other. A wide range of stimuli like drought, salinity, cold, mechanical perturbation, hormonal signals, symbiotic and pathogenic microorganisms trigger rapid and transient increases in intracellular Ca^{2+} concentration of plant cells that is transported from the apoplast as well as the intracellular compartments [43]. This transient increase in cytosolic Ca^{2+} initiates the stress signal transduction leading to salt adaptation. Ca^{2+} release is primarily from extracellular source (apoplastic space) and also from activation of PLC, leading to hydrolysis of PIP2 to IP3 and subsequent release of Ca^{2+} from intracellular Ca^{2+} stores [3,4]. The Ca^{2+}-binding proteins sense and relay the information downstream to initiate a phosphorylation cascade leading to regulation of gene expression [11]. Wu *et al.* (1996) [44] commenced a mutant screen for Arabidopsis plants, which were over-sensitive to salt stress. As a result of this screen, three genes SOS1, SOS2, and SOS3 (Salt Overlay-Sensitive) were identified. *SOS3* gene (also known as *AtCBL4*) encodes a calcineurin B-like protein (CBL, Ca^{2+} sensor), which is a Ca^{2+} binding protein and senses the change in cytosolic Ca^{2+} concentration and transduces the signal downstream. The SOS pathway results in the exclusion of excess Na^+ ions out of the cell via the plasma membrane Na^+/H^+ antiporter and helps in reinstating cellular ion homeostasis. The discovery of SOS genes paved the way for elucidation of a novel pathway linking the Ca^{2+} signaling in response to a salt stress [45,46]. SOS genes (*SOS1, SOS2 and SOS3*) were genetically confirmed to function in a common pathway of salt tolerance [47].

In the SOS pathway, the salinity stress signal is perceived by an unknown hypothetical plasma membrane sensor resulted in increased cytoplasmic Ca^{2+} perturbations, which is sensed by SOS3 followed by transduction of the signal to the downstream components. The myristoylation motif of SOS3 results in the recruitment of SOS3-SOS2 complex to the plasma membrane, where SOS2 phosphorylates and activates SOS1 [48]. The SOS1 is a Na^+/H^+ antiporter and sos1 mutant was hypersensitive to salt and showed impaired osmotic/ionic balance. The SOS pathway also seems to have other branches, which help to remove the excess of Na^+ ions out of the cell and thereby maintain the cellular ion homeostasis. In Arabidopsis, Na^+ entry into root cells during salt stress appears to be mediated by AtHKTI, a low affinity Na^+ transporter, which block the entry of Na^+ [3, 46]. SOS2 also interacts and activates NHX (vacuolar Na^+/H^+ exchanger) resulting in sequestration of excess Na^+ ions and pushing it into vacuoles and thereby further contributes to Na^+ ion homeostasis. Some other Ca^{2+}-binding proteins like calnexin and calmodulin (CaM) also sense the increased level of Ca^{2+} and can interact and activate the NHX. Over-expression of AtNHX1 antiporter substantially enhanced salt tolerance of Arabidopsis [49]. CAX1 (H^+/Ca^{2+} antiporter) has been identified as an additional target for SOS2 activity reinstating cytosolic Ca^{2+} homeostasis. This reflects that the components of SOS pathway may cross-talk and interact with other branching components to maintain cellular ion homeostasis, which helps in salinity tolerance. So far, the main avenue in breeding crops for salt tolerance has been to reduce Na^+ uptake and transport from roots to shoots. It has been demonstrated that retention of cytosolic K^+ could also be considered as another key factor in conferring salt tolerance in plants. Recently, Zepeda-Jazo *et al.* [50] have shown that the expression of *NORC* was significantly lower in salt-tolerant genotypes. As NORC is capable of mediating K^+ efflux coupled to Na^+ influx, which suggested that the restriction of its activity could be beneficial for plants under salt stress.

ABA is a phytohormone that regulates plant growth and development and also plays an important role in the plant's response to abiotic stresses including salinity stress [3,4,46,51]. The role of ABA in salinity stress was confirmed by a study of Zhu's group where they have shown that ABA-deficient mutants performed poorly under salinity stress [52]. ABA level is known to be induced under stress condition, which is mainly due to the induction of genes for enzymes responsible for ABA biosynthesis. The induction of osmotic stress responsive genes imposed by salinity is transmitted through either ABA-dependent or ABA-independent pathways, though some others are only partially ABA-dependent [53]. However, the components involved in these pathways often cross talk through Ca^{2+} in stress signaling pathways. The transcript accumulation of *RD29A* gene is reported to be regulated in both ABA-dependent and ABA-independent manner [54]. The proline accumulation in plants can be mediated by both ABA-dependent and ABA-independent signaling pathways [46]. The salinity stress-induced upregulation of transcript of pea DNA helicase 45 (PDH45) followed ABA-dependent pathway [20] while calcineurin B-like protein (CBL) and CBL-interacting protein kinase (CIPK) from pea followed the ABA-independent pathway [48, 55]. The role of Ca^{2+} in ABA-dependent induction of *P5CS* gene during salinity stress has been reported [56]. Overall, the ABA-dependent pathways are involved essentially in osmotic stress gene expression.

Transcriptional regulatory network of cis-acting elements and transcription factors involved in ABA and salinity stress responsive gene expression has been described [4]. The ABA-dependent salinity stress signaling activates basic leucine zipper transcription factors called AREB, which binds to ABRE element to induce the stress responsive gene *RD29A*. Transcription factors like DREB2A and DREB2B transactivate the DRE cis-element of osmotic stress genes and thereby are involved in maintaining the osmotic equilibrium of the cell. Some genes such as RD22 lack the typical CRT/DRE elements in their promoter suggesting their regulation by some other mechanism. The MYC/MYB transcription factors, RD22BP1 and AtMYB2, could bind MYCRS and MYBRS elements, respectively, and help in the activation of *RD22* gene [3, 4]. Overall, these transcription factors may also cross-talk with each other for their maximal response to stress tolerance.

Salt Stress Tolerance through Osmotic Adjustment

Salt stress represents major environmental constraints for sessile plants. To cope with these stresses, plants have developed adaptive strategies by expressing specific genes and synthesizing compatible solutes like proline, glycine betaine, trehalose and sugars [57]. Among these, proline and GB (glycinebetaine) are two major osmoprotectant osmolytes, which are synthesized by many plants (but not all) in response to salt stress [31]. In higher plants the amino acid proline is synthesized by glutamic acid by the actions of two enzymes, pyrroline-5-carboxylate synthetase (P5CS) and pyrroline-5-carboxylate reductase (P5CR). It is well documented that following salt, drought and metal stress there is a dramatic accumulation of Pro may be due to increased synthesis or decreased degradation. Free Pro has been proposed to act as an osmoprotectant, a protein stabilizer, a metal chelator, an inhibitor of lipid peroxidation, OH· and $^1O_2$ scavenger [58, 59]. The exogenous application of proline also provided the osmoprotection and facilitated the growth of salinity-stressed plants. Proline can also protect cell membranes from salinity-induced oxidative stress by upregulating activities of various antioxidants [60]. It is reported that the salt stress enhances proline utilization in the apical region of barley roots [61]. The function of proline is thought to be an osmotic regulator under water stress, and its transportation into cells is mediated by a proline transporter. In an interesting study, sorbitol, mannitol, myo-inositol and proline has been tested for OH· scavenging capacity and it has been found that proline appeared as an effective scavenger of OH· [62]. Therefore, proline is not only an important molecule in redox signaling, but also an effective quencher of ROS formed under salt, metal and dehydration stress conditions in all plants, including algae [63]. Enhanced synthesis of Pro under drought or salt stress has been implicated as a mechanism to alleviate cytoplasmic acidosis and maintain $NADP^+$:NADPH at values compatible with metabolism [64]. An additional advantage of the refilling of $NADP^+$ supply by proline synthesis may be to support redox cycling, which is especially important in plant antioxidant defense mechanisms during stress [65]. Recently, it has been noted that salt stress increased the accumulation of Pro in the leaves of two rice cultivars differing in salt tolerance [66]. Su and Wu [67] reported that both constitutive expression and stress-inducible expression of the *P5CS* cDNA in transgenic *O. sativa* have led to the accumulation of *P5CS* mRNA and proline which resulted in higher salt and water deficiency stress tolerance. *Triticum aestivum* plants overexpression of *Vigna aconitifolia* Δ^1-pyrroline-5-carboxylate synthetase (*P5CS*) cDNA under the control of a stress-induced promoter complex-AIPC resulted in enhanced proline accumulation under water deficit. The tolerance to water deficit in transgenic plants was mainly due to protection mechanisms against oxidative stress and not caused by osmotic adjustment. Overexpression of *P5CS* gene in transgenic tobacco resulted in increased production of proline and salinity/drought tolerance [68]. However, recently, Ueda *et al.* (2008) [69] have reported that altered expression of barley proline transporter (HvProT) causes different growth responses in Arabidopsis, as it leads to the reduction in biomass production and decreased proline accumulation in leaves. Impaired growth of *HvProT* transformed plants was restored by exogenously adding proline, which suggested that growth reduction was caused by a deficiency of endogenous proline.

Glycine betaine has been reported to protect higher plants against salt/osmotic stresses not only by maintaining osmotic adjustment but by also protecting the photosystem II (PSII) complex by stabilizing the connection of extrinsic PSII complex proteins in the presence of salt or under extremes of temperature or pH. Many important agronomical crops, such as rice, potato, tomato and tobacco cannot synthesize GB. Therefore, such plants overexpressing GB synthesizing genes can result in the production of enough amount of GB, which lead plants to tolerate stresses including salinity stress. GB is synthesized from choline by the action of choline monooxygenase and betaine aldehyde dehydrogenase enzymes. The overexpression of the genes encoding betaine aldehyde decarboxylase from halophyte *Suaeda liaotungensis* improved the salinity tolerance in tobacco plants. The choline dehydrogenase gene (*codA*) from *Arthrobacter globiformis* helped salinity tolerance in rice [31]. The overexpression of N-methyl transferase gene in cyanobacteria and Arabidopsis resulted in accumulation of GB in higher levels and improved salinity tolerance [70]. Further, the overexpression of *betA* from *E. coli* in *Triticum aestivum* resulted in better salt tolerance and accumulated higher level of GB [71]. It is also reported that foliar application of GB exogenously to low- or non-accumulating plants helped in improving the growth of plants in salinity-stress condition as reported in the *Zea mays* [72]. In plants, betaine is synthesized upon abiotic stress *via* choline oxidation, in which choline monooxygenase (CMO) is a key enzyme. Although it had been thought that betaine synthesis is well regulated to protect abiotic stress, however, recently, it has been shown that an exogenous supply of precursors such as choline, serine and glycine in the betaine-accumulating plant Amaranthus further enhances the accumulation of betaine under salt stress, but not under normal conditions [73]. Recently, Waditee *et al.* [74] have shown that expression of Aphanothece gene encoding 3-phosphoglycerate dehydrogenase (PGDH) in Arabidopsis plants enhances the levels of betaine by providing the precursor serine for both choline oxidation and glycine methylation pathways. It has also been found that overexpression of proline and GB biosynthetic pathway genes enhance the abiotic stress tolerance in transgenic plants (Table **2**).

Table 2: Transgenic plants overexpressing genes of proline and glycine betaine biosynthetic pathway for abiotic stress tolerance

Gene	Source	Target transgenic	Response in transgenic plants	Reference
P5CS (Δ^1-Pyrroline-5-carboxylate-synthetase)	*Vigna aconitifolia* L.	*T. aestivum* L.cv. CD200126	Drought tolerance	[188]
P5CS	*V. aconitifolia* L.	*Saccharum spp.* variety RB855156	Drought tolerance	[189]
P5CS	*Arabidopsis thaliana* L. and *Oryza sativa* L.	*Petunia hybrida* cv. Mitchell	Drought tolerance and high proline	[190]
P5CR (Δ^1-pyrroline-5-carboxylate reductase)	*T. aestivum*	*A. thaliana* L.	Salt tolerance	[191]
P5CR	*A. thaliana*	*Glycine max* L. Merr. cv. Ibis	Drought stress tolerance	[192]
HvBADH1(Betaine aldehyde dehydrogenase)	*H. vulgare* L. var. nudum Hook. f	*N. tabacum* L.	Salt tolerance	[193]
BADH-1	*Spinacia Oleracea* L.	*N. tabacum* L.	Salt tolerance	[194]
OsBADH1	*Oryza sativa* L. ssp. *indica* 'Homjan,'	*N. tabacum* L.	Salt tolerance	[195]
betA	*E.coli*	*Zea mays* (DH4866)	Drought tolerance	[196]
COX (Choline oxidase)	*Arthrobacter pascens*	*N. tabacum* cv Xanthi, Arabidopsis ecotype RLD, *Brassica napus* cv. Westar	Salinity, drought, and freezing tolerance	[197]
codA(Choline oxidase)	*A. globiformis*	*Lycopersicon esculentum* Mill. cv. Moneymaker	Chilling, high salt and oxidative stress tolerance	[198]
codA	*A. globiformis*	*O. sativa* L.	Salt stress recovery	[199]
codA	*A. globiformis*	*L. esculentum* Mill. cv. Moneymaker	Chilling tolerance	[200]

DROUGHT STRESS

According to United Nations Convention to combat desertification (UNCCD), drought means the naturally-occurring phenomenon that exits when precipitation has been significantly below normal recorded levels, causing serious hydrological imbalances that adversely affect land resource production systems. Water deficit conditions are known to cause drought stress, which reduces agricultural production mainly by disrupting the osmotic equilibrium and membrane structure of the cell. Climate models have indicated that drought stress will become more frequent because of the long-term effects of global warming, which indicate the urgent need to develop adaptive agricultural strategies for a changing environment. Actually the water stress within the lipid bilayer results in displacement of membrane proteins, which contributes to loss of membrane integrity, selectivity, disruption of cellular compartmentalization and a loss of membrane based enzymes activity. The high concentration of cellular electrolytes due to the dehydration of protoplasm may also cause disruption of cellular metabolism. To avoid drought stress plants close their stomata, repress cell growth and photosynthesis, activate respiration, reduce leaf expansion and start shedding older leaves to reduce the area that transpires [15]. The relative root growth may undergo enhancement, which facilitates the capacity of the root system to extract more water from deeper soil layers. The components of drought and salt stress cross-talk as both these stresses ultimately result in dehydration of the cell and osmotic imbalance. Overall, this stress signaling encompasses three important parameters [75].

- Reinstating osmotic as well as ionic equilibrium of the cell to maintain cellular homeostasis under the condition of stress.

- Control as well as repair of stress damage by detoxification signaling.

- Signaling to coordinate cell division to meet the requirements of the plant under stress.

As a consequence of drought stress many changes occur in the cell, which include change in the expression level of LEA/dehydrin-type genes, synthesis of molecular chaperones, which help in protecting the partner protein from degradation and proteinases that function to remove denatured and damaged proteins. This stress also leads to activation of enzymes involved in the production and removal of ROS [25, 76]. Overexpression of some genes has been reported to help plants in drought stress tolerance [10]. Overexpression of barley group 3 LEA gene *HVA1* in leaves and roots of rice and wheat lead to improved tolerance against osmotic stress as well as improved recovery after drought and salinity stress [77]. Dehydrins are also known to accumulate in response to both dehydration as well as low temperature stresses [78]. Overexpression of the vacuolar Na^+/H^+ antiporter and H^+-pyrophosphatase pump (H^+-PPase) has resulted in enhanced tolerance of plant to both salinity [79,80] and drought stress [81,82]. These results suggest that the enhanced vacuolar H^+-pumping in the transgenic plants provide additional driving force for vacuolar sodium accumulation via the vacuolar Na^+/H^+ antiporter. Brini *et al.* [83] have reported that the overexpression of wheat Na^+/H^+ antiporter *TNHX1* and H^+-pyrophosphatase *TVP1* improve salt- and drought-stress tolerance in *Arabidopsis thaliana* plants. Recently, Jung *et al.* [84] have shown that overexpression of *AtMYB44* enhanced stomatal closure and confer dehydration stress tolerance in transgenic Arabidopsis. Recently, Jia *et al.* [85] have shown that a Ca^{2+}-binding protein calreticulin (CRT) from wheat is involved in the plant response to drought stress. *TaCRT*-overexpressing tobacco (*Nicotiana benthamiana*) plants grew better and exhibited less wilt under drought stress.

Plants produce compounds in roots that are transported to shoots via the xylem sap. Some of these compounds are vital for signaling and adaptation to drought stress. Recently, several metabolomic and proteomic changes in the xylem sap of maize under drought stress [86]. The application of these new methodologies provides insights into the range of compounds in sap and how alterations in composition may lead to changes in development and signaling during adaptation to drought.

Drought Stress and Their Effect Plant Performance

The first response of plants to drought stress is the closure of stomata to prevent transpirational water loss [87]. The closure of stomata may result from direct evaporation of water from the guard cells with no metabolic involvement and is referred to as hydropassive closure. Stomatal closure may also be metabolically dependent and involve processes that result in reversal of the ion fluxes that cause stomatal opening. This process of stomatal closure, which requires ions and metabolites, is known as hydroactive closure. Plant growth and response to a stress condition is largely under the control of hormones. ABA promotes the efflux of K^+ ions from the guard cells, which results in the loss of turgor pressure leading to stomata closure. The closure of stomata does not always depend upon the perception of water deficit signals arising from leaves. In fact, stomata closure also responds directly to soil desiccation even before there is any significant reduction in leaf mesopyll turgor pressure. The fact that ABA can act as a long distance communication signal between water deficit roots and leafs, inducing the closure of stomata is about two decades old [88]. Stomatal closure in response to drought stress primarily results in a decline in the rate of photosynthesis. Severe drought condition was reported to decline rubisco activity, which leads to limited photosynthesis [89]. The photosystem II (PS II) has been reported to decline under drought conditions [90]. It has been shown that the decline in the rate of photosynthesis in drought stress is primarily due to CO_2 deficiency [91]. Decline in intracellular CO_2 levels also results in the over-reduction of components within the electron transport chain and the electrons get transferred to oxygen at photosystem I (PS I). This generates ROS including superoxide, H_2O_2 and hydroxyl radicals [92]. These ROS need to be scavenged by the plant as they may lead to photo-oxidation. Plant detoxifying systems, which include ascorbate and glutathione pools control the intracellular concentration of ROS. In a longer drought situation the plant cells can undergo shrinkage and lead to mechanical constraint on cellular membranes, which impairs the functioning of ions and transporters as well as membrane associated enzymes.

Drought Stress and Oxidative Stress

Various abiotic stresses including drought stress leads to the overproduction of toxic ROS ($O_2^{\cdot-}$, H_2O_2, $^1O_2$, and OH·) in plants (Fig. **3**) which are highly reactive and can cause damage to proteins, lipids, carbohydrates, DNA which ultimately results in oxidative stress. ROS accumulation as a result of various environmental stresses including drought is a major cause crop productivity loss worldwide. ROS can be efficiently scavenged by the plant's antioxidant machinery which comprised of SOD, CAT, APX, GR etc. Therefore, transgenic plants overexpressing one or more antioxidant enzymes may be a useful tool to get rid of the toxic effect of ROS following drought stress. Increase in SOD activity following drought stress was noted in three cultivars of bean [93], *Alternanthera philoxeroides* [94], *Oryza sativa* [95]. Wang and Li [96] studied the effect of polyethylene glycol (PEG)-induced water stress on the activities of total leaf SOD and chloroplast SOD (including thylakoid-bound SOD and stroma SOD) in white clover (*Trifolium repens* L.) Both leaf SOD and chloroplast SOD activities were markedly enhanced with increasing concentration of PEG stress. The enhanced chloroplastic SOD activity, especially thylakoid-bound SOD activity suggests that Fe-SOD located in chloroplasts play a more important role than cytosolic Cu/Zn containing SODs in scavenging $O_2^{\cdot-}$. Transgenic rice plants overexpressing *OsMT1a* showed increase in CAT activity and thus enhanced tolerance to drought [97]. It has also been noted that overexpression of APX in *Nicotiana tabacum* chloroplasts enhanced plant tolerance to salt and water deficit [98]. Yang *et al.* [97] correlated the enhanced tolerance of *OsMT1a* overexpressing transgenic rice plants to drought stress with the increase in APX activity. Overexpression of different antioxidants has been found to enhance abiotic stress tolerance in transgenic plants (Table **1**).

Drought Stress Tolerance through Osmotic Adjustment

Abiotic stress factors are limiting crop productivity worldwide. The normal plant growth, development and productivity are negatively affected by drought stress. To cope with drought stress plants need to do undergo osmotic adjustment where plants decrease their cellular osmotic potential by the synthesis/accumulation of solutes including proline, glutamate, glycine-betaine, carnitine, mannitol, sorbitol, fructans, polyols, trehalose, sucrose, oligosaccharides and inorganic ions like K^+. These compounds help the plant cells to maintain their hydrated state and therefore function to provide resistance against drought and cellular dehydration [99]. The hydroxyl group of sugar alcohols substitutes the OH group of water to maintain the hydrophilic interactions with the membrane lipids and proteins and therefore, help to maintain the structural integrity of the membranes. These stress-accumulated solutes do not intervene with cells normal metabolic processes. It has been reported that the accumulation of simple sugars such as glucose and fructose cause an increase in the invertase activity in the leaves of the drought challenged plants [100]. ABA has been implicated in enhancing the activity and expression of vacuolar invertase [100]. ABA biosynthesis is also reported to be directly controlled by glucose, as the transcript of several genes responsible for ABA synthesis was increased by glucose in Arabidopsis seedlings [101]. Cross-talk may exist between the sugars and plant hormones such as ABA and ethylene. Glucose and ABA signaling act in coordination for regulating plant growth and development. A high concentration of ABA and sugars can inhibit growth in a severe drought stress, while a low concentration can promote growth.

In response to drought stress the osmolytes in low accumulation function in protecting macromolecules either by stabilizing the tertiary structure of protein or by scavenging ROS [102]. However, higher accumulation of osmolytes in transgenic plants can impair the growth in the absence of any stress probably due to plant adaptation strategy to conserve water in acute stress [3]. Therefore controlled synthesis of osmolytes is the main concern in designing transgenic strategies for crop improvement. Oligosaccharides such as raffinose and galactinol are among the sugars synthesized in response to drought. These compounds seem to function as osmoprotectants rather than providing osmotic adjustment. Mannitol is one of the most widely distributed sugar alcohols in nature and functions to scavenge the ROS, hydroxyl radicals and it also stabilizes the macro molecular structure of enzymes [103]. Trehalose is a non-reducing disaccharide of glucose and has been shown to exert its positive influence during drought by stabilizing membranes and macromolecules. Trehalose over-expression helps in the maintenance of an elevated capacity for photosynthesis primarily due to increased protection of PS II against photooxidation [104]. The over-expression of *P5CS* (pyrroline-5-carboxylate synthase) gene from *Vigna aconitifolia* in tobacco, leads to increased levels of proline and consequently improved growth under drought stress [105].

Drought Stress and Signal Transduction in Plants

Lipids are important membrane components and are also major targets of environmental stresses including drought stress. The changes in the lipids composition may help to maintain membrane integrity and preserve cell compartmentalisation under water stress conditions. It has been shown that in response to drought, total leaf lipid contents decreased progressively [106]. However, for leaf relative water, content as low as 47.5%, total fatty acids still represented 61% of control contents. Lipid content of extremely dehydrated leaves rapidly increased after rehydration. In general, phospholipids from plant cell membrane constitute a dynamic system that generates a multitude of signaling molecules like IP3, DAG and PA [25]. In response to the stress the PLC get activated, which catalyzes the hydrolysis of PIP_2 into IP_3 and DAG, IP_3 releases Ca^{2+} from internal stores (Fig. **2**). Several studies have shown that in various plant systems IP_3 levels rapidly increase in response to hyperosmotic stress [3,107]. The IP_3 level was also reported to be increased upon treatment with exogenous ABA in *Vicia faba* guard cell protoplast [108] and in Arabidopsis seedlings [109]. An Arabidopsis PLC gene, *AtPLC*, is also induced by salt and drought stress [110]. In guard cells, IP_3 induced Ca^{2+} increase in the cytoplasm lead to stomatal closure and thus retention of water in the cells [111]. PLD was shown to be rapidly activated in response to drought stress in two plant species, i.e. *Craterostigma plantagineum* and Arabidopsis [112,113]. When drought stress-induced PLD activity was compared between drought-resistant and sensitive cultivars of cowpea, it was found that activity was higher in the drought sensitive cultivars [114].

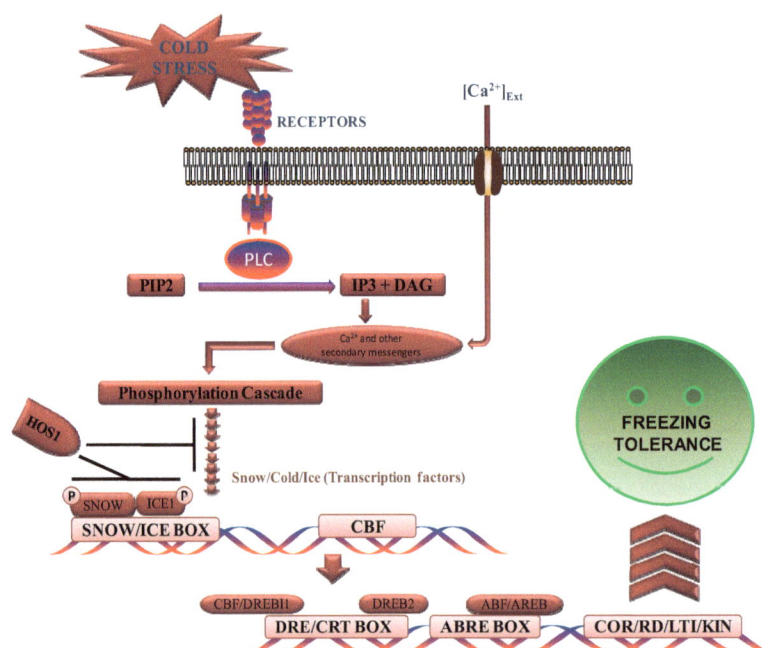

Figure 2: Generic pathway for plant response to cold stress.

COLD STRESS

Cold stress is one of the most important abiotic stress factors which limit the crop growth and productivity worldwide. Each plant has its own set of temperature requirements, which are optimum for its proper growth and development. Deviation from optimum temperature may lead to plant growth inhibition and yield loss. The cold stress experienced by plants can be classified into two types: those occurring at (a) temperatures below freezing and (b) low temperatures above freezing (non-freezing temperatures). In this section various aspects of cold stress are covered.

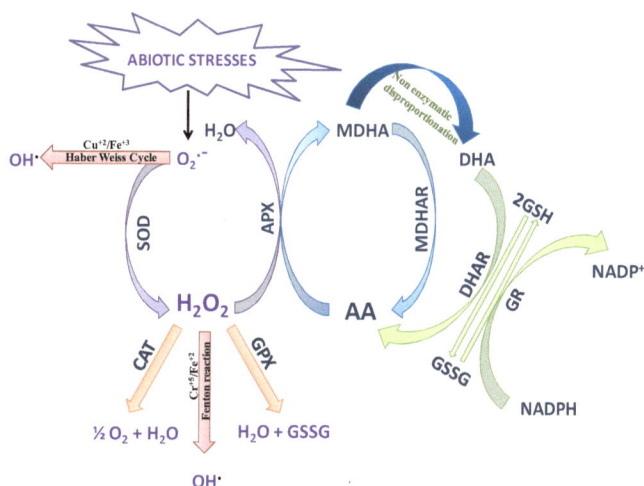

Figure 3: Generation of reactive oxygen species by abiotic stresses and their scavenging mechanism by antioxidants

Many plants like maize, soybean, cotton, tomato and banana are sensitive to non-freezing temperatures (10-15°C) and exhibit signs of injury [115,116]. Various phenotypic symptoms in response to chilling stress include reduced leaf expansion, wilting and chlorosis, which may lead to necrosis. Low temperature can also severely hamper the reproductive development of plants, as reported in rice [117]. Freezing temperatures can induce severe membrane damage, which is largely due to the acute dehydration associated with freezing [118,119] (Fig. **4**). The temperature at which a membrane changes from semi fluid state to a semi crystalline state is known as the transition temperature. Chilling sensitive plants usually have a higher transition temperature as compared to the chilling resistant plants, which have a lower transition temperature. An understanding of how freezing induces injurious effects on plants is essential for the development of frost tolerant crops. The real cause of freeze-induced injury to plants is the ice formation rather than low temperatures. Ice formation in plants begins in the apoplastic space as it has relatively lower solute concentration. This causes a mechanical strain on the cell wall and plasma membrane leading to cell rupture [120]. Freezing temperature exerts its effect largely by membrane damage due to severe cellular dehydration, but certain additional factors including ROS also contribute to damage induced by freezing. Overall, chilling ultimately results in loss of membrane integrity, which leads to solute leakage. The integrity of intracellular organelles is also disrupted leading to the loss of compartmentalization, reduction and impairing of photosynthesis, protein assembly and general metabolic processes. The primary environmental factors responsible for triggering increased tolerance against freezing, is the phenomenon known as 'cold acclimation'. It is the process where certain plants increase their freezing tolerance upon prior exposure to low non-freezing temperatures [3].

Cold Acclimation, Freezing Tolerance and Signal Transduction

Cold acclimation is the process whereby plants increase their freezing tolerance upon exposure to low temperature conditions. Considerable research has been done over the past 10 years to understand the molecular basis of cold acclimation and it was noted that low temperatures induce a number of alterations in cellular components, including the extent of unsaturated fatty acids, the composition of glycerolipids, changes in protein and carbohydrate composition and the activation of ion channels [3,121]. For cold acclimation the membranes have to be stabilized against freeze injury, which can be achieved through changes in the lipid composition and induction of other non-enzymatic proteins that alter the freezing point of water. Accumulation of sucrose and other simple sugars also contributes to the stabilization of membrane as these molecules can protect membranes against freeze-damage. Low temperatures activate a number of cold-inducible genes, such as those that encode dehydrins, lipid transfer proteins, translation elongation factors and the late-embryogenesis abundant (LEA) proteins [3,118,121]. Overall, cold acclimation results in protection and stabilization of the integrity of cellular membranes, enhancement of the antioxidative mechanisms, increased intercellular sugar levels as well as accumulation of other cryoprotectants including polyamines that protect intracellular proteins by inducing the genes encoding molecular chaperones. All these modifications help the plant withstand and surpass severe dehydration associated with freezing stress [3,121].

First, it is important to know the meaning of freezing tolerance, it is a multigenic trait and injury due to freezing in most plants and tissues results largely from the severe cellular dehydration that occurs upon ice formation and the cellular membrane systems are a primary site of freeze-induced injury. The Arabidopsis *FAD8* gene [122] encodes a fatty acid desaturase that contributes to freezing tolerance by altering the lipid composition. Cold-responsive genes encoding molecular chaperones including a spinach *hsp70* gene [123], and a *Brassica napus hsp90* gene [124], contribute to freezing tolerance by stabilizing proteins against freeze-induced denaturation. Many cold-responsive genes encoding various signal transduction and regulatory proteins have been identified and this list includes the mitogen-activated protein Kinase (*MAPK*) [125], MAP Kinase Kinase Kinase (*MAPKKK*) [126] and the genes for calmodulin-related proteins [127]. The largest class of cold-induced genes encodes polypeptides, that are homolog of LEA proteins and the polypeptides, are synthesized during the late embryogenesis phase, just prior to seed desiccation and also in the seedlings in response to dehydration stress [128]. Other examples of cold responsive genes include: *COR15a*, alfalfa *Cas15*, and wheat *WCS120* [3]. The expression of *COR* genes has been shown to be critical for both chilling tolerance and cold acclimation in plants [129]. Arabidopsis COR genes include: *COR78/RD29, COR47, COR15a, COR6.6* and the genes encoding LEA like proteins [129]. These genes are induced by cold, dehydration or ABA. The analysis of the promoter elements of *COR* genes revealed that they contain DRE (Dehydration Responsive Elements) or CRT (C-Repeats) and some of them contain ABRE (ABA-Responsive Element) as well [130,131]. Induction of the *COR* genes was accomplished by over-expression of transcription factor CBF (CRT/DRE binding factor) [131]. CBF binds to the CRT/DRE elements present in the promoter of the *COR* genes and other cold regulated genes. The over-expression of these regulatory elements not only resulted in increased freezing tolerance but also an increase to drought tolerance [132]. Lee and coworkers in 2001 [133] genetically analyzed *HOS1* (High Expression of Osmotically responsive genes) locus of Arabidopsis. *HOS1* gene encodes a ring finger protein and is constitutively expressed but is drastically down-regulated within 10 min of cold stress. Genetic analysis led to the identification of ICE1 (Inducer of CBF Expression 1) as an activator of CBF3 [134]. ICE1 encoded a transcription factor that specifically recognized MYC sequence on the CBF3 promoter. Transgenic lines overexpressing *ICE1* did not express CBF3 at warm temperature but showed a higher level of expression for CBF3 as well as RD29 and COR15a at low temperatures. This study suggests that cold induced modification of ICE1 is necessary for it to act as an activator of CBF3 in plant. Two CBF1-like cDNAs, *CaCBF1A* and *CaCBF1B*, have been cloned and characterized from hot pepper [3]. These were induced in response to low temperature stress (4°C) and not in response to wounding or ABA. The gene expression as well as protein accumulation of *OsCDPK13* (*Oryza sativa* CDPK 13) were up-regulated in response to cold. Cold-tolerant rice varieties exhibited higher expression of *OsCDPK13* than the cold sensitive ones [135].

Proline has been shown to be an effective cryoprotectant and this is also one of the major factors imparting freezing tolerance. The *eskimo 1* (*esk1*) gene is known to play an important role in freezing tolerance. The concentration of free proline [136] in *esk1* mutant was found to be 30 fold higher than in the wild-type plants. Sui *et al.* [137] have reported that overexpression of glycerol-3-phosphate acyltransferase gene improves chilling tolerance in tomato. Recently, Soybean *GmbZIP44, GmbZIP62* and *GmbZIP78* genes have been shown to function as negative regulator of ABA signaling and their overexpression confer salt and freezing tolerance in transgenic Arabidopsis [138]. Recently, *OsCSP* (*Oryza sativa* cold shock domain protein) transcripts are reported to be transiently up-regulated in response to low-temperature stress and rapidly return to a basal level of gene expression [139]. OsCSP1 and OsCSP2 (*Oryza sativa* CSD protein) encode putative proteins consisting of an N-terminal CSD and glycine-rich regions that are interspersed by 4 and 2 CX2CX4HX4C (CCHC) retroviral-like zinc fingers, respectively. *In vivo* functional analysis confirmed that OsCSPs could complement a cold-sensitive bacterial strain, which lacks four endogenous cold shock proteins.

The generic pathway for plant response to cold stress (Fig. **2**). The extracellular cold stress signal is first perceived by the membrane receptors/sensors and then activates large and complex signaling cascade intracellularly including the generation of secondary signal molecules. Increases in cytosolic free Ca^{2+} ($[Ca^{2+}]cyt$) are common to many stress-activated signalling pathways, including the response to cold environments. In Arabidopsis [127] and alfalfa [140] cytoplasmic Ca^{2+} levels increase rapidly in response to low temperature, largely due to an influx of Ca^{2+} from extracellular stores. Through the use of pharmacological and chemical reagents, it has been demonstrated that Ca^{2+} is required for the full expression of some of the cold induced genes including the *CRT/DRE* controlled *COR6* and *KIN1* genes of Arabidopsis [140]. Ca^{2+} release can be primarily from extracellular source (apoplastic space) as addition of EGTA or BAPTA was shown in many cases to block Ca^{2+} effects (Fig. **2**). Ca^{2+} release may also result from activation of PLC (phospholipase C), leading to hydrolysis of phosphatidyl inositol bisphosphate (PIP_2) to (inositol triphosphate (IP_3) and diacyl glycerol (DAG), which resulted in subsequent release of Ca^{2+} from intracellular Ca^{2+} stores [3,4]. Furthermore, Ca^{2+}-binding proteins (Ca^{2+} sensors) can provide an additional level of regulation in the Ca^{2+} signaling. These sensor proteins recognize and decode the information provided in the Ca^{2+} signatures, relay the information downstream to initiate a phosphorylation cascade leading to regulation of gene expression. The phosphorylation cascade leads to induction of the cis regulatory elements like *SNOW* and *ICE1*, which in turn regulates the level of cold binding factors which in turn induces the DRE/CRT and ABRE regulatory elements and these elements in turn up-regulate the level of cold responsive genes like *COR, KIN, LT1* and *RD* [3]. The product of these cold stress responsive genes can provide cold stress tolerance directly or indirectly (Fig. **2**). Overall, the cold stress response could be coordination action of many genes, which may cross-talk also with each other.

HEAVY METAL STRESS

Various abiotic and biotic stresses are adversely affecting the plant growth, productivity and genome stability. Among abiotic stresses, heavy metal contamination is a serious environmental problem that limits plant productivity and threatens human health [141]. The use of urban waste for the production of vegetables and other important crops in urbanized areas of Asia is a growing concern because of the contamination of soil, water and agricultural produce by heavy metals [142]. The agricultural soil may have toxic levels of heavy metals due to agricultural and industrial practices. The agricultural practices which lead to the accumulation of heavy metals like Cd, Cu,Zn, Ni, Co, Cr, Pb and As in agricultural soils are as follows:

- Uncontrolled long-term use of phosphatic fertilizers, pesticides and other chemical fertilizers

- Bad watering practices such as application of sewage sludge, waste water and industrial waste

- Precipitation from heavy coal combustion, and

- Smelter wastes and residues from metalliferous mining

Although many metals are required as structural and catalytic components of enzymatic proteins involved in various physiological processes, they can still be toxic to a plant if present at higher concentrations [143].

Heavy Metals and Their Effect On Plants

Over the past decades, the emission of toxic heavy metals into the ecosphere is so extensive that in some regions it overrides the adaptive potential of plants. Therefore, contamination of agricultural soil by heavy metals has become a critical environmental concern due to their potential adverse ecological effects. It has been reported that the regulatory limit of Cd in agricultural soil is 100 mg/kg soil [144], but anthropogenic activities is continuously exceeding this threshold limit. Cd reduces plant growth and biomass and causes plant death in extreme

cases [141]. It causes inhibition of shoot and root growth, disorganization of the grana structures and reduction in the biosynthesis of chlorophyll. It also interferes with photosynthesis, respiration and water relations. The effect of Cd on nitrate and sulfate assimilation has been studied in several plants showing an inhibition of the nitrate uptake rate and the activity of the enzymes involved in the nitrate assimilation pathway [6 and references therein]. Unlike Cd, Zn is an essential nutrient for living organisms. It has been noted that higher concentration of Zn in the agricultural fields leads to phytotoxicity. According to an estimation, it has been reported that metal polluted soils have Zn concentration over 150 mg/kg soil [145]. Such concentrations are toxic for normal plant growth and development and results in inhibited growth and productivity.

However, Cu is considered as a micronutrient for plants which plays important role in CO_2 assimilation and ATP synthesis. Cu is also an essential component of various proteins like plastocyanin of photosynthetic system and cytochrome oxidase of respiratory electron transport chain [146]. Various anthropogenic activities like industrial and mining activities contribute to high levels of Cu in the agricultural soils which cause cytotoxicity and injury to crop plants. Cu can occur in very high concentrations that are detrimental or even lethal to most plants. It is widely used as a pesticide in agriculture, and field runoff may easily reach concentrations of several micromolar. Photosynthetic reactions, both photochemical and biochemical ones, belong to the most important sites of inhibition by many heavy metals and in particular Cu.

Among all heavy metals, Hg is the most hazardous metal in the environment. Considerable amounts of Hg may be added to agricultural land with the application sludge, pesticides, fertilizers, lime and manures. The most important sources of contaminating agricultural soil have been the use of organic mercurials as a seed-coat dressing to prevent fungal diseases in seeds. Hg is a transition metal but able to cause oxidative stress in plants by triggering the generation ROS which results in lipid peroxidation and alteration of antioxidant enzyme activities [147,148]. The possible causal mechanisms of Hg toxicity are changes in the permeability of the cell membrane, reactions of sulphydryl (-SH) groups with cations, affinity for reacting with phosphate groups and active groups of ADP or ATP, and replacement of essential ions, mainly major cations. It has been reported that Hg binds with water channel proteins of root cells which results in the obstruction to the water flow. Furthermore, Hg accumulation in plants significantly decrease the chlorophyll and proteins content which can affect both photochemical and carbon reduction reactions of photosynthesis [149].

Among all heavy metals, Al is the most abundant metal in the earth's crust and its toxicity is a major factor limiting crop yields on acid soils, which cover about 50% of the world's potentially arable land surface [150]. There is considerable genotypic variation for Al tolerance in most common plant species. Al causes various adverse effects, such as disruption of signal transduction pathways, inhibition of cell division and ion fluxes, disruption of cytoskeletal dynamics, induced generation of ROS, and disturbance of plasma membrane stability and function.

Heavy Metals and Oxidative Stress

Exposure to high levels of heavy metals results in the overproduction of ROS. It has been found that metal exposure leads to generation of ROS either by Haber-Weiss reactions or indirectly leads to overproduction of ROS which results in oxidative stress (Fig. 3). As a whole, heavy induced ROS production causes oxidative damage to photosynthetic pigments, bio-molecules such as lipids, proteins and nucleic acids, leakage of electrolytes *via* lipid peroxidation resulting in dramatic reductions of growth and productivity and eventually causing the death of plants. The indirect mechanisms include the interference with antioxidant machinery, disrupting the electron transport chain and/or disturbing the metabolism of essential elements [151]. Plants adopt several strategies to avoid metal induced toxicity:

- Physically avoiding the metal-contaminated environment by making exudates of complexing agents into rhizosphere region,

- Binding metal ions in the cell wall,

- Effluxing the metal ions from the symplasm,

- Preventing the upward transport of metal ions to the above ground parts,

- Transporting metal-peptide/ligand complexes into the vacuole,

- Storing metal ions in the vacuoles by complexation with vacuolar peptides/ligands and or Forming metal-resistant enzymes or metabolites to minimize metal-induced severe internal metabolic injuries

Plants have evolved a complex array of mechanisms to maintain optimal metal levels and avoid the detrimental effects of excessively high concentrations [143]. When these homeostatic mechanisms are overwhelmed, plants suffer metal-induced damage and pro-oxidant conditions within cells. However, higher plants are very well equipped with antioxidant mechanisms [152,153,154,155]. Plant cells display an antioxidant network including numerous soluble and membrane compounds, particularly in mitochondria and in chloroplasts where respiratory and photosynthetic electron transfer chains, respectively, take place. Antioxidant enzymes are considered as those that either catalyze such reactions, or are involved in the direct processing of ROS [156]. Plants possess very efficient enzymatic (SOD, CAT, APX, GR, MDHAR, DHAR, GPX, GOPX and GST) and non-enzymatic (AsA, GSH, phenolic compounds, alkaloids, non-protein amino acids and α-tocopherols) antioxidant defense systems. Components of antioxidant defense systems control the cascades of uncontrolled oxidation [157] and protect plant cells from oxidative damage by scavenging of ROS.

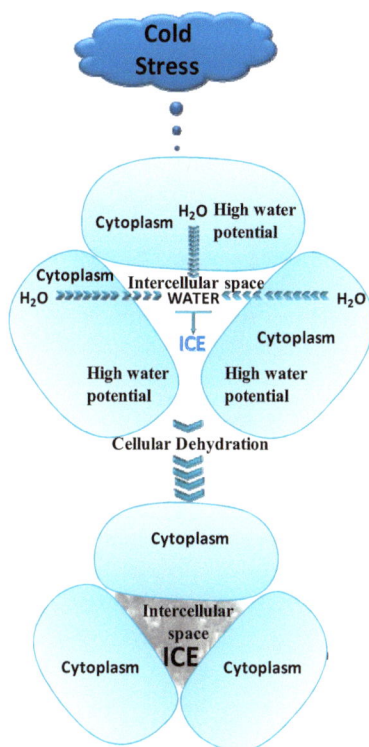

Figure 4: Mechanism of cellular dehydration at low temperature

Plants developed a number of strategies to resist metal toxicity, including active efflux, sequestration, binding of heavy metals inside the cells by strong ligands or by the activation of antioxidant defense machinery. Heavy metal exposed plants adopt the process of avoidance of the production of ROS as the first line of defence against oxidative stress. Once formed, ROS must be detoxified as efficiently as possible to minimize eventual damage. Thus, the detoxification mechanisms constitute the second line of defence against the detrimental effects of ROS [158]. In fact, compounds having the property of quenching the ROS without undergoing conversion to a destructive radical can be described as antioxidant. Antioxidant enzymes are considered as those that either catalyses such reactions, or are involved in the direct processing of ROS [156]. Hence, antioxidants (enzymatic and non-enzymatic) function to interrupt the cascades of uncontrolled oxidation [157]. Though, the expression for antioxidant enzymes is altered under stress conditions, their up regulation has a key role in combating the abiotic stress-induced oxidative stress. However, the level of up regulation is subject to type and magnitude of the stress.

Heavy Metals and Antioxidant Defense Machinery

Several antioxidants like SOD, APX and GR have been cloned from a variety of plants. The upregulation of these antioxidants is implicated in combating oxidative stress caused due to biotic and abiotic stress and have a critical role in the survival of plants under stresses environment. Significant increase in SOD activity has been detected following Cd treatment in many plants like *Hordeum vulgare*, *Arabidopsis thaliana*, *Oryza sativa*, *Triticum aestivum*, *Brassica juncea*, *Vigna mungo* and *Cicer arietinum* [6 and references therein). Simonovicova *et al.* [96] reported increase in SOD activity in barley (*Hordeum vulgare* L. cv. Alfor) root tips under Al stress at 72h. Li *et al.* [159] reported significant increase in SOD activity in two cultivars of *Brassica compestris* under Cu stress. A general induction in SOD activity in *Anabaena doliolum* under NaCl and Cu^{2+} stress has also been reported [160]. Calgaroto *et al.* [148] observed direct correlation of SOD activity with H_2O_2 concentration in *Pfaffia glomerata* plantlets treated with Hg. The variable response of CAT activity has been observed under metal stress. Its activity declined in *Glycine max*, *Phragmites australis*, *Capsicum annuum* and *Arabidopsis thaliana*, whereas, its activity increased in *Oryza sativa* leaves, *Brassica juncea*, *Triticum aestivum*, *Vigna mungo* roots and *Cicer arietinum* under Cd stress (6 and references therein). Hso and Kao [161] reported that pretreatment of rice seedlings with H_2O_2 under non-heat shock conditions resulted in an increase in CAT activity and protected rice seedlings from subsequent Cd stress. Srivastava *et al.* [160] reported a decrease in CAT activity in *Anabaena doliolum* under NaCl and Cu^{2+} stress. Hso and Kao [161] reported that pretreatment of *Oryza sativa* seedlings with H_2O_2 under non-heat shock conditions resulted in an increase in APX activity and protected rice seedlings from subsequent Cd stress. Pekker *et al.* [162] studied the expression of cAPX in leaves of de-rooted bean plants in response to iron overload and found that cAPX expression (mRNA and protein) was rapidly induced in response to iron overload. GR activity increased in the presence of Cd in *Capsicum annuum*, *Arabidopsis thaliana*, *Vigna mungo*, *Triticum aestivum* and *Brassica juncea* (6 and references therein). Srivastava *et al.* [160] reported decline in GR activity in *Anabaena doliolum* under Cu^{2+} stress but it increased under salt stress.

The combined expression of Cu/Zn-SOD and APX in transgenic *Festuca arundinacea* plants led to increased tolerance to MV, H_2O_2, and Cu, Cd and As [163]. Azpilicueta *et al.* [164] reported that incubation of *Helianthus annuus* leaf discs with 300 and 500 μM $CdCl_2$ under light conditions increased *CATA3* transcript level but this transcript was not induced by Cd in etiolated plants. Moreover, in roots of the transgenic CAT-deficient tobacco lines (*CAT 1AS*), the DNA damage induced by Cd was higher than in wild type tobacco roots [165]. A CAT gene from *Brassica juncea* (*BjCAT3*) was cloned and up-regulated in tobacco under Cd. CAT activity of transgenic plants was approximately two-fold higher than that of WT which was correlated with enhanced tolerance under Cd stress [166]. Bashir *et al.* [167] studied the expression patterns and enzyme activities of GR in graminaceous plants under Fe-sufficient and Fe-deficient conditions by isolating cDNA clones for chloroplastic GR (*HvGR1*) and cytosolic GR (*HvGR2*) from barley. Both proteins showed *in vitro* GR activity, and the specific activity for *HvGR1* was 3 times higher than *HvGR2*. The expression patterns of *GR1* and *GR2* in rice, wheat, barley, and maize was examined by northern blot analysis. Upregulation of *HvGR1*, *HvGR2*, and *TaGR2* was found in response to Fe-deficient conditions than Fe-sufficient. Overexpression of a eukaryotic GR from *Brassica campestris* (BcGR) and *E. coli* GR (EcGR) was studied in *E. coli* in pET-18a. It was found that BcGR overproducing *E. coli* showed better growth and survival rate than the control but far better growth was noted in *E. coli* strain transformed with the inducible EcGR in the presence of paraquat, SA and Cd. Finally, it was found that BcGR functions in a prokaryotic system by providing protection against oxidative damages in *E. coli* [168].

It has also been found that DHAR overexpression also enhance plant tolerance against heavy metal stress. In a study, under Al stress, the role of MDAR or DHAR in AsA regeneration has been studied in transgenic tobacco plants overexpressing *A. thaliana* cytosolic DHAR (*DHAR-OX*) or MDAR (*MDAR-OX*). It was found that *DHAR-OX* transgenic plants showed higher levels of AsA with or without Al, whereas, *MDAR-OX* plants only showed higher AsA level in the absence of Al in comparison to WT. Significantly higher levels of AsA and APX in *DHAR-OX* plants showed better tolerance under Al stress but not *MDAR-OX* plants. It is clear that plants overexpressing DHAR showed tolerance to Al stress by maintaining high AsA level [169]. Leisinger *et al.* [170] reported the upegulation of a GPX homologous gene (*Gpxh* gene) in *Chlamydomonas reinhardtii* following oxidative stress. It was noted that *Gpxh* gene showed strong induction by the $^1O_2$-generating photosensitizers neutral red, methylene blue and rose bengal. It was also noted that Gpxh showed transcriptionally up-regulation by $^1O_2$ photosensitizers when *Gpxh* promoter fusions with the arylsulfatase reporter gene [170]. There have been many reports of the production of heavy metal stress tolerant transgenic plants having different antioxidant enzymes (Table **1**).

ACKNOWLEDGEMENTS

We thank Department of Biotechnology (DBT), and Department of Science and Technology (DST), Government of India grants for partial support. I apologize if some references could not be cited due to space constraint.

REFERENCES:

[1]   FAO, "More people than ever are victims of hunger", Press release, FAO, Rome, 2009; www.fao.org/fileadmin/user_upload/newsroom/docs/Press%20release%20june-en.pdf

[2]   Tester M, Langridge P. Breeding Technologies to Increase Crop Production in a Changing World. Science 2010; 327: 818-822.

[3] Mahajan S, Tuteja N. Cold, salinity and drought stress: an overview. Arch Biochem Biophys 2005; 444: 139-158.

[4] Tuteja N. Mechanisms of high salinity tolerance in plants. Meth Enzymol 2007; 428: 419-438.

[5] Khan NA, Singh S. (Eds.) Abiotic Stress and Plant Responses. IK International Publishing House Pvt. Ltd., New Delhi, 2008; pp. 1-299.

[6] Gill SS, Khan NA, Anjum NA, Tuteja N. Amelioration of Cadmium Stress in Crop Plants by Nutrient Management: Morphological, Physiological and Biochemical Aspects. Plant Stress (Spl issue) 2010; (In press)

[7] Bray EA, Bailey-Serres J, Weretilnyk E. In: Gruissem W, Buchannan B, Jones R (Eds.) Biochemistry and Molecular Biology of Plants, Responses to abiotic stresses, American Society of Plant Biologists, Rockville, MD, 2000; pp. 1158-1249.

[8] Nakashima K, Ito Y, Yamaguchi-Shinozaki K. Transcriptional Regulatory Networks in Response to Abiotic Stresses in Arabidopsis and Grasses. Plant Physiol 2009; 149: 88-95.

[9] Jones HG, Jones MB. In: Jones HG, Flowers TJ, Jones MB (Eds.), Plants under stress. Introduction: Some terminology and common mechanisms, Cambridge: Cambridge university Press, 1989; pp. 1-10.

[10] Tuteja N, Sopory SK. Plant Signaling in Stress: G-Protein Coupled Receptors, Heterotrimeric G-Proteins and Signal Coupling *via* Phospholipases. Plant Sig Beh 2008; 3: 79-86.

[11] Tuteja N, Mahajan S. Calcium signaling network in plants: an overview. Plant Sig Beh 2007; 2: 79-85.

[12] Mishra NS, Tuteja R, Tuteja N. Signaling through MAP kinase networks in plants. Arch Biochem Biophys 2006; 452: 55-68.

[13] Seki M *et al.* Monitoring the expression profiles of 7000 Arabidopsis genes under drought, cold and high-salinity stresses using a full-length cDNA microarray. Plant J 2002; 31: 279-292.

[14] Seki M *et al.* Monitoring the expression pattern of around 7,000 Arabidopsis genes under ABA treatments using a full-length cDNA microarray. Funct Integ Gen 2002; 2: 282-291.

[15] Shinozaki K, Yamaguchi-Shinozaki K. Gene networks involved in drought stress response and tolerance. J Exp Bot 2007; 58: 221-227.

[16] Bhatnagar-Mathur P, Vadez V, Sharma KK. Transgenic approaches for abiotic stress tolerance in plants: retrospect and prospects. Plant Cell Rep 2008; 27: 411-424.

[17] Li A, Wang X, Leseberg CH, Jia J, Mao L. Biotic and abiotic stress responses through calcium-dependent protein kinase (CDPK) signaling in wheat (*Triticum aestivum* L.). Plant Sig Beh 2008; 3(9): 654-660.

[18] FAO. FAO land and plant nutrition management service. Available online at: http://www.fao.org/ag/agl/agll/ spush/. Accessed 25 April 2008; 2008.

[19] Ma S, Gong Q, Bohnert HJ. Dissecting salt stress pathways. J Exp Bot 2006; 57: 1097-1107.

[20] Sanan-Mishra N, Phan XH, Sopory SK, Tuteja N. Pea DNA helicase 45 overexpression in tobacco confers high salinity tolerance without affecting yield. Proc Natl Acad Sci USA 2005; 102: 509-514.

[21] Misra S, Wu Y, Venkataraman G, Sopory SK, Tuteja N. Heterotrimeric G-proteins complex and GPCR from a legume (*Pisum sativum*): role in salinity and heat stress and cross talk with PLC. Plant J 2007; 51: 656-669.

[22] Ge LF, Chao DY, Shi M, Zhu MZ, Gao JP, Lin HX. Overexpression of the trehalose-6-phosphate phosphatase gene OsTPP1 confers stress tolerance in rice and results in the activation of stress responsive genes. Planta 2008; 228: 191-201.

[23] Wang X, Liu Z, He Y. Responses and tolerance to salt stress in bryophytes. Plant Sig Beh 2008; 3: 516 - 518.

[24] Wang W, Vinocur B, Altman A. Plant responses to drought, salinity and extreme temperatures: towards genetic engineering for stress tolerance. Planta 2003; 218: 1-14.

[25] Zhu JK. Salt and drought stress signal transduction in plants. Annu Rev Plant Biol 2002; 53: 247-273.

[26] Niu X, Bressan RA, Hasegawa PM, Pardo JM. Ion homeostasis in NaCl stress environments. Plant Physiol 1995; 109: 735-742.

[27] Flowers TJ, Troke PF, Yeo AR. The mechanisms of salt tolerance in halophytes. Annu Rev Plant Physiol 1977; 28: 89-121.

[28] Munns R, James RA, Lauchli A. Approaches to increasing the salt tolerance of wheat and other cereals. J Exp Bot 2006; 57: 1025-1043.

[29] Shoji T *et al.* Salt Stress Affects Cortical Microtubule Organization and Helical Growth in Arabidopsis. Plant Cell Physiol 2006; 47: 1158-1168.

[30] Kovtun Y, Chiu WL, Tena G, Sheen J. Functional analysis of oxidative stress-activated mitogen-activated protein kinase cascade in plants. PNAS USA 200; 97: 2940–2945.

[31] Vinocur B, Altman A. Recent advances in engineering plant tolerance to abiotic stress: achievements and limitations. Curr Opin Biotech 2005; 16: 123-132.

[32] Harinasut P, Poonsopa D, Roengmongkol K, Charoensataporn R. Salinity effects on antioxidant enzymes in mulberry cultivar. Sci Asia 2003; 29: 109–113.

[33] Kukreja S *et al.* Plant water status, H_2O_2 scavenging enzymes, ethylene evolution and membrane integrity of *Cicer arietinum* roots as affected by salinity. Biol Plant 2005; 49: 305–308.

[34] Gapińska M, Skłodowska M, Gabara B. Effect of short- and long-term salinity on the activities of antioxidative enzymes and lipid peroxidation in tomato roots. Acta Physiol Plant 2008; 30: 11-18.

[35] Eyidogan F, Oz MT. Effect of salinity on antioxidant responses of chickpea seedlings. Acta Physiol Plant 2005; 29: 485–493.

[36] Pan Y, Wu LJ, Yu ZL. Effect of salt and drought stress on antioxidant enzyme activities and SOD isozymes of liquorice (*Glycyrrhiza uralensis* Fisch.). Plant Growth Regul 2006; 49: 157-165.

[37] Yang Z, Wu Y, Li Y, Ling H-Q, Chu C. OsMT1a, a type 1 metallothionein, plays the pivotal role in zinc homeostasis and drought tolerance in rice. Plant Mol Biol 2009; 70: 219-229.

[38] Melchiorre M, Robert G, Trippi V, Racca R, Lascano HR. Superoxide dismutase and glutathione reductase overexpression in wheat protoplast: photooxidative stress tolerance and changes in cellular redox state. Plant Growth Regul 2009; 57: 57–68.

[39] Wang Y, Ying Y, Chen J, Wang XC. Transgenic arabidosis overexpressing Mn SOD enhanced salt-tolerance. Plant Sci 2004; 167: 671-677.

[40] Wang Y *et al.* Ectopic expression of Mn-SOD in Lycopersicon esculentum leads to enhanced tolerance to salt and oxidative stress. J Appl Horticul 2007; 9: 3-8.

[41] Lu ZQ, Liu D, Liu SK. Two rice cytosolic ascorbate peroxidases differentially improve salt tolerance in transgenic Arabidopsis. Plant Cell Rep 2007; 26: 1909-1917.

[42] Eltayeb AE *et al.* Overexpression of monodehydroascorbate reductase in transgenic tobacco confers enhanced tolerance to ozone, salt and polyethylene glycol stresses. Planta 2007; 225: 1255-1264.

[43] Knight H, Trewavas AJ, Knight MR. Calcium signalling in Arabidopsis thaliana responding to drought and salinity. Plant J 1997; 12: 1067-1078.

[44] Wu SJ, Lei D, Zhu JK. SOS1, a genetic locus essential for salt tolerance and potassium acquisition. Plant Cell 1996; 8: 617-627.

[45] Mahajan S, Pandey G, Tuteja N. Calcium and Salt Stress Signaling in Plants: Shedding Light on SOS Pathway. Arch Biochem Biophys 2008; 471: 146-158.

[46] Zhu JK. Salt and drought stress signal transduction in plants. Annu Rev Plant Physiol Plant Mol Biol 2002; 53: 247-273.

[47] Zhu JK, Liu J, Xiong L. Genetic analysis of salt tolerance in Arabidopsis. Evidence for a critical role of potassium nutrition. Plant Cell 1998; 10: 1181-1191.

[48] Quintero FJ, Ohta M, Shi H, Zhu JK, Pardo JM. Reconstitution in yeast of the Arabidopsis SOS signaling pathway for Na^+ homeostasis. Proc Natl Acad Sci USA 2002; 99: 9061-9066.

[49] Apse MP, Aharon GS, Snedden WA, Blumwald E. Salt tolerance conferred by over expression of a vacuolar Na^+/H^+ antiport in Arabidopsis. Science 1999; 285: 1256-1258.

[50] Zepeda-Jazo I, Shabala S, Chen Z, Pottosin II. Na^+-K^+ transport in roots under salt stress. Plant Sig Beh 2008; 3: 401-403.

[51] Tuteja N. Abscisic acid and abiotic stress signalling. Plant Sig Beh 2007; 2: 135-138.

[52] Xiong L., Ishitini M., Lee H., Zhu J-K. The Arabidopsis LOS5/ABA3 locus encodes a molybdenum cofactor sulfurase and modulates cold stress- and osmotic stress-responsive gene expression. Plant Cell 2001; 13: 2063-2083.

[53] Shinozaki K, Yamaguchi-Shinozaki K. Gene Expression and Signal Transduction in Water-Stress Response. Plant Physiol 1997; 115: 327-334.

[54] Yamaguchi-Shinozaki K, Shinozaki K. Characterization of the expression of a desiccation-responsive rd29 gene of Arabidopsis thaliana and analysis of its promoter in transgenic plants. Mol Gen Genet 1993; 236: 331-340.

[55] Mahajan S, Sopory SK, Tuteja N. Cloning and characterization of CBL-CIPK signalling components from a legume (*Pisum sativum*). FEBS J 2006b; 273: 907-925.

[56] Knight H, Trewavas AJ, Knight M. Calcium signalling in *Arabidopsis thaliana* responding to drought and salinity. Plant J 1997; 12: 1067-1078.

[57] Djilianov D *et al.* Improved abiotic stress tolerance in plants by accumulation of osmoprotectants–gene transfer approach. Biotechnol Biotechnol Eq 2005; 19: 63-71.

[58] Ashraf M, Foolad MR. Roles of glycinebetaine and proline in improving plant abiotic stress tolerance. Environ Exp Bot 2007; 59: 206–216.

[59] Trovato M, Mattioli R, Costantino P. Multiple roles of proline in plant stress tolerance and development. Rendiconti Lincei – Scienze Fisiche e Naturali 2008; 19: 325-346.

[60] Yan H, Gong LZ, Zhao CY, Guo WY. Effects of exogenous praline on the physiology of soyabean plantlets regenerated from embryos *in vitro* and on the ultrastructure of their mitochondria under NaCl stress. Soybean Sci 2000; 19: 314-319.

[61] Ueda A, Yamamoto-Yamane Y, Takabe T. Salt stress enhances proline utilization in the apical region of barley roots. Biochem Biophys Res Commun 2007; 355: 61-66.

[62] Smirnoff N, Cumbes QJ. Hydroxyl radical scavenging activity of compatible solutes. Phytochemistry 1989; 28: 1057–1060.

[63] Alia P. Pardha Saradhi. Proline accumulation under heavy metal stress. J Plant Physiol 1991; 138: 554–558.

[64]   Hare PD, Cress WA. Metabolic implications of stress-induced proline accumulation in plants. Plant Growth Regul 1997; 21: 79–102.

[65]   Babiychuk E, Kushnir S, Belles-Boix E, Van Montagu M, Inzé D. *Arabidopsis thaliana* NADPH oxidoreductase homologs confer tolerance of yeast toward the thiol-oxidizing drug diamide. J Biol Chem 1995; 270: 26224-26231.

[66]   Demiral T, Türkan I. Comparative lipid peroxidation, antioxidant defense systems and proline content in roots of two rice cultivars differing in salt tolerance. Environ Exp Bot 2005; 53: 247–257.

[67]   Su J, Wu R. Stress-inducible synthesis of proline in transgenic rice confers faster growth under stress conditions than that with constitutive synthesis. Plant Sci 2004 ; 166 : 941–948.

[68]   Kishor PBK, Hong Z, Miao PB, Hu CAA, Verma DPS. Overexpression of D1-pyrroline-5-carboxylate synthetase increases proline production and confers osmotolerance in transgenic plants. Plant Physiol. 1995; 108: 1387-1394.

[69]   Ueda A, Shi W, Shimada T, Miyake H, Takabe T. Altered expression of barley y-transporter causes different growth responses in Arabidopsis. Planta 2008; 227: 277-286.

[70]   Waditee R *et al.* Genes for direct methylation of glycine provide high levels of glycine betaine and abiotic-stress tolerance in Synechococcus and Arabidopsis. Proc Natl Acad Sci USA 2005; 102: 1318-1323.

[71]   He C, Yang A, Zhang W, Gao Q, Zhang J. Improved salt tolerance of transgenic wheat by introducing betA gene for glycine betaine synthesis. Plant Cell Tiss Organ Cult 2010; 101: 65-78.

[72]   Yang X, Lu C. Photosynthesis is improved by exogenous glycinebetaine in salt stressed maize plants. Physiol Plant 2005; 124: 343-352.

[73]   Bhuiyan NH, Hamada A, Yamada N, Rai V, Hibino T, Takabe T. Regulation of betaine synthesis by precursor supply and choline monooxygenase expression in *Amaranthus tricolor*. J Exp Bot 2007; 58: 4203-4212.

[74]   Waditee R, Bhuiyan NH, Hirata E, Hibino T, Tanaka Y, Shikata M, Takabe T. Metabolic Engineering for Betaine Accumulation in Microbes and Plants. J Biol Chem 2007; 282: 34185-34193.

[75]   Liu J, Zhu JK. Calcium sensor homolog required for plant salt tolerance. Science 1998; 280: 1943-1945.

[76]   Cushman JC, Bohnert HJ. Genomic approaches to plant stress tolerance. Curr Opin Plant Biol 2000; 3: 117-124.

[77]   Sivamani E, Bahieldin A, Wraith JM, Al-Niemi T, Dyer WE, Ho THD, Qu R. Improved biomass productivity and water use efficiency under water deficit conditions in transgenic wheat constitutively expressing the barley HVA1 gene. Plant Sci 2000, 155, 1-9.

[78]   Close TJ. Dehydrins: a commonality in the response of plants to dehydration and low temperature. Physiol Plant 1997; 100: 291-296.

[79]   Apse MP, Aharon GS, Sneddon WA, Blumwald E. Salt tolerance conferred by over expression of a vacuolar Na+/H+ antiport in Arabidopsis. Science 1999; 285: 1256-1258.

[80]   Zhang HX, Blumwald E. Transgenic salt-tolerant tomato plants accumulate salt in foliage but not in fruit. Nature Biotechnol 2001; 19: 765-768.

[81]   Gaxiola RA *et al.* Drought-and salt-tolerant plants result from overexpression of the AVP1 H^+-pump. PNAS USA 2001; 98: 11444-11449.

[82]   Park S *et al.* Up-regulation of a H^+-pyrophosphatase (H^+-PPase) as a strategy to engineer drought-resistant crop plants. PNAS USA 2005; 102: 18830-18835.

[83]   Brini F, Hanin M, Mezghani I, Berkowitz GA, Masmoudi KJ. Overexpression of wheat Na+/H+ antiporter TNHX1 and H+-pyrophosphatase TVP1 improve salt- and drought-stress tolerance in Arabidopsis thaliana plants. Exp Bot 2007; 58: 301-308.

[84]   Jung C *et al.* Overexpression of AtMYB44 Enhances Stomatal Closure to Confer Abiotic Stress Tolerance in Transgenic Arabidopsis. Plant Physiol. 2008; 146: 623-635.

[85]   Jia XY *et al.* Molecular cloning and characterization of wheat calreticulin (CRT) gene involved in drought-stressed responses. J Exp Bot 2008; 59: 739-751.

[86]   Alvarez S, Marsh EL, Schroeder SG, Schachtman DP. Metabolomic and proteomic changes in the xylem sap of maize under drought. Plant Cell Environ 2008; 31: 325-340.

[87]   Mansfield TJ, Atkinson CJ. In: Alscher RG, Cumming JR (Eds.), Stress responses in Plants: Adaptation and acclimation Mechanisms, Stomatal behaviour in water stressed plants. New York: Wiley-Liss, 1990; pp. 241-264.

[88]   Blackman PG, Davies WJ. Root-to-shoot communication in maize plants of the effects of soil drying. J Exp Bot 1985; 36: 39-48.

[89]   Bota J, Flexas J, Medrano H. Is photosynthesis limited by decreased Rubisco activity and RuBP content under progressive water stress? New Phytol 2004; 162: 671-681.

[90]   Loreto F, Tricoli D, Di Marco G. On the relationship between electron transport rate and photosynthesis in leaves of the C_4 plant Sorghum bicolor exposed to water stress, temperature changes and carbon metabolism inhibition. Aust J Plant Physiol 1995; 22: 885-892.

[91]   Meyer S, Genty B. Mapping intercellular CO_2 mole fraction (Ci) in Rosa rubiginosa leaves fed with abscisic acid by using chlorophyll fluorescence imaging: significance of Ci estimated from leaf gas exchange. Plant Physiol 1998; 116: 947-957.

[92]   Zlatev ZS, Lidon FC, Ramalho JC, Yordanov IT. Comparison of resistance to drought of three bean cultivars. Biol Plant 2006; 50: 389-394.

[93]   Wang L, Zhou Q, Ding L, Sun Y. Effect of cadmium toxicity on nitrogen metabolism in leaves of *Solanum nigrum* L. as a newly found cadmium hyperaccumulator. J Hazardous Metals 2008; 154: 818-825.

[94]   Sharma P, Dubey RS. Modulation of nitrate reductase activity in rice seedlings under aluminium toxicity and water stress: Role of osmolytes as enzyme protectant. J Plant Physiol 2005; 162: 854-864.

[95]   Wang CQ, Li RC. Enhancement of superoxide dismutase activity in the leaves of white clover (Trifolium repens L.) in response to polyethylene glycol-induced water stress. Acta Physiol Plant 2008; 30: 841–847.

[96]   Simonovicova M, Tamás L, Huttová J, Mistrík I. Effect of aluminium on oxidative stress related enzymes activities in barley roots. Biol Plant 2004; 48: 261-266.

[97]   Yang Z, Wu Y, Li Y, Ling H-Q, Chu C. OsMT1a, a type 1 metallothionein, plays the pivotal role in zinc homeostasis and drought tolerance in rice. Plant Mol Biol 2009; 70: 219-229.

[98]   Badawi GH, Yamauchi Y, Shimada E, Sasaki R, Kawano N, Tanaka K, Tanaka K. Enhanced tolerance to salt stress and water deficit by overexpressing superoxide dismutase in tobacco (*Nicotiana tabacum*) chloroplasts. Plant Sci 2004; 166: 919-928.

[99]   Ramanjulu S, Bartels D. Drought- and desiccation-induced modulation of gene expression in plants. Plant Cell Environ 2002; 25: 141-151.

[100]   Trouverie J, Thevenot C, Rocher JP, Sotta B, Prioul JL. The role of abscisic acid in the response of a specific vacuolar invertase to water stress in the adult maize leaf. J Exp Bot 2003; 54: 2177-2186.

[101]   Cheng WH *et al.* A unique short-chain dehydro-genase/reductase in Arabidopsis glucose signaling and abscisic acid biosynthesis and functions. Plant Cell 2002; 14: 2732-2743.

[102]   Zhu JK. Plant salt tolerance. Trends Plant Sci 2001; 6: 66-71.

[103]   Shen B, Jensen RG, Bohnert HJ. Mannitol protects against oxidation by hydroxyl radicals. Plant Physiol 1997; 115: 527-532.

[104]   Garg AK *et al.* Trehalose accumulation in rice plants confers high tolerance levels to different abiotic stresses. PNAS USA 2002; 99: 15898-15903.

[105]   Kishor KPB, Hong Z, Miao GH, Hu CAA, Verma DPS. Overexpression of D1-pyrroline-5-carboxylate synthetase increases proline production and confers osmotolerance in transgenic plants. Plant Physiol 1995; 108: 1387-1394.

[106]   Gigon A, Matos AR, Laffray D, Zuily-Fodil Y, Pham-Thi AT. Effect of Drought Stress on Lipid Metabolism in the Leaves of Arabidopsis thaliana (Ecotype Columbia). Ann Bot (Lond) 2004; 94: 345-351.

[107   DeWald DB *et al.* Rapid accumulation of phosphatidylinositol 4,5-bisphosphate and inositol 1,4,5-trisphosphate correlates with calcium mobilization in salt-stressed Arabidopsis. Plant Physiol 2001; 126: 759-769.

[108]   Lee Y *et al.* Abscisic acid-induced phosphoinositide turnover in guard cell protoplasts of *Vicia faba*. Plant Physiol 1996; 110: 987-996.

[109]   Xiong L, Ishitani M, Lee H, Zhu JK. The Arabidopsis LOS5/ABA3 locus encodes a molybdenum cofactor sulfurase and modulates cold stress- and osmotic stress-responsive gene expression. Plant Cell 2001; 13: 2063-2083.

[110]   Hirayama T, Ohto C, Mizoguchi T, Shinozaki K. A gene encoding a phosphatidylinositol-specific phospholipase C is induced by dehydration and salt stress in Arabidopsis thaliana. Proc Natl Acad Sci USA 1995; 92: 3903-3907.

[111]   Sanders D, Brownlee C, Harper JF. Communicating with calcium. Plant Cell 1999; 11: 691-706.

[112]   Frank W, Munnik T, Kerkmann K, Salamini F, Bartels D. Water deficit triggers phospholipase D activity in the resurrection plant *Craterostigma plantagineum*. Plant Cell 2000; 12: 111-123.

[113]   Katagiri T, Takahashi S, Shinozaki K. Hyperosmotic stress induced a rapid and transient increase in inositol 1,4,5-trisphosphate independent of abscisic acid in Arabidopsis cell culture. Plant J 2001; 26: 595-605.

[114]   El Maarouf H, Zuily-Fodil Y, Gareil M, d'Arcy-Lameta A, Pham-Thi AT. Enzymatic activity and gene expression under water stress of phospholipase D in two culitivars of *Vigna unguiculata* L. Walp. Differing in drought tolerance. Plant Mol Biol 1999; 39: 1257-1265.

[115]   Lynch DV. In: Katterman F (Ed.), Environmental injury to plants. Chilling injury in plants: the relevance of membrane lipids, New York: Academic Press, 1990; pp.17-34.

[116]   Hopkins WG. In: Introduction to Plant Physiology, The physiology of plants under stress, John Wiley and Sons, Inc. New York, 2nd edition, 1999; pp 451-475.

[117]   Jiang QW, Kiyoharu O, Ryozo I. Two Novel Mitogen-Activated Protein Signaling Components, OsMEK1 and OsMAP1, Are Involved in a Moderate Low-Temperature Signaling Pathway in Rice. Plant Physiol 2002; 129: 1880-1891.

[118] Guy CL. Cold Accelimation and Freezing Stress Tolerance: Role of Protein Metabolism. Annu Rev Plant Physiol 1990; 41: 187-223.

[119] Steponkus PL. Role of the plasma membrane in freezing injury and cold acclimation. Annu Rev Plant Physiol 1984; 35: 543-84.

[120] Olien CR, Smith MN. Ice adhesions in relation to freeze stress. Plant Physiol 1977; 60: 499-503.

[121] Jones PG, Inouye M. The cold shock response-a hot topic. Mol Microbiol 1994; 11: 811-818.

[122] Gibson S, Arondel V, Iba K, Somerville C. Cloning of a temperature - regulated gene encoding a chloroplast Omega -3 desaturase from *Arabidopsis thaliana*. Plant Physiol 1994; 106: 1615-1621.

[123] Anderson JV, Li QB, Haskell DW, Guy CL. Structural organization of the spinach endoplasmic reticulum-luninal 70-kilodalton heat-shock cognate gene and expression of 70-kilodalton heat shock genes during cold acclimation. Plant Physiol. 1994, 104, 1359-70.

[124] Krishna P, Sacco M, Cherutti JF, Hill S. Cold-induced accumulation of hsp 90 transcripts in *Brasscia napus*. Plant Physiol 1995; 107; 915-923.

[125] Mizoguchi T, Hayashida N, Yamaguchi-Schinozaki K, Kamada H, Shinozaki K. Shinozaki, AtMPKs: a gene family of plant MAP Kinases in *Arabidopsis thaliana*. FEBS Lett 1993; 336: 440-444.

[126] Mizoguchi T *et al.* A gene encoding a mitogen activated protein kinase kinase kinase is induced simultaneously with genes for a mitogen-activated protein kinase and an S6 ribosomal protein kinase by touch, cold and water stress in *Arabidopsis thaliana*. Proc Natl Acad Sci USA 1996; 93: 765-769.

[127] Polisensky DH, Braam J. Cold shock regulation of the Arabidopsis TCH genes and the effects of modulating intracellular Calcium levels. Plant Physiol 1996; 111: 1271-1279.

[128] Ingram J, Bartels D. The molecular basis of dehydration tolerance in plants. Annu Rev Plant Physiol Plant Mol Biol 1996; 47: 377-403.

[129] Thomashow MF. Plant cold acclimation: Freezing tolerance genes and regulatory mechanism. Annu Rev Plant Physiol Plant Mol Biol 1999; 50: 571-599.

[130] Yamaguchi-Shinozaki K, Shinozaki K. A novel cis-acting element in an Arabidopsis gene is involved in responsiveness to drought, low-temperature, or high-salt stress. Plant Cell 1994; 6: 251-264.

[131] Stockinger EJ, Gilmour SJ, Thomashow MF. Arabidopsis thaliana CBF1 encodes an AP2 domain-containing transcription activator that binds to the C-repeat/DRE, a cis-acting DNA regulatory element that stimulates transcription in response to low temperature and water deficit. PNAS USA 1997; 94: 1035-1040.

[132] Liu Q *et al.* Two transcription factors, DREBI and DREB2, with an EREBP/AP2 DNA binding domain separate two cellular signal transduction pathways in drought and low temperature responsive gene expression respectively, in Arabidopsis. Plant Cell 1998; 10: 1391-1406.

[133] Lee H, Xiong L, Gong Z, Ishitani M, Stevenson B, Zhu JK. The Arabidopsis HOS1 gene negatively regulates cold signal transduction and encodes a RING finger protein that displays cold-regulated nucleo-cytoplasmic partitioning. Genes Dev 2001; 15: 912-924.

[134] Chinnusamy V *et al.* ICE1: a regulator of cold-induced transcriptome and freezing tolerance in Arabidopsis. Genes Dev 2003; 17: 1043-1054.

[135] Abbasi F, Onodera H, Toki S, Tanaka H, Komatsu S. OsCDPK 13, a calcium-dependent protein kinase gene from rice, is induced by cold and gibberellin in rice leaf sheath. Plant Mol Biol 2004; 55: 541-552.

[136] Xin Z, Browse J. Eskimo1 mutants of Arabidopsis are constitutively freezing tolerant. Proc Natl Acad Sci USA 1998; 95: 7799-7804.

[137] Sui N, Li M, Zhao S-J, Li F, Liang H, Meng Q-W. Overexpression of glycerol-3-phosphate acyltransferase gene improves chilling tolerance in tomato. Planta 2007; 226: 1097-1108.

[138] Liao Y *et al.* Soybean GmbZIP44, GmbZIP62 and GmbZIP78 genes function as negative regulator of ABA signaling and confer salt and freezing tolerance in transgenic Arabidopsis. Planta 2008; 228: 225-240.

[139] Chaikam V, Karlson D. Functional characterization of two cold shock domain proteins from *Oryza sativa*. Plant Cell Environ 2008; 31: 995-1006.

[140] Monroy AF, Dhindsa RS. Low-Temperature Signal Transduction: Induction of Cold Acclimation-Specific Genes of Alfalfa by Calcium at 25[deg]C. Plant Cell 1995; 7: 321-331.

[141] Sanita di Toppi L, Gabbrielli R. Response to cadmium in higher plants. Env Exp Bot 1999; 41: 105-130.

[142] Simmons RW, Ahmad W, Noble AD, Blummel M, Evans A, Weckenbrock P. Effect of long-term un-treated domestic wastewater reuse on soil quality, wheat grain and straw yields and attributes of fodder quality. Irrigation and Drainage Systems 2009; (in press)

[143] Clemens S. Molecular mechanisms of plant metal tolerance and homeostasis. Planta 2001; 212: 475-486

[144] Salt DE, Blaylock M, Kumar NPBA, Dushenkov V, Ensley BD, Chet I, Raskin I. Phytoremediation: a novel strategy for the removal of toxic metals from the environment using plants. Biotechnology 1995; 13: 468-474.

[145] Warne MS *et al.* Modeling the toxicity of copper and zinc salts to wheat in 14 soils. Environmental Toxicology and Chemistry 2008; 27: 786–792.

[146] Demirevska-kepova K, Simova-Stoilova L, Stoyanova Z, Holzer R, Feller U. Biochemical changes in barely plants after excessive supply of copper andmanganese. Environ Exp Bot 2004; 52: 253–266.

[147] Zhou ZS, Huang SQ, Guo K, Mehta SK, Zhang PC, Yang ZM. Metabolic adaptations to mercury-induced oxidative stress in roots of *Medicago sativa* L. J Inorg Biochem 2007; 101: 1–9.

[148] Calgaroto NS *et al.* Antioxidant system activation by mercury in Pfaffia glomerata plantlets. BioMetals 2010; 23:295-305.

[149] Cargnelutti D *et al.* Mercury toxicity induces oxidative stress in growing cucumber seedlings. Chemosphere 2006; 65: 999–1006.

[150] Maron LG, Kirst M, Mao C, Milner MJ, Menossi M, Kochian LV. Transcriptional profiling of aluminum toxicity and tolerance responses in maize roots. New Phytol 2008; 179: 116–128.

[151] Yadav SK. Heavy metals toxicity in plants: An overview on the role of glutathione and phytochelatins in heavy metal stress tolerance of plants. South Afric J Bot 2010; 76: 167-179.

[152] Mittler R, Vanderauwers S, Gollery M, Van Breusegem F. Reactive oxygen gene network of plants. Trends Plant Sci 2004; 9: 490-498

[153] Gratao PL, Gomes-Junior RA, Delite FS, Lea PJ, Azevedo RA. In: Khan NA, Samiullah (Eds) Cadmium Toxicity and Tolerance in Plants, Antioxidant stress responses of plants to cadmium. Narosa Publishing House, New Delhi, 2006; pp 1-34

[154] Singh S, Anjum NA, Khan NA, Nazar R. In: Khan NA, Singh S (Eds) Abiotic stress and plant responses, Metal-binding peptides and antioxidant defence system in plants: significance in cadmium tolerance. IK International, New Delhi, 2008; pp 159-189

[155] Singh S, Khan NA, Nazar R, Anjum NA. Photosynthetic traits and activities of antioxidant enzymes in blackgram (*Vigna mungo* L. Hepper) under cadmium stress. Am J Plant Physiol 2008; 3: 25-32.

[156] Medici LO, Azevedo RA, Smith RJ, Lea PJ. The influence of nitrogen supply on antioxidant enzymes in plant roots. Funct Plant Biol 2004; 31: 1-9.

[157] Noctor G, Foyer CH. Ascorbate and glutathione: Keeping active oxygen under control. Ann Rev Plant Physiol Mol Biol 1998; 49: 249-279.

[158] Moller IM. Plant mitochondria and oxidative stress: electron transport, NADPH turnover, and metabolism of reactive oxygen species. Ann Rev Plant Physiol Mol Biol 2001; 52: 561-591.

[159] Li Y, Song Y, Shi G, Wang J, Hou X, Response of antioxidant activity to excess copper in two cultivars of Brassica campestris ssp. chinensis Makino. Acta Physiol Plant 2009; 31: 155-162.

[160] Srivastava AK, Bhargava P, Rai LC, Salinity and copper-induced oxidative damage and changes in antioxidative defense system of Anabaena doliolum. Microb Biotechnol 2005; 22: 1291–1298.

[161] Hsu YT, Kao CH. Heat shock-mediated H_2O_2 accumulation and protection against Cd toxicity in rice seedlings. Plant Soil 2007; 300: 137-147.

[162] Pekker I, Telor E, Mittler R. Reactive oxygen intermediates and glutathione regulate the expression of cytosolic ascorbate peroxidase during iron-mediated oxidative stress in bean. Plant Mol Biol 2002; 49: 429–438.

[163] Lee SH *et al.* Simultaneous overexpression of both CuZn superoxide dismutase and ascorbate peroxidase in transgenic tall fescue plants confers increased tolerance to a wide range of abiotic stresses. J Plant Physiol 2007; 164: 1626-1638.

[164] Azpilicueta CE, Benavides MP, Tomaro ML, Gallego SM. Mechanism of CATA3 induction by cadmium in sunflower leaves. Plant Physiol Biochem 2007; 45: 589-595.

[165] Gichner T, Patkova Z, Szakova J, Demnerova K. Cadmium induces DNA damage in tobacco roots, but no DNA damage, somatic mutations or homologous recombinations in tobacco leaves. Mut Res Genet Toxicol Environ Mut 2004; 559: 49-57.

[166] Guan ZQ, Chai TY, Zhang YX, Xu J, Wei W. Enhancement of Cd tolerance in transgenic tobacco plants overexpressing a Cd-induced catalase cDNA. Chemosphere 76 (2009) 623-630.

[167] Bashir K *et al.* Expression and enzyme activity of glutathione reductase is upregulated by Fe-deficiency in graminaceous plants. Plant Mol Biol 2007; 65: 277-284.

[168] Yoon HS, Lee IA, Lee H, Lee BH, Jo J. Overexpression of a eukaryotic glutathione reductase gene from Brassica campestris improved resistance to oxidative stress in *Escherichia coli*. Biochem Biophys Res Comm 2005; 326: 618-623.

[169] Yin L *et al.* Overexpression of dehydroascorbate reductase, but not monodehydroascorbate reductase, confers tolerance to aluminum stress in transgenic tobacco. Planta 2010 ; 231 : 609-621.

[170] Leisinger U, Rüfenacht K, Fischer B, Pesaro M, Spengler A, Zehnder AJB, Eggen RIL. The glutathione peroxidase homologous gene from *Chlamydomonas reinhardtii* is transcriptionally up-regulated by singlet oxygen. Plant Mol Biol 2001; 46: 395–408.

[171] Wang YC *et al.* Enhanced salt tolerance of transgenic poplar plants expressing a manganese superoxide dismutase from Tamarix androssowii. Mol Biol Rep 2010; 37: 1119–1124.

[172] Prashanth SR, Sadhasivam V, Parida A. Over expression of cytosolic copper/zinc superoxide dismutase from a mangrove plant Avicennia marina in indica Rice var Pusa Basmati-1 confers abiotic stress tolerance. Transgenic Res 2008; 17: 281–291.

[173] Rubio MC *et al.* Effects of water stress on antioxidant enzymes of leaves and nodules of transgenic alfalfa overexpressing superoxide dismutases. Physiol Plant 2002; 115: 531–540.

[174] Lee SH *et al.* Simultaneous overexpression of both CuZn superoxide dismutase and ascorbate peroxidase in transgenic tall fescue plants confers increased tolerance to a wide range of abiotic stresses. Plant Physiol 2007; 164: 1626-1638.

[175] Guan ZQ, Chai TY, Zhang YX, Xu J, Wei W. Enhancement of Cd tolerance in transgenic tobacco plants overexpressing a Cd-induced catalase cDNA. Chemosphere 2009; 76: 623-630.

[176] Al-Taweel K, Iwaki T, Yabuta Y, Shigeoka S, Murata N, Wadano A. A bacterial transgene for catalase protects translation of d1 protein during exposure of salt-stressed tobacco leaves to strong light. Plant Physiol 2007; 145: 258–265.

[177] Xu WF, Shi WM, Ueda A, Takabe T. Mechanisms of Salt Tolerance in Transgenic Arabidopsis thaliana Carrying a Peroxisomal Ascorbate Peroxidase Gene from Barley. Pedosphere 2008; 18: 486-495.

[178] Kim YH, Kim CY, Song WK, Park DS, Kwon SY, Lee HS, Bang JW, Kwak SS. Overexpression of sweetpotato *swpa4* peroxidase results in increased hydrogen peroxide production and enhances stress tolerance in tobacco. Planta 2008; 227: 867–881.

[179] Wang Y, Wisniewski M, Meilan R, Cui M, Fuchigami L. Transgenic tomato (*Lycopersicon esculentum*) that overexpress cAPX exhibits enhanced tolerance to UV-B and heat stress. J Appl Horticult 2006; 8: 87-90.

[180] Melchiorre M, Robert G, Trippi V, Racca R, Lascano HR. Superoxide dismutase and glutathione reductase overexpression in wheat protoplast: photooxidative stress tolerance and changes in cellular redox state. Plant Growth Regul 2009; 57: 57–68.

[181] Kornyeyev D, Logan BA, Payton P, Allen RD, Holaday AS. Elevated chloroplastic glutathione reductase activities decrease chilling-induced photoinhibition by increasing rates of photochemistry, but not thermal energy dissipation, in transgenic cotton. Functional Plant Biol 2003; 30: 101–110.

[182] Eltayeb AE *et al.* Overexpression of monodehydroascorbate reductase in transgenic tobacco confers enhanced tolerance to ozone, salt and polyethylene glycol stresses. Planta 2007; 225: 1255-1264.

[183] Ushimaru T *et al.* Transgenic Arabidopsis plants expressing the rice dehydroascorbate reductase gene are resistant to salt stress. J Plant Physiol 2006; 163: 1179-1184.

[184] Zhao F, Zhang H. Salt and paraquat stress tolerance results from co-expression of the Suaeda salsa glutathione S-transferase and catalase in transgenic rice. Plant Cell Tiss Organ Cult 2006; 86: 349-358.

[185] Roxas VP, Lodhi SA, Garrett DK, Mahan JR, Allen RD. Stress Tolerance in Transgenic Tobacco Seedlings that Overexpress Glutathione S-Transferase/Glutathione Peroxidase. Plant Cell Physiol 2000; 41: 1229-1234.

[186] Yoshimura K *et al.* Enhancement of stress tolerance in transgenic tobacco plants overexpressing Chlamydomonas glutathione peroxidase in chloroplasts or cytosol. Plant J 2004; 37: 21–33.

[187] Gaber A *et al.* Glutathione peroxidase-like protein of Synechocystis PCC 6803 confers tolerance to oxidative and environmental stresses in transgenic Arabidopsis. Physiol Plant 2006; 128: 251-262.

[188] Vendruscolo ECG *et al.* Stress-induced synthesis of proline confers tolerance to water deficit in transgenic wheat. J Plant Physiol 2007; 164: 1367-1376.

[189] Molinari HBC *et al.* Evaluation of the stress-inducible production of proline in transgenic sugarcane (Saccharum spp.): osmotic adjustment, chlorophyll fluorescence and oxidative stress. Physiol Plant 2007; 130: 218-229.

[190] Yamada M *et al.* Effects of free proline accumulation in petunias under drought stress. J Exp Bot 2005; 56: 1975-1981.

[191] Ma L, Zhou E, Gao L, Mao X, Zhou R, Jia J. Isolation, expression analysis and chromosomal location of P5CR gene in common wheat (*Triticum aestivum* L.). S African J Bot 2008; 74: 705-712.

[192] Simon-Sarkadi L, Kocsy G, Várhegyi Á, Galiba G, De Ronde JA. Stress-induced changes in the free amino acid composition in transgenic soybean plants having increased proline content. Biol Plant 2006; 50 793-796.

[193] Zhao YW, Hao JG, Bu HY, Wang YJ, Jia JF. Cloning of HvBADH1 Gene from Hulless Barley and its Transformation to Tobacco. Acta Agronomica Sinica 2008; 34: 1153-1159.

[194] Yang X, Liang Z, Wen X, Lu C. Genetic engineering of the biosynthesis of glycinebetaine leads to increased tolerance of photosynthesis to salt stress in transgenic tobacco plants. Plant Mol Biol 2008; 66:73–86.

[195] Hasthanasombut S, Ntui V, Supaibulwatana K, Mii M, Nakamura I. Expression of Indica rice OsBADH1 gene under salinity stress in transgenic tobacco. Plant Biotech Rep 2010; 4: 75–83.

[196] Quan R, Shang M, Zhang H, Zhao Y, Zhang J. Improved chilling tolerance by transformation with betA gene for the enhancement of glycinebetaine synthesis in maize. Plant Sci 2004; 166: 141-149.

[197] Huang J *et al.* Genetic engineering of glycinebetaine production toward enhancing stress tolerance in plants: metabolic limitations. Plant Physiol 2000; 122: 747–756.

[198] Park EJ, Jeknić Z, Chen THH, Murata N. The codA transgene for glycinebetaine synthesis increases the size of flowers and fruits in tomato. Plant Biotech J 2007; 5: 422-430.

[199] Mohanty A *et al.* Transgenics of an elite indica rice variety Pusa Basmati 1 harboring the *codA* gene are highly tolerant to salt stress. Theor Appl Genet 2002; 106: 51–57.

[200] Park EJ *et al.* Genetic engineering of glycinebetaine synthesis in tomato protects seeds, plants, and flowers from chilling damage. Plant J 2004; 40: 474–487.

CHAPTER 4

Genomic Overview of Ion Transporters in Plant Salt Tolerance

Faïçal Brini, Kaouther Feki, Habib Khoudi, Moez Hanin and Khaled Masmoudi*

Plant Molecular Genetic Laboratory, Centre of Biotechnology of Sfax (CBS) Route Sidi Mansour Km 6, B.P "1177" 3018 Sfax –Tunisia

Abstract: Salinity is one of the most severe environmental stresses affecting plant productivity worldwide. In many plant species, salt sensitivity is associated with the accumulation of sodium (Na^+) in photosynthetic tissues. Adaptation of plants to salt stress (i.e. resumption of growth after exposure to high soil salinity) requires cellular ion homeostasis. To prevent the accumulation of Na^+ in the cytoplasm, plants have developed three mechanisms that function in a cooperative manner, i.e restriction of Na^+ influx, active Na^+ extrusion at the root-soil interface, and subsequent vacuolar compartmentalization without toxic ion accumulation in the cytosol. Sodium ions can enter the cell through several low- and high-affinity K^+ carriers. Voltage-independent, non selective cation channels (NSCC) provide a pathway for the entry of Na^+ into plant cells. Some members of the HKT family (High Affinity K^+ transporter) function as sodium transporter and contribute to Na+ removal from the ascending xylem sap and recirculation from the leaves to the roots *via* the phloem vasculature. Sodium extrusion is presumed to be of critical importance for ion homeostasis and salt tolerance of glycophytes. Na^+ sequestration into the vacuole depends on expression and activity of Na^+/H^+ antiporter that is driven by electrochemical gradient of protons generated by the vacuolar H^+-ATPase and the H^+-PPase. However, we have a limited molecular understanding of the overall control of Na^+ accumulation, the role of each transporter and salt stress tolerance at the whole plant level. Genomics and functional genomics provide a new opportunity in addressing the multigenicity of the plant abiotic stress response through genome sequences, stress-specific transcript collections, protein and metabolite profiles, their dynamic changes and protein interactions. In this review, we analyze available data related to omics and plant abiotic stress responses in order to enhance our understanding about how salinity and other abiotic stresses affect the most fundamental processes of cellular function which have a substantial impact on plant growth development.

PLANT ADAPTATION TO SALT STRESS

The increasingly agricultural and environmental problems caused by soil salinity on a worldwide scale have received much attention. Nearly 20% of the world's cultivated land and nearly half of all irrigated lands are affected by salinity [1]. To understand how plants tolerate and acclimate to saline environments is of great importance for genetic modification and plant selection. In general, salinity causes direct and indirect stresses on plant tissues, e.g (1) a reduced water availability, due to the high soluble salts in the soil; (2) ion-specific effects, resulting from the excessive accumulation of toxic ions in plants (Na^+ and Cl^-); and, (3) the salt-induced secondary stress, e.g oxidative bursts caused by over-production of ROS (reactive oxygen species) [1,2,3,4]. Ion accumulation contributes to decreasing cellular osmotic potentials but this makes plants confront ion toxicity and oxidative stress. Therefore, whether plants could survive salinity is dependent to a large extent on the ability to retain the ionic homeostasis under saline conditions. Among the altered ion relations, the maintenance of low Na^+ and Cl^-, as well as keeping high concentrations of nutrition elements, especially K^+ homeostasis, are crucial traits for plant salt adaptation [1-3].

The ability of plants to grow in high NaCl concentrations is associated with the ability of the plants to transport, compartmentalize, extrude, and mobilize Na^+ ions. While the influx and efflux at the roots establish the steady state rate of entry of Na^+ into the plant, the compartmentation of Na^+ into the cell vacuoles and the radial transport of Na^+ to the stele and its loading into the xylem establish the homeostatic control of Na^+ in the cytosol of the root cells. Removal of Na^+ from the transpirational stream, its distribution within the plant and its progressive accumulation in the leaf vacuoles, will determine the ability to deal with the toxic effects of Na^+. In addition to the inhibition of Na^+ influx, there are two major strategies to prevent Na^+ toxicity in the cytosol: (1) improving vacuolar Na^+

*Address correspondence to: Dr. Khaled Masmoudi, Plant Molecular Genetic Laboratory, Centre of Biotechnology of Sfax (CBS) Route Sidi Mansour Km 6, B.P "1177" 3018 Sfax–Tunisia; E-mail: khaled.masmoudi@cbs.rnrt.tn

compartmentation *via* tonoplast Na^+/H^+ antiporters, and (2) increasing active Na^+ extrusion to the external environment through Na^+/H^+ antiporters located in the plasma membrane (PM) [1,2].

ION HOMEOSTASIS

Since NaCl is the principal soil salinity stress, a research focus has been given to the transport determinants that are necessary for the control of Na^+ and Cl^- homeostasis in saline environments and the utilization of Na^+ as an osmotic solute [3,5,6]. Although all the transport systems necessary for ion homeostasis have not been verified by molecular genetic tools, huge data has been acquired about how Na^+ homeostasis is achieved, including the function of the plasma membrane, the tonoplast H^+ pumps, Na^+/H^+ antiporters and Na^+ influx system [7,8]. Much less is known about how intracellular Cl^- homeostasis is achieved [6]. Table **1** lists the genes that have been identified either by loss- or gain of function approaches as salt tolerance determinants.

Table 1: Plant ion transporters involved in salt tolerance identified by functional analysis

Name	Source species	Gene product	Function	Identification method	Target species	References
			Sodium influx			
AtHKT1	*A. thaliana*	Na^+ transporter	Na^+/K^+ homeostasis	Mutation	Arabidopsis	[104]
HKT1	*T. aestivum*	Na^+/K^+ transporter	K^+/Na^+ homeostasis	Overexpression	Wheat	[105]
			Sodium efflux			
AtSOS1	*T. aestivum*	Plasma membrane Na^+/H^+ antiporter	Na^+ detoxification	Mutation / Overexpression	Arabidopsis	[106,36]
TaSOS1	*T. aestivum*	Plasma membrana Na^+/H^+ antiporter	Na^+ detoxification	Mutation / Overexpression	Arabidopsis	[39]
			Sodium compartmentation			
AtNHX1	*A. thaliana*	Vacuolar Na^+/H^+ antiporter	Na^+ vacuolar sequestration	Overexpression	Arabidopsis, Cabbage, tomato	[44,45]
TNHX1	*T. aestivum*	Vacuolar Na^+/H^+ antiporter	Na^+ vacuolar sequestration	Overexpression	Arabidopsis	[62,63]
AgNHX1	*A. gmelini*	Vacuolar Na^+/H^+ antiporter	Na^+ vacuolar sequestration	Overexpression	Rice	[107]
AVP1	*A. thaliana*	Vacuolar H^+-PPase	H^+ transport, vacuolar acidification	Overexpression	Arabidopsis	[108]
TVP1	*T. aestivum*	Vacuolar H^+-PPase	H^+ transport, vacuolar acidification	Overexpression	Arabidopsis	[62,63]
			Regulatory genes			
SOS2	*A. thaliana*	Serine/ threonine Protein kinase	SOS1 regulator	Mutation	Arabidopsis	[109]
SOS3	*A. thaliana*	Ca^{++} binding protein	Ca^{++} sensor/SOS2 activator	Mutation	Arabidopsis	[110]

Sodium Influx: The HKT and NSCC Families of Ion Transporters

A high K^+/Na^+ ratio in the cytosol is essential for normal cellular functions of plants. Na^+ competes with K^+ uptake through Na^+-K^+ cotransporters, and may also block the K^+ specific transporters of root cells under salinity [8]. This results in accumulation of toxic levels of sodium as well as insufficient K^+ concentration for enzymatic reactions and osmotic adjustment. Under salinity, sodium gains entry into root cell cytosol through cation channels or transporters (selective and nonselective) or into the root xylem stream *via* an apoplastic pathway depending on the plant species. The transport systems named HKT upon first identification, for High Affinity K^+ Transporters, are active at the plasma membrane and have been shown to function as Na^+/K^+ symporters and as Na^+-selective uniporters [9,10]. Phylogenetic analyses of available HKT sequences revealed two major subfamilies named 1 and 2 [11]. It was suggested that subfamily 1 (HKT1;x) would correspond to HKT transporters permeable to Na^+ only and subfamily 2 (HKT2;y) to transporters permeable to both Na^+ and K^+. In *Arabidopsis*, the HKT family comprises a single member, AtHKT1;1, which is permeable to Na^+ only [12] and contributes to Na^+ removal from the ascending xylem sap and recirculation from the leaves to the roots *via* the phloem vasculature [13,14]. Interestingly, the HKT family comprises a much larger number of members in rice, with 7 to 9 genes depending on the cultivar. OsHKT1 specifically mediates Na^+ uptake in rice roots when the plants are K^+ deficient [10]. Functional analyses in *Xenopus laevis* oocytes revealed striking diversity. OsHKT1;1 and OsHKT1;3, shown to be permeable to Na^+ only, are strongly different in terms of affinity for this cation and direction of transport (inward only or reversible). OsHKT2;1 displays a diverse permeation modes, Na^+-K^+ symport, Na^+ uniport, or inhibited states, depending on external Na^+ and K^+ concentration within the physiological concentration range [15]. Horie *et al.* [16] showed that OsHKT2;1 is the central transporter for nutritional Na^+ uptake into K^+ starved rice roots. High-affinity Na^+ uptake mediated by HKT transporters seem not to be universal for plants. In the moss *Physcomitrella patens*, high-affinity Na^+ uptake is not mediated by PpHKT1 [17]. Several QTLs responsible for variation of K^+ and Na^+ content were mapped to HKT family genes. SKC1 gene, member of the HKT-type transporters was mapped to rice QTL that maintained K^+ homeostasis in the salt-tolerant variety under salt stress [18]. A major QTL for Na^+ exclusion, Nax2, which removes Na^+ from the xylem in the roots and leads to a high K^+- to Na^+ ratio in the leaves was mapped to chromosome 5AL in durum wheat. TmHKT1;5-A is a candidate gene for Nax2 [19]. HKT transporters are strongly expressed in tissues (root epidermis and cortex, xylem and phloem, vascular bundle regions) playing crucial roles in ion uptake, or long distance transport and redistribution in the plant. The subfamily 1 transporters (AtHKT1;1 from *Arabidopsis*, OsHKT1;5 from rice and McHKT1;1 from *Mesembryanthemum crystallinum* have been constantly reported to be expressed in the plant vasculature, and rarely in other tissues. Expression patterns of the subfamily 2 transporters (TaHKT2;1 from wheat, OsHKT2;1 and OsHKT2;2 from rice) have always been found to include root periphery cells, and often tissues in or close to the plant vasculature.

Ion channel transporters are likely candidates for mediating the passive transport of Na^+ into the cells. In the last few years, evidence has been presented supporting the existence of weakly voltage-dependent NSCC that are the main pathway for Na^+ entry into the roots, at high soil NaCl concentrations [9]. Although there are many candidate genes in the databases that could encode these NSCC channels, their identity remains elusive. Two families of NSCC, CNGCs (cyclic nucleotide-gated channels) [20], and GLRs (glutamate-activated channels) [21] have been suggested to be candidate NSCC channels [22].

Sodium efflux: The Plasma Membrane Na$^+$/H$^+$ Antiporter (SOS1)

Sodium efflux from root cells prevents accumulation of toxic levels of Na^+ in the cytosol and transport of Na^+ to the shoot. Molecular genetic analysis in *Arabidopsis* SOS mutants have led to the identification of a plasma membrane Na^+/H^+ antiporter, SOS1, which plays a crucial role in sodium extrusion from root epidermal cells under salinity. The Na^+/H^+ antiporter localized to the plasma membrane (SOS1) is the only Na^+ efflux protein from plants characterized so far. The expression of SOS1 is ubiquitous, but stronger in epidermal cells surrounding the root-tip, as well as parenchyma cells bordering the xylem. Thus, SOS1 functions as a Na^+/H^+ antiporter on the plasma membrane and plays a crucial role in sodium efflux from root cells and the long distance Na^+ transport from root to shoot [23]. In *Arabidopsis thaliana*, three salt overly sensitive genes (SOS1, SOS2 and SOS3) have been found to function in a common pathway [24,25]. Mutants of *Arabidopsis* lacking SOS1 are extremely salt sensitive and have combined defects in Na^+ extrusion and in controlling long-distance Na^+ transport from roots to shoots [26,23]. Under salt or oxidative stress, SOS1 interacts through its predicted cytoplasmic tail with RCD1 (radical induced cell death), a regulator of oxidative stress responses, and functions in oxidative stress tolerance in *Arabidopsis* [27]. Na^+ efflux with

SOS1 exchanger is regulated through protein phosphorylation by SOS2/SOS3 protein kinase complex [26,28]. This model acts primarily in roots under salt stress. By contrast, the SOS3 homolog SOS3-like calcium binding protein8 (SCABP8)/calcineurin B-like10 functions mainly in the shoot response to salt toxicity [29,30]. SOS3 is a myristoylated calcium-binding protein capable of sensing Ca^{2+} elicited by salt stress. SOS2 is a serine/threonine protein kinase belonging to the SNF1-related kinase (SnRK3) family [31,32]. SOS2 is activated by Ca-SOS3 and subsequently phosphorylates the ion transporter SOS1 to bring about cellular ion homeostasis under salt stress [33,28,34]. The crystal structure of the binary complex of Ca-SOS3 with the C-terminal regulatory part of SOS2 resolves crucial questions regarding the dual function of SOS2 as a kinase and a phosphatase-binding protein [35].

Transgenic *Arabidopsis* plants overexpressing SOS1 have lower Na^+ in the xylem transpirational stream and in shoots compared with wild-type plants. These plants also show enhanced salt tolerance, measured in terms of their growth, ability to bolt and flower at high salt concentrations (50-200 mM NaCl), while control plants became necrotic and have failed to bolt [36]. Salt tolerance of transgenic plants overexpressing SOS1, SOS2 and SOS3 from *Arabidopsis* showed similar tolerance to plants overexpressing either SOS1 or SOS3 [37]. Besides the AtSOS1 from *Arabidopsis*, OsSOS1 from rice and TaSOS1 from bread wheat are the most characterized plant plasma membrane Na^+/H^+ antiporters reported in literature [38,39].

Sodium Compartmentation: The Vacuolar Na^+/H^+ Antiporter and the H^+-PPase

The compartmentation of Na^+ ions into vacuoles provides an efficient mechanism to avert the toxic effects of Na^+ in the cytosol. The transport of Na^+ into the vacuoles is mediated by cation/H^+ antiporters that are driven by the electrochemical gradient of protons generated by the vacuolar H^+-translocating enzymes, the H^+-ATPase and the H^+-PPase. Although the activity of these cation/H^+ antiporters was demonstrated more than 20 years ago [40], their molecular characterization was only possible after the *Arabidopsis* Genome-sequencing project. The *Arabidopsis* Na^+/H^+ antiporter AtNHX1 was identified based on its sequence similarity to the *Saccharomyces cerevisiae* NHX1 [41]. NHX antiporters in *Arabidopsis* are encoded by a gene family comprising of six members AtNHX1–6 [42,43]. AtNHX1 was shown to mediate both Na^+/H^+ and K^+/H^+ exchange in plant vacuoles [44,45,46], and in vacuoles from yeast expressing AtNHX1 [47]. Since then, the existence of NHX-like proteins has been demonstrated in all plants tested; whether gymnosperm or angiosperm, monocots or dicots [48]. Vacuolar NHX transporters have been shown to play significant roles in endosomal pH regulation [49], cellular K^+ homeostasis and cell expansion [50], vesicular trafficking and protein targeting [51,52,53]. A correlation between the expression of genes encoding NHX antiporters in salt-tolerant cultivars and their salt tolerance was shown in cotton [54]. Similar results were observed in wheat suggesting that the higher expression of endogenous vacuolar Na^+/H^+ antiporters in roots and shoots of the salt-resistant wheat genotypes facilitated Na^+ exclusion from the cytosol, improving salt tolerance [55]. The key role of NHX genes has been emphasized with the generation of salt-tolerant transgenic plants through the overexpression of NHX genes in a wide variety of species [22,56]. It was shown that overexpression of NHX proteins in various plants improve salt tolerance, indicating a role for the protein in vacuolar Na^+ accumulation [44,45]. However, it was shown that the encoded proteins also catalyze K^+/H^+ exchange with similar activity [46]. Based on localization and ion specificity, it was proposed that endosomal NHX isoforms are essential to set the pH of endosomal compartments, which is believed to be fundamental for proper functioning of vesicle sorting in the secretory and endocytic membrane system [48]. Since NHXs transport Na^+ and K^+ with similar affinities and in some cases K^+ is the only exchanged cation, as demonstrated for tomato LeNHX2 by Venema *et al.* [57], there is no evidence for an effective Na^+ sequestration into the vacuole mediated by NHX. Overexpression of LeNHX2 antiporter in *A. thaliana* plants increases salt tolerance, but reduces growth at suboptimal K^+ concentration [58].

The H^+-PPases are considered to form a multigene family. Two cDNA clones (OVP1 and OVP2) encoding vacuolar H^+-PPases isolated from rice were reported [59]. Indeed, there are two genes in *Arabidopsis* annotated as inorganic pyrophosphatase H^+-PPase (AVP1, AVP3) and a third loci encoding a pyrophosphatase-like (AVP2=AVPL1), more than five isoforms in rice, and at least three isoforms in barley [60,61]. It has been previously demonstrated that overexpression of the *Arabidopsis* H^+-pyrophosphatase (AVP1) confers salt tolerance to yeast strains only when the yeast contains a functional vacuolar Na^+/H^+ exchanger (*Nhx1*) [41]. Thus, the Na^+/H^+ exchanger acts in concert with the vacuolar H^+-PPase (and ATPase) to sequester cations in the vacuole (and prevacuolar compartment). Heterologous expression in yeast mutants allowed for an indirect characterization of their functions. Functional characterization of wheat Na^+/H^+ antiporter TNHX1 and vacuolar H^+-PPase pump TVP1 was reported by Brini *et al.*

[62]. Transgenic *Arabidopsis* plants overexpressing the wheat vacuolar Na^+/H^+ antiporter TNHX1 and H^+-PPase TVP1 are much more resistant to high concentrations of NaCl and to water deprivation than the wild-type strains. These transgenic plants grow well in the presence of 200 mM NaCl and also under a water-deprivation regime, while wild-type plants exhibit chlorosis and growth inhibition [63].

FUNCTIONAL GENOMICS AND STRESS RESPONSE IN PLANTS

Transcript Profiling

Understanding the mechanisms involved in the response of plants to adverse environmental conditions is the first step in the generation of crops with higher tolerance to stress. Research at the level of genes (genomics), proteins (proteomics) and metabolites (metabolomics) has been fundamental in the current understanding of the response of plants to stress.

Genome analysis has been mostly limited to model plants that fulfill some specific requirements such as: (1) small genome size, (2) short generation time, (3) small size to enable growth in limited space, and (4) availability of gene manipulation technologies [64]. In particular, two of the most important model species are *Arabidopsis thaliana* and rice (*Oryza sativa* L.) for dicotyledonous and monocotyledonous plant species, respectively. Besides its importance as a crop, rice has a high degree of synteny with genomes of other cereals plants, such as maize, wheat, barley and other grasses because their genomes share a considerable similarity in their organization, as well as sequence similarity [65,66,67]. Great advances in the comparison of genomes and transcriptomes of different organisms have contributed to the development of comparative genomics as one of the most promising fields in the area [68]. In this context, finding variations in the genome or the transcriptome from the current model species related to interesting agronomic traits is very important for crop biotechnology [69].

Gene expression profiling has allowed the identification of hundreds of genes induced when plants are exposed to stress [70,71,72,73]. The availability of the complete genome sequence of some model plants, such as *O. sativa* and *A. thaliana*, has allowed the development of whole genome tiling microarrays. This constitutes a new powerful technology that has already made possible the identification of several unannotated transcripts responsive to abiotic stress [74,75]. However, finding a gene responsive to stress does not necessarily guarantee its participation in tolerance to this condition. Identification and sequencing allow assigning a putative function to a sequence when a significant homology with genes of known function is found. These results are then usually complemented with a proper validation by the use of transgenics. This approach has been especially important in the discovery of several candidate genes in crops in the last decade and, in some cases, it has led to significant improvements in tolerance to stress. Several contributions have reported the changes to transcript profiles under cold, drought, and high-salinity conditions, mainly in *A. thaliana* and rice [76,77,78]. AtNHX1, a vacuolar cation/proton antiporter of Arabidopsis, plays an important role in salt tolerance, ion homeostasis and development. Sottosanto *et al.* [52] used the T-DNA insertional mutant of AtNHX1 (nhx1 plants) and Affymetrix ATH1 DNA arrays to assess differences in transcriptional profiles. They demonstrate that the nhx1 transcriptome was differentially affected when the plants were grown in the absence or presence of salt. In addition to the known role of AtNHX1 on ion homeostasis, the vacuolar cation/proton antiporter plays a significant role in intracellular vesicular trafficking, protein targeting, and other cellular processes. SOS signal pathway that mediates ion homeostasis and salt tolerance controls expression of only a few salt stress-specific tolerance determinant genes among the numerous genes that are regulated in plant response to salt treatment. Analysis of salt responsive gene expression profiles in wild type and the salt-hypersensitive mutant

SOS3 seedlings of *Arabidopsis* revealed six of 89 genes that were expressed differentially [79]. Comparative genomics in salt tolerance between *Arabidopsis* and *Arabidopsis*-related halophyte salt cress (*Thellungiella halophyla*) using a full-length *Arabidopsis* cDNA microarray revealed that only 6 genes were strongly induced in response to high salinity stress in salt cress, whereas 40 genes were identified as salt stress-inducible genes in *Arabidopsis* [80].

In addition, profiling studies have been directed towards understanding transcriptional regulation during nutrient stresses [81,82] or on probing the effects of stresses on the transcriptome by eliminating regulatory genes that are involved in stress perception or signalling [83]. The consolidation and integration of individual transcript-profiling results (for which the establishment of data standards will be necessary) into a species-wide and later, a plant wide database would provide control and a high level of confidence. The transcript profiles for *A. thaliana* collated in AtGenExpress (http://www.arabidopsis.org/ info/expression/ATGenExpress.jsp) form one such database. Nonetheless, we have insufficient information about stress-dependent, cell-specific expression patterns (for example, in guard cells, cells in the vasculature or shoot and root meristems) and current technologies do not readily distinguish the presence of different splicing isoforms or tell us how resident transcripts might be utilized in translation [84].

Despite similarities among different plants, it must not be forgotten that species such as wheat and barley, with far less characterized genomes compared to model plants, may offer unique and interesting features. Their high level of abiotic tolerance and diversity may provide important resources for validation of candidate genes and accelerate important breeding programs [85]. The currently available number of EST sequences in plantGDB is increasing 3 times per year with 19,875 ESTs in durum wheat, 1,076,665 ESTs in bread wheat, 522,561 ESTs in barley, and 2,099,800 ESTs in maize (www.plantgdb.org). The large number of ESTs and the diversity of cDNA libraries that have been used to generate sequences have made 'electronic Northerns' a useful method for assessing gene expression and this provides a good first measure of transcript abundance. Several microarray and macroarray platforms have been generated for the cereals. For wheat and barley, there are a number of proprietal arrays such as the 10 000 cDNA array reported by Leader [86]. More recently, Affymetrix arrays have been developed for both wheat and barley [87]. Currently, there are few published reports on the use of barley or wheat chips for studying altered gene expression in response to abiotic stress. Transcript profiling of stressed rice and maize has been conducted. For example, Hazen *et al.* [88] screened a 21000 rice Affymetrix array with RNA from drought stressed rice to identify 662 differentially expressed genes. A further important resource for transcript profiling can be found at the Rice MPSS site (massively parallel signature sequencing) [89]. This contains the results of an extensive screen of rice transcript signatures for different developmental stages and for cold-, drought- and salt-stressed rice plants. The MPSS data gives a very broad picture of transcript profiles in the target tissues.

Recently, the development of novel high-throughput DNA sequencing methods has provided a new method for both mapping and quantifying transcriptomes. This method, termed RNA-Seq (RNA sequencing), has clear advantages over existing approaches and is expected to revolutionize the manner in which eukaryotic transcriptomes are analysed. RNA-Seq is the first sequencing based method that allows the entire transcriptome to be surveyed in a very high-throughput and quantitative manner [90, 91,92,4]. This method offers both single-base resolution for annotation and 'digital' gene expression levels at the genome scale, often at a much lower cost than either tiling arrays or large-scale Sanger EST sequencing. Currently, the most widely used systems to generate RNASeq data are those developed by Illumina (formerly Solexa), Applied Biosystems (AB), and 454 Life Sciences (Roche).

Proteomics and Metabolomics

The importance of protein profiling has long been acknowledged in plant abiotic stress studies. Previous studies have provided useful information on individual enzymes or transporters, measuring their stress-dependent changes in quantity and activity. The results of such studies have formed the current hypotheses on stress responsive networks, in which protein modifications, protein–protein interactions, stress-dependent protein movements, de novo synthesis, and controlled degradation play significant roles, literally spanning the time and space of the stress response. Consequently, large scale high-throughput proteome analyses must be integrated with transcriptome and metabolome analyses to improve our understanding of the stress response [93].

Improvements in proteomic technology regarding protein separation and detection, as well as mass spectrometry based protein identification, have an increasing impact on the study of plant responses to salinity stress [94,95,96]. New insights have been obtained into salinity stress responses by comparative proteome studies of salt-stressed roots from *Arabidopsis* and rice. A proteomic study of drought- and salt-stressed rice plants found that around 3000 proteins could be detected in a single gel and over 1000 could be quantified [97]. This study found 42 proteins that changed in abundance or position in response to stress. Several of the key proteins were identified and are the subject of further studies. The identification of novel protein candidates associated with salinity stress [98,99], detection of alterations in protein phosphorylation patterns [100] and recently, the location of a salinity stress-responsive protein to the rice root apoplast with a putative function in stress signalling [101] indicate the importance of ion uptake and transport, and regulation of water status and signal transduction processes in the root.

The importance of metabolite changes during plant responses to abiotic stress suggests that detailed metabolite profiling may provide valuable insights into stress response mechanisms. Metabolomics is a relatively new area of research and there are no published reports on its application to stress tolerance in cereals. However, a recent report on rice found that 88 main metabolites could be successfully quantified from the extract of rice leaves [102]. The compounds identified covered pathways of sugar and amino acid metabolism; hence these types of analyses should prove valuable for assessing stress responses.

CONCLUSIONS AND FUTURE PROSPECTS

Recent progress in the elucidation of salt stress signalling and effector output determinants that mediate ion homeostasis has uncovered some potential biotechnology tactics that may be used to obtain salt tolerant crop plants, i.e. enhance the yield stability under salinity. Two basic strategies are feasible; regulate the salt stress signal pathway that controls tolerance effectors or modulate effector's activity or efficacy. The recent demonstration that a constitutively activated SOS2 kinase can be achieved by deletion of the auto-inhibitory domain or by site-specific modifications to the catalytic domain of the protein kinase [33] offers an approach to regulate stress signaling that controls ion homeostasis. Although the overexpression of AtNHX1 enhances plant salt tolerance and facilitates growth in the saline environment [44,45], there is no evidence for an effective Na^+ sequestration into the vacuole mediated by NHX. Perhaps regulating net Na^+ influx across the plasma membrane would enhance salt tolerance efficacy achieved by overexpressing the vacuolar antiporter. Control of net Na^+ flux across the plasma membrane should be achieved by modulating the expression or activation of SOS1 (Na^+ efflux) and/or HKT1 (Na^+ influx), or by expressing more efficient forms of the Na^+ transport proteins. For example, mutant variant forms of HKT1 transport more K^+ at the expense of Na^+ and succeed to greater salt tolerance [103].

Gene expression profiling constitutes an exciting tool to unveil mechanisms involved in the response of plants to environmental stress. Its application in crop research is just starting as technologies are becoming more accessible and cost-effective and are expected to fuel huge advances in agriculture in the coming decades. Currently, the importance of biotechnology is being acknowledged by breeding programs around the world and is resulting in the development of new techniques and approaches to increase crop tolerance to stress.

REFERENCES

[1]   Zhu JK. Plant salt tolerance. Trends Plant Sci 2001; 6: 66-71.

[2]   Munns R, Tester M. Mechanisms of salinity tolerance. Annu Rev Plant Biol 2008; 59:651-81.

[3]   Blumwald E, Aharon GS, Apse MP. Sodium transport in plant cells. Biochemica et Biophysica Acta 2000; 1465: 140-151.

[4]   Wang R *et al.* Ionic homeostasis and reactive oxygen species control in leaves and xylem sap of two poplars subjected to NaCl stress. Tree Physiol 2008; 28: 947-57.

[5]   Hasegawa PM, Bressan RA, Zhu JK, Bohnert HJ. Plant cellular and molecular responses to high salinity. Annu Rev Plant Physiol Plant Mol Biol 2000; 51: 463-499.

[6]   Niu X, Bressan RA, Hasegawa PM, Pardo JM. Ion homeostasis in NaCl stress environments. Plant Physiol 1995; 109: 735-742.

[7]   Blumwald E. Engineering salt tolerance in plants. Biotechnol Gen Eng Rev 2003; 20: 261-275.

[8]   Zhu JK. Regulation of ion homeostasis under salt stress. Curr Opin Plant Biol 2003; 6: 441–445.

[9]   Horie T, Schroeder JI. Sodium transporters in plants. Diverse genes and physiological functions. Plant Physiol 2004; 136: 2457–2462.

[10]   Garciadeblas B, Senn ME, Banuelos MA, Rodrıguez-Navarro A. Sodium transport and HKT transporters: the rice model. Plant J 2003; 34: 788–801.

[11]   Platten JD *et al.* Nomenclature for HKT transporters, key determinants of plant salinity tolerance. Trends Plant Sci 2006; 11: 372–374.

[12]   Uozumi N *et al.* The Arabidopsis HKT1 gene homolog mediates inward Na^+ currents in Xenopus laevis oocytes and Na^+ uptake in Saccharomyces cerevisiae. Plant Physiol 2000; 122: 1249-1259.

[13]   Berthomieu P *et al.* Functional analysis of AtHKT1 in Arabidopsis shows that Na^+ recirculation by the phloem is crucial for salt tolerance. EMBO J 2003; 22: 2004–2014.

[14]   Sunarpi Q *et al.* Enhanced salt tolerance mediated by AtHKT1 transporter induced Na^+ unloading from xylem vessels to xylem parenchyma cells. Plant J 2005; 44: 928–938.

[15]   Jabnoune M *et al.* Diversity in expression patterns and functional properties in the rice HKT transporter family. Plant Physiol 2009; 150(4): 1955-1971.

[16]   Horie T *et al.* Rice OsHKT2;1 transporter mediates large Na$^+$ influx component into K$^+$-starved roots for growth. EMBO J 2007; 26: 3003-3014.

[17]   Haro R, Banuelos MA, Rodriguez-Navarro A. High-affinity sodium uptake in land plants. Plant Cell Physiol 2010; 51 (1): 68-79.

[18]   Ren ZH *et al.* A rice quantitative trait locus for salt tolerance encodes a sodium transporter. Nat Gen 2005; 37:1141-60.

[19]   Byrt CS *et al.* HKT1;5-like cation transporters linked to Na$^+$ exclusion loci in wheat, Nax2 and Kna1. Plant Physiol 2007; 143:1918-1928.

[20]   Leng Q, Mercier RW, Hua BG, Fromm H, Berkowitz GA. Electrophysiological analysis of cloned cyclic nucleotide- gated ion channels. Plant Physiol 2002; 128: 400–410.

[21]   Demidchik B, Essah PA, Tester M. Glutamate activates cation currents in the plasma membrane of Arabidopsis root cells. Planta 2004; 219: 167–175.

[22]   Tester M, Davenport R. Na$^+$ tolerance and Na$^+$ transport in higher plants. Ann Bot 2003; 91: 503–527.

[23]   Shi H, Quintero FJ, Pardo JM, Zhu JK. The putative plasma membrane Na$^+$/H$^+$ antiporter SOS1 controls long-distance Na$^+$ transport in plants. Plant cell 2002; 14:465–477.

[24]   Wu SJ, Ding L, Zhu JK. SOS1, a genetic locus essential for salt tolerance and potassium acquisition. Plant Cell 1996; 8: 617–627.

[25]   Zhu JK. Genetic analysis of plant salt tolerance using Arabidopsis. Plant Physiol 2000; 124: 941-948.

[26]   Qiu QS, Guo Y, Dietrich MA, Schumaker KS, Zhu JK. Regulation of SOS1, a plasma membrane Na$^+$/H$^+$ exchanger in Arabidopsis thaliana, by SOS2 and SOS3. PNAS USA 2002; 99:8436-41.

[27]   Agarwal M *et al.* A R2R3 type MYB transcription factor is involved in the cold regulation of CBF genes and in acquired freezing tolerance. J Biol Chem 2006; 281(49): 37636-37645.

[28]   Quintero FJ, Ohta M, Shi H, Zhu JK, Pardo JM. Reconstitution in yeast of the Arabidopsis SOS signalling pathway for Na$^+$ homeostasis. Proc Natl Acad Sci USA 2002; 99: 9061-9066.

[29]   Quan R *et al.* SCABP8/CBL10, a putative calcium sensor, interacts with the protein kinase SOS2 to protect Arabidopsis shorts from SALT stress. The Plant Cell 2007; 9: 1415-1431.

[30]   Kim BG *et al.* The calcium sensor CBL10 mediates salt tolerance by regulating ion homeostasis in Arabidopsis. Plant J 2007; 52: 473-484.

[31]   Gong JM *et al.* Microarray-based rapid cloning of an ion accumulation deletion mutant in Arabidopsis thaliana. Proc Natl Acad Sci USA 2004; 101: 15404–15409.

[32]   Kolukisaoglu U, Weinl S, Blazevic D, Batistic O, Kudla J. Calcium sensors and their interacting protein kinases: genomics of the Arabidopsis and rice CBL-CIPK signaling networks. Plant Physiol 2004; 134(1):43-58.

[33]   Guo Y, Halfter U, Ishitani M, Zhu JK. Molecular characterization of functional domains in the protein kinase SOS2 that is required for plant salt tolerance. Plant Cell 2001; 13: 1383-400.

[34]   Sánchez-Barrena MJ, Martínez-Ripoll M, Zhu JK, Albert A. The structure of the Arabidopsis thaliana SOS3: molecular mechanism of sensing calcium for salt stress response. J Mol Biol 2005; 345(5):1253-64.

[35]   Sánchez-Barrena MJ, Fujii H, Angulo I, Martinez-Ripoll M, Zhu JK, Albert A. The structure of the C-terminal domain of the protein kinase AtSOS2 bound to the calcium sensor AtSOS3. Mol Cell 2007; 26 (3): 427-35.

[36]   Shi H, Lee BH, Wu SJ, Zhu JK. Overexpression of a plasma membrane Na$^+$/H$^+$ antiporter gene improves salt tolerance in *Arabidopsis thaliana*. Nat Biotech 2003; 21: 81–85.

[37]   Yang Q *et al.* Overexpression of SOS (Salt Overly Sensitive) Genes Increases Salt Tolerance in Transgenic Arabidopsis. Molecular Plant 2009; 2(1): 22-31.

[38]   Martinez-Atienza J *et al.* Conservation of the salt overly sensitive pathway in rice. Plant Physiol 2007; 143: 1001–1012.

[39]   Xu H *et al.* Functional characterization of a wheat plasma membrane Na$^+$/H$^+$ antiporter in yeast. Arch Biochem Biophys 2008; 473 (1): 8-15.

[40]   Blumwald E, Poole RJ. Na$^+$/H$^+$ antiport in isolated tonoplast vesicles from storage tissue of *Beta vulgaris*. Plant Physiol 1985; 78: 163-167.

[41]   Gaxiola RA, Rao R, Sherman A, Grisafi P, Alper SL, Fink GR. The Arabidopsis thaliana proton transporters, AtNhx1 and Avp1, can function in cation detoxification in yeast. Proc Natl Acad Sci USA 1999; 96: 1480-1485.

[42]   Yokoi S. Differential expression and function of Arabidopsis thaliana NHX Na$^+$/H$^+$ antiporters in the salt stress responses. Plant J 2002; 30(5): 529-539.

[43]   Aharon R, Shahak Y, Wininger S, Bendov R, Kapulnik Y, Galili G. Overexpression of a plasma membrane aquaporin in transgenic tobacco improves plant vigor under favorable growth conditions but not under drought or salt stress. Plant Cell 2003; 15(2):439-447.

[44]   Apse MP, Aharon GS, Snedden WA, and Blumwald E. Salt tolerance conferred by overexpression of a vacuolar Na$^+$/H$^+$ antiport in Arabidopsis. Science 1999 ; 285: 1256-1258.

[45]   Zhang HX, Blumwald E. Transgenic salt-tolerant tomato plants accumulate salt in foliage but not in fruit. Nat Biotechnol 2001; 19: 765-768.

[46]   Venema L, Quintero FJ, Pardo JM, Donaire JP. The Arabidopsis Na$^+$/H$^+$ exchanger catalyzes low affinity Na$^+$ and K$^+$ transport in reconstituted vesicles. J Biol Chem 2002 ; 277 : 2413–2418.

[47]   Darley CP, van Wuytswinkel OCM, van der Woude K, Mager WH, de Boer AH. Arabidopsis thaliana and Saccharomyces cerevisiae NHX1 genes encode amiloride sensitive electroneutral Na$^+$/H$^+$ exchangers. Biochem J 2000; 351: 241–249.

[48]   Pardo JM, Cubero B, Leidi EO, Quintero FJ. Alkali cation exchangers: roles in cellular homeostasis and stress tolerance. J Exp Bot 2006; 57: 1181–1199.

[49]   Yamaguchi T *et al.* Genes encoding the vacuolar Na$^+$/H$^+$ exchanger and flower coloration. Plant Cell Physiol 2001; 142: 451–461.

[50]   Apse MP, Sottosanto J, Blumwald E. Ion homeostasis in Arabiodpsis thaliana: role of the vacuolar Na$^+$/H$^+$ antiporters. Plant J 2003; 36: 229–239.

[51]   Bowers K, Levi BP, Patel FI, Stevens TH. The sodium/proton exchanger Nhx1p is required for endosomal protein trafficking in the yeast Saccharomyces cerevisiae. Mol. Biol. Cell 2000; 11: 4277–4294.

[52]   Sottosanto JB, Gelli A, Blumwald E. DNA array analyses of Arabidopsis thaliana lacking a vacuolar Na$^+$/H$^+$ antiporter: impact of AtNHX1 on gene expression. Plant J 2004; 40, 752–771.

[53]   Brett CL, Tukaye DN, Mukherjee S, Rao R. The yeast endosomal Na$^+$(K$^+$)/H$^+$ exchanger Nhx1 regulates cellular pH to control vesicle trafficking. Mol Biol Cell 2005; 16, 1396–1405.

[54]   Wu YY, Chen GD, Meng QW, Zheng CC. The cotton GhNHX1 gene encoding a novel putative tonoplast Na$^+$/H$^+$ antiporter plays an important role in salt stress. Plant Cell Physiol 2004; 45: 600–607.

[55]   Saqib M, Zorb C, Rengel Z, Schubert S. The expression of the endogenous vacuolar Na$^+$/H$^+$ antiporters in roots and shoots correlates positively with the salt tolerance of wheat (*Triticum aestivum* L.). Plant Sci 2005; 169: 959–965.

[56]   Yamaguchi T, Blumwald E. Developing salt tolerant crop plants: challenges and opportunities. Trends Plant Sci 2005; 12: 615–620.

[57]   Venema K, Belver A, Marin-Manzano MC, Rodriguez-Rosales MP, Donaire JP. A novel intracellular K$^+$/H$^+$ antiporter related to Na$^+$/H$^+$ antiporters is important for K$^+$ ion homeostasis in plants. J Biol Chem 2003; 278: 22453-22459.

[58]   Rodriguez-Rosales MP, Jiang X, Galvez FJ, Aranda MN, Cubero B, Venema K. Overexpression of the tomato K$^+$/H$^+$ antiporter LeNHX2 confers salt tolerance by improving potassium compartmentalization. New Phytologist 2008; 179: 366-377.

[59]   Sakakibara Y, Kobayashi H, Kasamo K. Isolation and characterization of cDNAs encoding vacuolar H$^+$-pyrophosphatase isoforms from rice (*Oryza sativa* L.), Plant Mol Biol 1996; 31 1029–1038.

[60]   Tanaka Y, Chiba K, Maeda M, Maeshima M. Molecular cloning of cDNA for vacuolar membrane proton-translocating inorganic pyrophosphatase in *Hordeum vulgare*. Biochem Biophys Res Commun 1993; 190 :1110–1114.

[61]   Maeshima M. Vacuolar H$^+$-pyrophosphatase. Biochim Biophys Acta 2000; 1465: 37-51.

[62]   Brini F, Gaxiola RA, Berkowitz GA, Masmoudi K. Cloning and characterization of a wheat vacuolar cation/proton antiporter and pyrophophatase proton pump. Plant Physiol Biochem 2005; 43: 347-354.

[63]   Brini F, Hanin M, Mezghanni I, Berkowitz G, Masmoudi K. Overexpression of wheat Na$^+$/H$^+$ antiporter TNHX1 and H$^+$-pyrophosphatase TVP1 improve salt and drought stress tolerance in Arabidopsis thaliana plants. J Exp Bot 2007; 58(2): 301–308.

[64]   Tabata S. Impact of genomics approaches on plant genetics and physiology. J Plant Res 2002; 115:271-275.

[65]   Gale MD, Devos KM. Comparative genetics in the grasses. Proc Natl Acad Sci USA 1998; 95:1971-1974.

[66]   Bowers JE *et al.* Comparative physical mapping links conservation of micro-synteny to chromosome structure and recombination in grasses. Proc Natl Acad Sci USA 2005; 102: 13206-13211.

[67]   Paterson AH, Freeling M, and Sasaki T. Grains of knowledge: Genomics of model cereals. Genome Res 2005; 15: 1643-1650.

[68]   Caicedo AL, Purugganan MD. Comparative plant genomics. Frontiers and prospects. Plant Physiol 2005; 138: 545-547.

[69]   Van de Mortel JE, Aarts MGM. Comparative transcriptomics - model species lead the way. New Phytol 2006; 170: 199-201.

[70]   Kreps JA, Wu Y, Chang HS, Zhu T, Wang X, Harper JF. Transcriptome changes for Arabidopsis in response to salt, osmotic, and cold stress. Plant Physiol 2002; 130: 2129-2141.

[71]   Oono Y *et al.* Monitoring expression profile of Arabidopsis genes during cold acclimation and deacclimation using DNA microarrays. Funct Integr Genomics 2006; 6: 212-234.

[72]   Jianping PW, Suleiman SB. Monitoring of gene expression profiles and identification of candidate genes involved in drought responses in *Festuca mairei*. Mol Genet Genomics 2007; 277: 571-587.

[73] Mantri N, Ford R, Coram T, Pang E. Transcriptional profiling of chickpea genes differentially regulated in response to high-salinity, cold and drought. BMC Genomics 2007; 8: 303.

[74] Gregory BD, Yazaki J, Ecker JR. Utilizing tiling microarrays for whole-genome analysis in plants. Plant J 2008; 53: 636-644.

[75] Matsui A *et al.* Arabidopsis transcriptome analysis under drought, cold, high-salinity and ABA treatment conditions using a tiling array. Plant Cell Physiol 2008; 49:1135-1149.

[76] Gong Q, Li P, Ma S, Indu Rupassara S, Bohnert HJ. Salinity stress adaptation competence in the extremophile Thellungiella halophila in comparison with its relative *Arabidopsis thaliana*. Plant J 2005; 44(5): 826-39.

[77] Rabbani MA *et al.* Monitoring expression profiles of rice genes under cold, drought, and high-salinity stresses and abscisic acid application asing cDNA microarray and RNA gel-blot analyses. Plant Physiol 2003; 133: 1755-1767.

[78] Buchanan CD *et al.* Sorghum bicolor's transcriptome response to dehydration, high salinity and ABA. Plant Mol Biol 2005; 58: 699-720.

[79] Gong Z *et al.* Genes that are uniquely stress regulated in salt overly sensitive (sos) mutants. Plant Physiol 2001; 126: 363-375.

[80] Taji T *et al.* Comparative genomics in salt tolerance between Arabidopsis and Arabidopsis-related halophyte salt cress using Arabidopsis microarray. Plant Physiol 2004; 135:1697-1709.

[81] Maathuis FJ *et al.* Transcriptome analysis of root transporters reveals participation of multiple gene families in the response to cation stress. Plant J 2003; 35(6): 675-92.

[82] Misson L, Panek JA, Goldstein AH. A comparison of three approaches to modeling leaf gas exchange in annually drought-stressed ponderosa pine forests. Tree Physiol 2005; 24(5): 529-41.

[83] Zhu J *et al.* HOS10 encodes an R2R3-type MYB transcription factor essential for cold acclimation in plants. Proc Natl Acad Sci USA 2005; 102 (28): 9966-71.

[84] Kawaguchi R, Girke T, Bray EA, Bailey-Serres J. Differcntial mRNA translation contributes to gene regulation under non-stress and dehydration stress conditions in *Arabidopsis thaliana*. Plant J 2004; 38(5): 823-839.

[85] Langridge P, Paltridge N, Fincher G. Functional genomics of abiotic stress tolerance in cereals. Brief Funct Genom Proteom 2006; 4: 343-354.

[86] Leader DJ. Transcriptional analysis and functional genomics in wheat. J Cereal Sci 2005; 41: 149–63.

[87] Close TJ *et al.* A new resource for cereal genomics: 22K barley Gene Chip comes of age. Plant Physiol 2004; 134: 960–968.

[88] Hazen SP *et al.* Expression profiling of rice segregating for drought tolerance QTLs using a rice genome array. Funct Integr Genomics 2005; 5: 104-116.

[89] Meyers B. Rice MPSS site, http://mpss.udel.edu/rice/, University of Delaware [25 October 2005, date last accessed], 2005.

[90] Li B, WWe A, Song C, Li N, Zhang J. Heterologous expression of the TsVP gene improves the drought resistance of maize. Plant Biotech J 2008; 6: 146-159.

[91] Marioni J, Mason C, Mane S, Stephens M, Gilad Y. RNA-seq: an assessment of technical reproducibility and comparison with gene expression arrays. Genome Res 2008; doi:10.1101/gr.079558.108

[92] Pan Q, Shai O, Lee LJ, Frey BJ, Blencowe BJ. Deep surveying of alternative splicing complexity in the human transcriptome by high-throughput sequencing. Nat Genet 2008; 40: 1413–1415.

[93] Bohnert HJ, Gong Q, Li P, Ma S. Unraveling abiotic stress tolerance mechanisms-getting genomics going. Curr Opin Plant Biol 2006; 9(2): 180-188.

[94] Parker R, Flowers TJ, Moore AL, Harpham NV. An accurate and reproducible method for proteome profiling of the effects of salt stress in the rice leaf lamina. J Exp Bot 2006; 57(5): 1109-18.

[95] Qureshi MI, Qadir S, Zolla L. Proteomics-based dissection of stress-responsive pathways in plants. J Plant Physiol 2007; 164(10): 1239-1260.

[96] Caruso G, Cavaliere C, Guarino C, Gubbiotti R, Foglia P, Lagana A. Identification of changes in Triticum durum L. leaf proteome in response to salt stress by two-dimensional electrophoresis and MALDI-TOF mass spectrometry. Analyt Bioanal Chem 2008; 391: 381–390.

[97] Salekdeh GH *et al.* A proteomic approach to analyzing drought- and salt-responsiveness in rice. Field Crops Res 2002; 76: 199–219.

[98] Yan Y, Weaver VM, Blair IA. Analysis of protein expression during oxidative stress in breast epithelial cells using a stable isotope labeled proteome internal standard. J Proteome Res 2005; 4(6): 2007-14.

[99] Jiang Y, Yang B, Harris NS, Deyholos MK. Comparative proteomic analysis of NaCl stress-responsive proteins in Arabidopsis roots. J Exp Bot 2007; 58(13): 3591-607.

[100] Chitteti BR, Peng Z. Proteome and phosphoproteome dynamic change during cell dedifferentiation in Arabidopsis. Proteomics 2007; 7(9): 1473-500.

[101] Zhang L, Tian LH, Zhao JF, Song Y, Zhang CJ, Guo Y. Identification of an apoplastic protein involved in the initial phase of salt stress response in rice root by two-dimensional electrophoresis. Plant Physiol 2009; 149: 916–928.

[102] Sato S *et al.* Simultaneous determination of the main metabolites in rice leaves using capillary electrophoresis mass spectrometry and capillary electrophoresis diode array detection. Plant J 2004; 40: 151–63.

[103] Rubio F, Schwarz M, Gassmann W, Schroeder JI. Genetic selection of mutants in the high affinity K^+ transporter HKT1 that define functions of a loop site for reduced Na^+ permeability and increased Na^+ tolerance. J Biol Chem 1999; 274: 6839-6847.

[104] Rus A *et al.* AtHKT1 is a salt tolerance determinant that controls Na^+ entry into plant roots. Proc Natl Acad Sci USA 2001; 98: 14150-14155.

[105] Laurie S, Feeney KA, Maathuis FJM, Heard PJ, Brown SJ, Leigh RA. A role for HKT1 in sodium uptake by wheat roots. Plant J 2002; 32: 139–149.

[106] Shi H, Ishitani M Kim C, Zhu JK. The Arabidopsis thaliana salt tolerance gene SOS1 encodes a putative Na^+ /H^+ antiporter. Proc Natl Acad Sci USA 2000; 97: 6896-6901.

[107] Gaxiola RA, Li J, Undurraga S, Dang LM, Allen GJ, Alper SL, Fink GR. Drought- and salt tolerant plants result from over expression of the AVP1 H^+-pump. Proc Natl Acad Sci USA 2001; 98: 11444-11449.

[108] Ohta M *et al.* Introduction of a Na^+/H^+ antiporter gene from Atriplex gemelini confers salt tolerance to rice. FEBS Lett 2002; 532: 279-282.

[109] Liu J, Ishitani M, Halfter U, Kim CS, Zhu JK. The Arabidopsis thaliana SOS2 gene encodes a protein kinase that is required for salt tolerance. Proc Natl Acad Sci USA 200; 97: 3730-3734.

[110] Liu J, Zhu JK. An Arabidopsis mutant that requires increased calcium for potassium nutrition and salt tolerance. Proc Natl Acad Sci USA 1997; 94: 14960-14964.

CHAPTER 5

Transcriptomics Identifies Cold Stress Determinants in *Arabidopsis*

John Einset*

Department of Plant and Environmental Sciences, P. O. Box 5003, Norwegian University of Life Sciences (UMB), 1432 Aas, Norway

Abstract: During the last forty years, researchers have had the working hypothesis that genes upregulated by low temperatures protect plants from cold stress. When cDNA cloning technologies were developed about twenty years ago, cold-responsive genes such as genes for CBF/DREB transcription factors could be identified in *Arabidopsis*. The next step was to use microarray technologies to identify cold-regulated genes as a way to improve our understanding of cold tolerance. Unfortunately, there have been several problems with this approach; 1) upregulation of mRNA levels is only one of many mechanisms for the control of gene expression in plants, 2) published microarray results have not always been repeatable by other labs, 3) there has been a heavy emphasis on cold-regulated transcription factor genes to the exclusion of other important determinants, 4) the cold treatments used in several laboratory studies have not always been comparable to natural stress conditions and 5) there has often been a lack of follow-up research, using mutants to prove through functional genomics that specific genes are actually involved in cold tolerance. An alternative transcriptomic approach to identify genes for cold tolerance is chemical genetics based on glycine betaine treatments, using transcriptomics followed by functional studies with mutants. This approach was validated first for the RabA4c GTPase involved in membrane trafficking and has also identified a bZIP transcription factor and FRO2 ferric reductase. In conclusion, although transcriptomics has identified some determinants of cold stress tolerance, there still exists large gaps in our knowledge of this important process.

BACKGROUND ON GENE EXPRESSION/COLD TOLERANCE RESEARCH

Plant responses to low temperatures are complex and they are crucial to plant distribution worldwide in nature as well as to plant production in agriculture. Temperatures between $0-10^0$ C cause 'chilling stress' while sub-zero temperatures lead to 'freezing stress'. While it is often assumed that chilling stress differs from freezing stress in terms of the mechanisms involved, one could just as easily speculate that plants are adapted to respond to low temperatures by activating processes that protect against both chilling stress and freezing stress simultaneously. A key question in research on cold stress has been: What are the protective processes that plants use to cope with cold stress?

The idea that gene expression changes during cold lead to changes in physiology to protect plants from cold stress is nearly forty years old [1,2]. Given the methods that became available for transcriptional analysis twenty years ago, starting in the early 1990s several research groups used cDNA cloning methods to identify genes upregulated by cold temperatures. Using the term cold-responsive genes (COR genes), Thomashow and coworkers identified several COR genes in *Arabidopsis*, designated as COR6.6, COR15a, COR47 and COR78 [3,4]. Sequencing of these COR genes led to the finding that they coded for several different types of proteins, including some highly hydrophilic proteins such as the proteins coded by COR6.6 and COR15 [5]. Subsequent functional studies with transgenic plants overexpressing the COR15a protein showed a 2°C enhancement of

freezing tolerance in chloroplasts from transgenics compared to wild type plants; i.e. a cryoprotective effect [6]. Nevertheless, COR15a overexpression seemed to have no effect on the survival of frozen plants [7]. At the same time that COR genes were being identifed in *Arabidopsis via* cDNA cloning, other researchers were using similar methods to identify cold-responsive genes in Arabidopsis and in other plants such as *Brassica napus, Hordeum vulgare, Triticum vulgare* and *Medicago sativa*, to name a few species [reference 8 and http//:stress-genomics.org]. Upregulated genes for a wide variety of enzymes were reported in these studies, including genes for reactive oxygen species (ROS) metabolism such as glutathione peroxidase, ascorbate peroxidase and catalase as well as genes for dehydrins, chaperones, heat shock proteins and transcription factors.

*Address correspondence to: **Dr. John Einset,** Department of Plant and Environmental Sciences, P. O. Box 5003, Norwegian University of Life Sciences (UMB), 1432 Aas, Norway; E-mail: john.einset@umb.no

Narendra Tuteja, Sarvajeet Singh Gill and Renu Tuteja [Eds.]

GENES FOR TRANSCRIPTION FACTORS SUCH AS CBF AND DREB HAVE POTENTIAL FOR CROP IMPROVEMENT

When one evaluates the overall success of the studies during the cDNA cloning era of the 1990s, the discovery of the CBF1 transcription factor in Arabidopsis as a cold-responsive gene has to be considered as one of the major achievements having practical consequences. Having isolated and sequenced genomic clones corresponding to several COR genes, a 5-bp core squence of CCGAC, designated the C-repeat, was identified as a common component in the promoters of several COR genes [9]. Parallel studies by Liu *et al.* [10] identified the same sequence in promoter regions of cold-regulated as well as dehydration stress-regulated genes. They designated this common promoter sequence as DRE. Stockinger *et al.* [11] screened cDNA clones in yeast, looking for genes that could use the common COR gene promoter sequence (C-repeat/DRE) to cause the expression of a reporter gene. By this procedure, three clones were selected which were all shown to contain an identical 1.8 kb cDNA coding for a 24 kDa polypeptide with DNA binding ability to the C-repeat/DRE ??. They designated the polypeptide as CBF1 and the corresponding gene as CBF1. At about the same time as this report, Liu *et al.* [10] used similar methods to isolate the genes for the DRE-binding proteins DREB1A and DREB2A from *Arabidopsis*.

In two landmark papers, Jaglo-Ottosen *et al.* [12] and Kasuga *et al.* [13] demonstrated that *Arabidopsis* transgenics overexpressing C-repeat/DRE transcription factors had significantly improved tolerance for freezing stress (CBF1) or drought, salt and freezing stresses (DREB1A). There are good reasons to expect that transgenic technologies developed as a result of this research will have significant impacts on agriculture in the future; e.g. Mendel Biotechnology Inc. WeatherGard and DroughtGard technologies. There is also a growing body of information on CBF/DREB genes in other plants [14, 15].

MICROARRAY TECHNOLOGIES WERE A LOGICAL NEXT STEP

By the late 1990s, Brown and Botstein [16] introduced microarray technologies for conducting genome-wide transcriptomic studies. For plant researchers, these technologies have resulted in several publications in relation to cold-regulated genes. As with the cDNA studies earlier, the assumption has been that manipulation of cold-regulated genes has the potential to be instrumental in protecting plants from chilling and freezing stress. Of course, there is no reason to believe that this assumption is true in all cases. For example, some cold-regulated genes may only protect against chilling stress while others may protect only against freezing stress. A third possibility is that some cold-regulated genes have nothing at all to do with cold stress tolerance. Further problems with this approach have been that 1) upregulation of mRNA levels is only one of many mechanisms for the control of gene expression in plants in addition to control at pre-mRNA processing, mRNA stability, mRNA translation, protein degradation or protein modification, 2) published microarray results have not always been repeatable by other labs, 3) there has been an emphasis on cold-regulated genes for transcription factors to the exclusion of other important determinants, 4) the cold treatments used in several laboratory studies have not always been comparable to natural stress conditions [17, 18] and, finally, 5) there has often been a lack of follow-up research, using mutants to prove through functional genomics that specific genes are actually involved in cold tolerance.

Despite many reservations in terms of the interpretation of microarray data, these studies have provided a starting point for further research to identify cold stress determinants. Unfortunately, the possible avenues that present themselves in terms of number of cold-responsive genes makes decisions about which genes to pursue further somewhat problematic [references 19-27 and AtGenExpress at http://jsp.weigelworld.org/ expviz/expviz.jsp]. For example, in one of the first microarray studies of cold-responsive genes, Seki *et al.* [28] used a full-length cDNA microarray for 1300 *Arabidopsis* genes to identify 19 COR genes, nine of which had been reported earlier. Newly identified genes included genes for ferritin, a nodulin-like protein, LEA protein and glyoxalase. A subsequent report by Fowler and Thomashow [29] using a microarray for 8000 genes identified 306 COR genes; i.e. 218 upregulated and 88 downregulated. They could confirm most of Seki *et al.* [28] designations but they failed to confirm upregulation of either the LEA protein or the glyoxalase. In terms of the CBF1 regulon, Fowler and Thomashow [29] concluded that 45 of the 306 COR genes were under control of CBF1. A recent review [30] summarizes research on CBF1/DREB1-induced genes in Arabidopsis, concluding that more than 40 target genes have been identified so far, including genes for transcription factors, dehydrins, phospholipase C, an RNA-binding protein, a sugar transport protein, a desaturase, carbon-metabolism proteins, osmoprotectant biosynthesis proteins and protease inhibitors.

CHEMICAL GENETICS REPRESENTS AN ALTERNATIVE APPROACH USING TRANSCRIPTOMICS

With the exceptions of several papers on CBFs, a report on a cold-shock protein [31] as well as a report on dehydrins [32], very few COR or CBF-regulon genes identified using microarrays have been tested in mutant and/or transgenic experiments. There are many reasons for this; e.g. 1) given the success of the CBF/DREB work, emphasis has been on the discovery of transcription factors as master regulators that control gene clusters, 2) the assumption has been that cold stress is dependent on several genes and that single gene changes would not show effects in mutants or transgenics, 3) it is difficult to decide which COR genes, among many possibilities, might be the most important ones in determining cold tolerance.

Based on these considerations, it was of interest to develop alternative approaches for discovering cold stress determinants. One approach has been mutant screening which led, for example, to the identification of the Eskimo1 gene [33]. Another approach has been chemical genetics based on pioneering research from the laboratory of Professor Norio Murata during the last 10 years, demonstrating that naturally occurring glycine betaine (GB) can confer tolerance to several types of stress at low concentrations, including cold stress, either after application to plants or in transgenics engineered to overproduce GB. GB is widely distributed in plants and it has been demonstrated to improve stress tolerance in agriculturally important crops. Examples of transgenics with improved stress tolerance that have been engineered to produce GB include rice, tobacco, tomato, cotton, maize, *Brassica napus*, *Brassica juncea* and *Diospyros kaki* [34]. Based on the fact that GB levels in transgenic plants are often quite low, it was hypothesized that GB might confer stress tolerance, at least in part, via effects on gene expression [35]. If this could be shown, then a new approach for identifying cold stress determinants in plants would be established, using GB in a chemical genetic screen plus transcriptomics.

Application of this approach using transcriptomics and chemical genetics was first demonstrated by Einset *et al.* [35] with the gene for the membrane trafficking RabA4c GTPase (At5g47960). In this report, a microarray with probes for >26,000 *Arabidopsis* genes was used to identify GB-upregulated genes and then upregulation of selected genes was confirmed by Northern blot analysis. Focusing on a gene for RabA4c GTPase, a knockout mutant for this gene was then tested in a chilling sensitivity bioassay with or without a GB-pretreatment. Although *Arabidopsis* is usually defined as chilling-resistant because it shows no obvious signs of chilling injury, chilling does have an effect because chilled plants show inhibited root growth upon transfer back to normal temperatures. Remarkably, when wild-type plants were pretreated with GB, root growth rates after chilling were comparable to non-chilled plants. By contrast, the RabA4c GTPase knockout showed no GB response in the chilling test, proving the requirement for a functional RabA4c GTPase gene for GB's effect. Similar proof has since been provided [36,37] for genes for FRO2 ferric reductase (At1g01580) and for a bZIP transcription factor (At3g62420). Models for how several GB upregulated genes might function in causing improved stress tolerance in leaves [38] and roots [39,40] have also been presented. A recent review on GB action [34] concludes, 'Taken together, the various observations discussed above indicate that GB, either applied exogenously or synthesized in transgenic plants, seems to be capable of activating specific genes. Identification of GB-inducible genes and the functions of their products will advance our understanding of GB-enhanced stress tolerance in plants.'

CONCLUSIONS

Given the success of the CBF approach, it is not surprising that there has been an emphasis on transcription factor discovery during the past years. Meanwhile, we still have very little understanding about what genes and processes actually protect plants from cold stress. If one looks at the genes and processes that transcriptomics has uncovered in relation to cold stress, there seems to be no obvious relationship to traditional areas of investigation. Examples of traditional cold stress research problems include discovery of supercooling mechanisms [41], identification of ice nucleation factors [42-44] as well as studies of changes in carbohydrate metabolism enzymes [45,46] and membrane components [47,48]. Finally, chemical genetics is a new approach having potential to identify new genes or processes that may or may not be regulated by cold but that are, just the same, possible targets to be used to develop plants with improved cold tolerance.

Overall a review of research on cold stress determinants leads us to several conclusions; 1) there are still substantial knowledge gaps in relation to our understanding of cold stress, 2) in relation to studies based on transcriptomics, what is needed are gene expression studies combined with functional genomics to prove that specific genes are involved in cold tolerance and, finally, 3) we need to recognize that transcriptomics alone won't tell us everything.

REFERENCES

[1]   Weiser CJ. Cold resistance and injury in woody plants. Science 1970; 169: 1269-1278.

[2]   Guy CL, Haskell D. Induction of freezing tolerance in spinach is associated with the synthesis of cold acclimation induced proteins. Plant Physiol 1987; 84: 872-878.

[3]   Hajela RK, Horvath DP, Gilmour SJ, Thomashow MF. Molecular cloning and expression of cor [cold-regulated] genes in *Arabidopsis thaliana*. Plant Physiol 1990; 93: 1246-1252.

[4]   Gilmour SJ, Artus N, Thomashow MF. cDNA sequence analysis and expression of two cold-regulated genes of *Arabidopsis thaliana*. Plant Mol Biol 1992; 18: 13-21.

[5]   Lin C, Thomashow MF. DNA sequence analysis of a cDNA for cold-regulated *Arabidopsis* gene cor15 and characterization of the COR15 polypeptide. Plant Physiol 1992; 99: 519-525.

[6]   Artus NN, Uemura M, Steponkus PL, Gilmour SJ, Lin C, Thomashow MF. Constitutive expression of the cold-regulated *Arabidopsis thaliana* COR15a gene affects both chloroplast and protoplast freezing tolerance. Proc Natl Acad Sci USA 1996; 93: 13404-13409.

[7]   Thomashow MF. In: E. Meyerowitz, Somerville C Eds. Arabidopsis. *Arabidopsis thaliana* as a model for studying mechanisms of plant cold tolerance. Cold Spring Harbor Laboratory Press, Plainview, NY, 1994; pp. 807-834.

[8]   Thomashow MF. Plant cold acclimation: freezing tolerance genes and regulatory mechanisms. Ann Rev Plant Physiol Plant Mol Biol 1999; 50: 571-599.

[9]   Baker SS, Wilhelm KS, Thomashow MF. The 5'-region of *Arabidopsis thaliana* cor15a has cis-acting elements that confer cold-, drought- and ABA-regulated gene expression. Plant Mol Biol 1994; 24: 701-713.

[10]   Liu Q, Kasuga M, Sakuma Y, Abe H, Miura S, Yamaguchi-Shinozaki K, Shinozaki K. Two transcription factors, DREB1 and DREB2, with an
EREBP/AP2 DNA binding domain separate two cellular signal transduction pathways in drought- and low-temperature-responsive gene expression, respectively, in Arabidopsis. Plant Cell 1998; 10: 1391-1406.

[11]   Stockinger EJ, Gilmour SJ, Thomashow MF. *Arabidopsis thaliana* CBF1 encodes an AP2 domain-containing transcriptional activator that binds to the C-repeat/DRE, a cis-acting DNA regulatory element that stimulates transcription in response to low temperature and water deficit. Proc Natl Acad Sci USA 1997; 94: 1035-1040.

[12]   Jaglo-Ottosen KR, Gilmour SJ, Zarka DG, Schabenberger O, Thomashow MF. *Arabidopsis* CBF1 overexpression induces COR genes and enhances freezing tolerance. Science 1998; 280: 104-106.

[13]   Kasuga M, Liu Q, Miura S, Yamaguchi-Shinozaki K, Shinozaki K. Improving plant drought, salt, and freezing tolerance by gene transfer of a single stress-inducible transcription factor. Nat Biotechnol 1999; 17: 287-291.

[14]   Stockinger EJ. In: Gusta L, Wisniewski M, Tanino K Eds. Plant cold hardiness: from the laboratory to the field. Winter hardiness and the CBF genes in the Triticeae. CABI Publishers, 2009; pp. 119-130.

[15]   Nassuth A, Siddiqua M. In: Gusta L, Wisniewski M, Tanino K Eds. Plant cold hardiness: from the laboratory to the field. Regulation of stress-responsive signalling pathway by eudicot CBF/DREB1 genes. CABI Publishers, 2009; pp. 131-139.

[16]   Eisen MB, Spellman PT, Brown PO, Botstein D. Cluster analysis and display of genome-wide expression patterns. Proc Natl Acad Sci USA 1998; 95: 14863-14868.

[17]   Gusta LV, Wisniewski M, Nesbitt NT, Tanino KT. Factors to consider in artificial freeze tests. Acta Hort [ISHS] 2003; 618: 493-507.

[18]   Gusta LV, Wisneiwski ME, Trischuk RG. In: Gusta L, Wisniewski M, Tanino K, Eds. Plant cold hardiness: from the laboratory to the field. Patterns of freezing in plants: the influence of species, environment and experiential procedures. CABI Publishers, 2009; pp.214-225.

[19]   Seki M *et al.* Monitoring the expression profiles of 7000 *Arabidopsis* genes under drought, cold and high-salinity stresses using a full-length cDNA microarray. Plant J 2002; 31: 279–292.

[20]   Kreps JA, Wu YJ, Chang HS, Zhu T, Wang X, Harper JF. Transcriptome changes for *Arabidopsis* in response to salt, osmotic, and cold stress. Plant Physiol 2002; 130: 2129-2141.

[21]   Zarka DG, Vogel JT, Cook D, Thomashow MF. Cold induction of *Arabidopsis* CBF genes involves multiple ICE [inducer of CBF expression] promoter elements and a cold-regulatory circuit that is desensitized by low temperature. Plant Physiol 2003; 133: 910–918.

[22]   Maruyama K *et al.* Identification of cold-inducible downstream genes of the *Arabidopsis* DREB1A/CBF3 transcriptional factor using two microarray systems. Plant J 2004; 38: 982–993.

[23]   Vogel JT, Zarka DG, Van Buskirk HA, Fowler SG, Thomashow MF. Roles of the CBF2 and ZAT12 transcription factors in configuring the low temperature transcriptome of *Arabidopsis*. Plant J 2005; 41: 195–211.

[24]   Lee BH, Henderson DA, Zhu JK. The *Arabidopsis* cold responsive transcriptome and its regulation by ICE1. Plant Cell 2005; 17: 3155–3175.

[25]   Kilian J *et al.* The AtGenExpress global stress expression data set: protocols, evaluation and model data analysis of UV-B light, drought and cold stress responses. Plant J 2007; 50, 347-363.

[26]   Matsui A *et al. Arabidopsis* transcriptome analysis under drought, cold, high-salinity and ABA treatment conditions using a tiling array. Plant Cell Physiol 2008; 49: 1135 - 1149.

[27]   Maruyama K *et al.* Metabolic pathways involved in cold acclimation identified by integrated analysis of metabolites and transcripts regulated by DREB1A and DREB2A. Plant Physiol 2009; 150: 1972-1980.

[28]   Seki M *et al.* Monitoring the expression pattern of 1300 *Arabidopsis* genes under drought and cold stresses by using a full-length cDNA microarray. Plant Cell 2001; 13, 61–72.

[29]   Fowler S, Thomashow MF. *Arabidopsis* transcriptome profiling indicates multiple regulatory pathways are activated during cold acclimation in addition to the CBF cold-response pathway. Plant Cell 2002; 14: 1675-1690.

[30]   Nakashima K, Ito Y, Yamaguchi-Shinozaki K. Transcriptional regulatory networks in response to abiotic stresses in *Arabidopsis* and grasses. Plant Physiol 2009; 149: 88-95.

[31]   Kim M-H, Sasaki K, Imai R. Cold shock domain protein 3 regulates freezing tolerance in Arabidopsis thaliana. J Biol Chem 2009; 284: 23454-23460.

[32]   Puhakainen T, Hess MW, Mäkelä P, Svensson J, Heino P, Palva ET. Overexpression of multiple dehydrin genes enhances tolerance to freezing stress in *Arabidopsis*. Plant Molecular Biol 2004; 54: 743-753.

[33]   Xin Z, Browse J. eskimo1 mutants of *Arabidopsis* are constitutively freezing-tolerant. Proc Natl Acad Sci USA 1998; 95: 7799–7804.

[34]   Chen THH, Murata N. Glycinebetaine: an effective protectant against abiotic stress in plants. Trends Plant Sci 2008; 13: 499-505.

[35]   Einset J *et al.* Membrane trafficking RabA4c involved in the effect of glycine betaine on recovery from chilling stress in *Arabidopsis*. Physiol Plant 2007; 130: 511-518

[36]   Einset J, Winge P, Bones A, Connolly EL. The FRO2 ferric reductase is required for glycine betaine's effect on chilling tolerance in *Arabidopsis* roots. Physiol Plant 2008; 134: 334-341.

[37]   Einset J. In: Gusta L, Wisniewski M, Tanino K, Eds. Plant cold hardiness: from the laboratory to the field. Chemical genetics identifies new chilling stress determinants in *Arabidopsis*. CABI Publishers, 2009; pp. 262-268

[38]   Einset J. An extracellular mechanism of light protection in plants identified using a chemical genetic screen. Acta Hortic 2006; 711: 339-344.

[39]   Einset J, Winge P, Bones A. ROS signaling pathways in chilling stress. Plant Sig Beh 2007; 2: 365-367.

[40]   Einset J, Connolly EL. Glycine betaine enhances extracellular processes blocking ROS signaling during stress. Plant Sig Beh 2009; 4: 197-199.

[41]   Wisniewski ME, Gusta LV, Fuller MP, Karlson D. In: Gusta L, Wisniewski M, Tanino K, Eds. Plant cold hardiness: from the laboratory to the field. Ice nucleation, propagation and deep supercooling: the lost tribes of freezing studies. CABI Publishers, 2009; pp. 1-11.

[42]   Takata N, Kasuga J, Takezawa D, Arakawa K, Fujikawa S. Gene expression associated with increased supercooling capability in xylem parenchyma cells of larch [*Larix kaempferi*]. J Exp Bot 2007; 58: 3731-3742.

43]   Griffith M, Yaish MWF. Antifreeze proteins in overwintering plants: a tale of two activities. Trends in Plant Science 2004; 9: 399-405.

[44]   Kasuga J, Hashidoko Y, Nishioka A, Yoshiba M, Arakawa K, Fujikawa S. Deep supercooling xylem parenchyma cells of katsura tree [*Cercidiphyllum japonica*] contain flavonol glycosides exhibiting ant-ice nucleation activity. Plant Cell Environ 2008; 31: 1335-1348.

[45]   Bravo LA, Bascunan-Godoy L, Perez-Torres E, Corcuera LJ. In: Gusta L, Wisniewski M, Tanino K Eds. Plant cold hardiness: from the laboratory to the field. Cold hardiness in Antarctic vascular plants CABI Publishers, 2009; pp. 198-213.

[46]   Renaut J, Planchon S, Oufir M, Hausman J-F, Hoffmann L, Evers D Identification of proteins from potato leaves submitted to chilling temperature. In: Gusta L, Wisniewski M, Tanino K, Eds. Plant cold hardiness: from the laboratory to the field. CABI Publishers, 2009; pp. 279-292.

[47] Minami A, Kawamura Y, Yamazaki T, Furuto A, Uemura M. In: Gusta L, Wisniewski M, Tanino K Eds. Plant cold hardiness: from the laboratory to the field. Plasma membrane and plant freezing tolerance: possible involvement of plasma membrane microdomains in cold acclimation. CABI Publishers, 2009; pp. 62-71.

[48] Minami A *et al.* Alterations in detergent-resistant plasma membrane microdomains in Arabidopsis thaliana during cold acclimation. Plant Cell Physiol 2009; 50: 341-359.

CHAPTER 6

Transcriptome Analysis of Polyamine Overproducers Reveals Activation of Plant Stress Responses and Related Signalling Pathways Tolerance in Plants

F. Marco[1,4], T. Altabella[2], R. Alcázar[3], J. Cuevas[2], C. Bortolotti[2], M.E. González[3], OA Ruiz[3], A.F. Tiburcio[2] and P. Carrasco[4]*

[1]Fundación CEAM. Parque Tecnológico c/ Charles Darwin 14, 46980. Paterna. Spain; [2]Unitat de Fisiologia Vegetal. Facultat de Farmàcia. Universitat de Barcelona. Diagonal 643. 08028 Barcelona. Spain; [3]IIB-INTECH, Camino Circ. Laguna km 6, (B7130IWA) Chascomús, Buenos Aires, Argentina and [4]Departament de Bioquímica i Biologia Molecular. Universitat de València. Facultat de Ciències Biològiques. Dr. Moliner 50, 46100 Burjassot, València, Spain

Abstract: Polyamines have been proposed to regulate multiple aspects of plant development and stress responses. In most of the cases, they interact with other signaling pathways to exert their action. Here we describe transcriptomic analyses that have lead to the identification of physiologically relevant interactions between polyamines and other hormones, such as abscisic acid, during the response to abiotic stress. Overall, these studies reinforce the premise that polyamines are regulatory signaling molecules which mediate by intricate cross-talks with hormonal pathways in response to different environmental stimuli.

POLYAMINES ARE INVOLVED IN PLANT RESPONSES TO ENVIRONMENTAL STIMULI

Plant development and productivity are negatively affected by environmental stresses. It is predicted that, during this century, global effects of climate changes will dramatically reduce crop yield. Therefore, identification of novel stress-regulatory genes and molecules, and development of strategies to obtain stress-tolerant plants are currently one of the major topics in plant research. Polyamines (PAs), putrescine (*Put*), spermidine (*Spd*), and spermine (*Spm*), are small polycationic compounds known to participate in plant development and stress responses. PA biosynthesis is a relevant metabolic route in the intermediary nitrogen metabolism that responds to environmental fluctuations in higher plants [1, 2]. Drastic changes in plant polyamine metabolism occur in response to several abiotic stresses such as: nutrient deficiency [3], osmotic stress [4], atmospheric pollutants [5] or heavy metal contamination [6]. Despite its potential importance, the physiological meaning of the massive polyamine accumulation under these challenging conditions has remained ill-defined. The relevance of the PA metabolic pathway in the plant response to abiotic stress has just recently been approached with the use of Arabidopsis as a model plant since it allows global approaches to study the function of PA metabolism [7,8,9,10] The complete sequencing of the Arabidopsis genome [11] has allowed the characterization of genes involved in the PA metabolic pathway. The first step of PA biosynthesis in Arabidopsis is decarboxylation of arginine, catalyzed by arginine decarboxylase (*ADC*), followed by two successive reactions catalyzed by agmatine iminohydrolase (*AIH*) and N-carbamoylputrescine aminohydrolase (*CPA*) to yield the diamine *Put* (Fig. **1**). Higher molecular weight PAs *Spd* and *Spm* are formed by sequential addition of aminopropyl groups to *Put* and *Spd*, respectively by the activity of spermidine synthase (*SPDS*) and spermine synthase (*SPMS*) respectively. The aminopropyl groups are generated by decarboxylation of S-adenosylmethionine (SAM) catalyzed by SAM decarboxylase (*SAMDC)* (Fig. **1**). The Arabidopsis genome contains two genes encoding ADC (*ADC1* and *ADC2*) [12, 13], and one for each AIH and CPA [14, 15]. There are also two genes for SPDS, *SPDS1* and *SPDS2* [16] one coding for *SPMS* [17] and at least four coding for SAMDC, *SAMDC1, SAMDC2, SAMDC3* and *SAMDC4* [18].

More recently, the accessibility of Arabidopsis genome has also allowed the development of global "omic" approaches to study the function of these compounds in response to abiotic stresses [10]. A summarized version of how expression of PA biosynthetic genes is affected by abiotic stress can be obtained in the microarray heat map

*****Address correspondence to: Dr. Pedro Carrasco,** Departament de Bioquímica i Biologia Molecular. Universitat de València. Facultat de Ciències Biològiques. Dr. Moliner 50, 46100 Burjassot, València, Spain; Phone: +34-96-354 4868, Fax: +34-96-354 4635, E-mail: pedro.carrasco@uv.es

Electronic-Northern of stress experiments from the Botany Array Resource (BAR) [19]. Water stress induces expression of *ADC2*, *SPDS1* and *SPMS* [20]; *ADC1*, *ADC2* and *SAMDC2* expression are also induced by cold [21, 22, 23]. However, despite the involvement of PAs in abiotic stress responses has been documented [1], a systematic analysis of the different PAs integrated in a global plant stress response has not yet been accomplished.

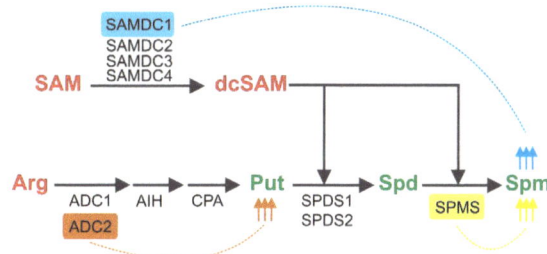

Figure 1: Polyamine biosynthetic pathway in *Arabidopsis thaliana*. Effects of *ADC*, *SAMdC* or *SPMS* over-expression: over-expressed genes are boxed; arrows of the same colour indicate the main effects of over expression on polyamine levels. *ADC*: arginine decarboxylase; AIH: agmatine iminohydrolase; CPA: N-carbamoylputrescine amidohydrolase; dcSAM: decarboxylated S-adenosylmethionine; SAM: S-adenosylmethionine; *SAMDC*: S-adenosylmethionine decarboxylase; *SPDS*: spermidine synthase; *SPMS*: spermine synthase.

TRANSCRIPTOMIC ANALYSIS OF ARABIDOPSIS PLANTS WITH INCREASED PUT LEVELS

Homologous over-expression of *ADC2* using a constitutive 35S promoter in Arabidopsis gave rise to plants with elevated Put levels [24] (Fig. **1**). *Put* accumulation by homologous over-expression of *ADC1* enhances freezing tolerance in Arabidopsis [25]. Likewise, elevated levels of *Put* by over-expressing *ADC2* produce drought tolerance in Arabidopsis [26]. On the contrary, deletion of Arabidopsis *ADC2* in T-DNA insertion mutants results in host plants defective in *Put* and more sensitive to salt stress [27]. Transcriptomic analysis revealed 1608 up-regulated and 2653 down-regulated genes in *ADC2* over-expressing plants when compared wild type (Col0) plants using the ATH1 microarray, and the Affymetrix Microarray Suite 5.0 statistical algorithms. Categorization of the over-expressed genes using the functional enrichment analysis tool GENECODIS 2.0 [28, 29] and the Mapman tool [30, 31] showed changes in expression of an important number of genes quoted as involved in both biotic and abiotic stress responses (Fig. **2**). Moreover, expression of a number of hormone signalling stress related genes was significantly altered in these plants (Fig. **2**). Among them, genes of the Indol acetic acid (IAA) biosynthetic pathway (*NIT2, GH3.3, GH3.5, YDK1, IAR3, ILL6*), auxin transport (*PIN3*) and genes coding for up to 20 auxin responsive proteins, (Fig. **2**). Moreover, genes coding for several ethylene-responsive transcription factors as well as for ethylene biosynthetic enzymes are up- and down- regulated (Fig. **2**). In addition, a set of genes coding for some of the ABA biosynthetic enzymes (ABA1, NCDE3, NCDE4) and several ABA responsive genes such as transcription factors *ABI1*, *ABF3*, *ABF4* appear down-regulated, while other ABA putative responsive proteins are up-regulated (Fig. **2**). Concerning biotic stress, some jasmonate-induced proteins (putative jacalin lectin family proteins) were under-expressed, but two JA pathway genes (*OPR2* and *OPR3*) were over-expressed (Fig. **2**). From a different point of view, application of MAPMAN analysis looking for specific categories in each set of genes (Fig. **2**) reveals that significant changes of expression in *ADC2* overexpressing plants correspond to signalling-related genes, as well as stress-related transcription factors. Finally, GENECODIS enrichment analysis applied to these particular gene sub-sets showed that some stress-related biological process appeared overrepresented.

TRANSCRIPTOME ANALYSIS OF ARABIDOPSIS PLANTS WITH INCREASED SPM LEVELS

Transgenic plants with altered *Spm* levels can be obtained either by homologous over expression of the *SAMDC1* gene or by homologous over expression of the *SPMS* gene (Fig. **1**). Elevated *Spm* levels produced by homologous over expression of *SAMDC1* in Arabidopsis enhances tolerance to drought and saline stress [Busó *et al.* in preparation]. In contrast, Arabidopsis *acl5/spms* double mutant plants are unable to produce *Spm* and show hypersensitivity to salt and drought stresses when compared to wild type [32, 33]. Addition of exogenous *Spm* supresses these stress hypersensitive phenotypes. Transcriptomic profiles obtained using Affymetrix ATH1 microarray and analyzed with Significance Analysis of Microarrays software (SAM 3.0) [34] for *SAMDC1* or *SPMS* over-expressing plants differ with respect to the one of Col0 wild type plants. GENECODIS functional enrichment

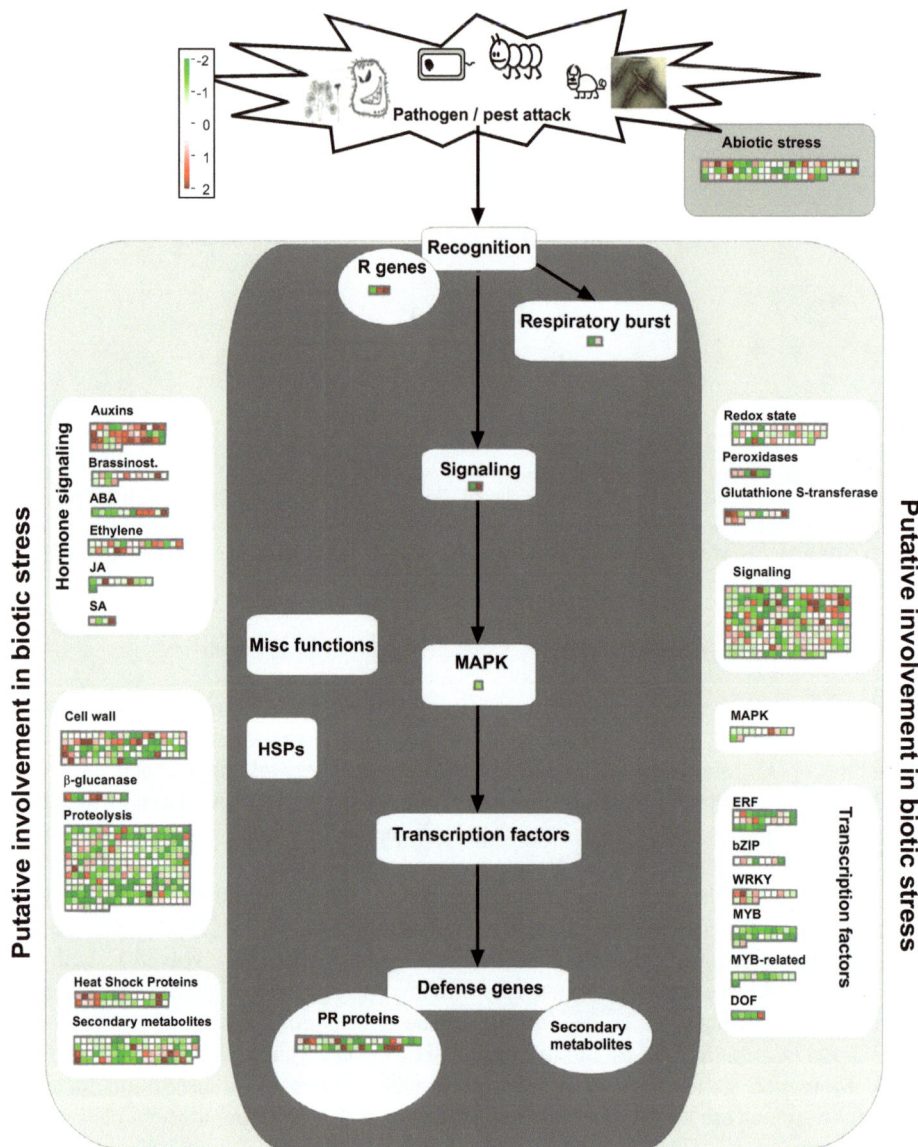

Figure 2: MAPMAN analysis of differentially expressed genes in *Arabidopsis* plants over-expressing the *ADC2* gene. Only genes involved in biotic and abiotic stress are shown. The signal from *ADC2* over-expressing plants is expressed as a ratio relative to the signal in wild type Col0 plants converted to a log2 scale, and displayed. The scale is shown in the figure.

shows that in both types of *Spm* overproducer plants, the respective sets of over expressed genes appear as enriched in annotation categories involved in defence-related processes from both biotic and abiotic stresses (data not shown). These correlations were confirmed after a deeper analysis of the distribution of over-expressed genes using the Mapman tool (Fig. **3**). A number of genes directly involved in some of the different events triggered in biotic stress appeared over-expressed in *Spm* accumulating plants. Moreover, a great number of signalling genes with a putative role in biotic stress appeared with high levels of expression. Some groups, although containing a minor number of genes, had majority of over-expressed transcripts, like ABA-related, JA-related or mitogen activated protein kinases (MAPKs). Those differences are observed for both types of *Spm* overproducer plants, although the number of genes and extent of the variation in expression were less pronounced for 35S-*SPMS* transgenic plants (Fig. **3**). From this picture, it is clear that elevated *Spm* levels have an important effect on plant stress response mechanisms. Although both *SAMDC1* and *SPMS* over expressing plants accumulate *Spm*, transcriptomic responses respect to wild type plants is only partially common. Thus, only 234 up-regulated and 333 down-regulated genes are present in both types of plants. Functional enrichment analysis of the up-regulated genes indicates that the most predominant biological processes were related with defence to biotic and abiotic stresses, as well as JA biosynthesis and JA and SA responses (data not shown). Moreover, two biosynthetic JA genes are up-regulated in both types of plants, as well as expression of the *NCDE3* gene, regulator of the ABA biosynthetic pathway (data not shown). Even though similar expression patterns for most of the categories presented are found in both *SAMDC1* and *SPMS* over-expressing plants, the differences with wild type plants are more pronounced in the case of *SAMDC1* over expressor lines.

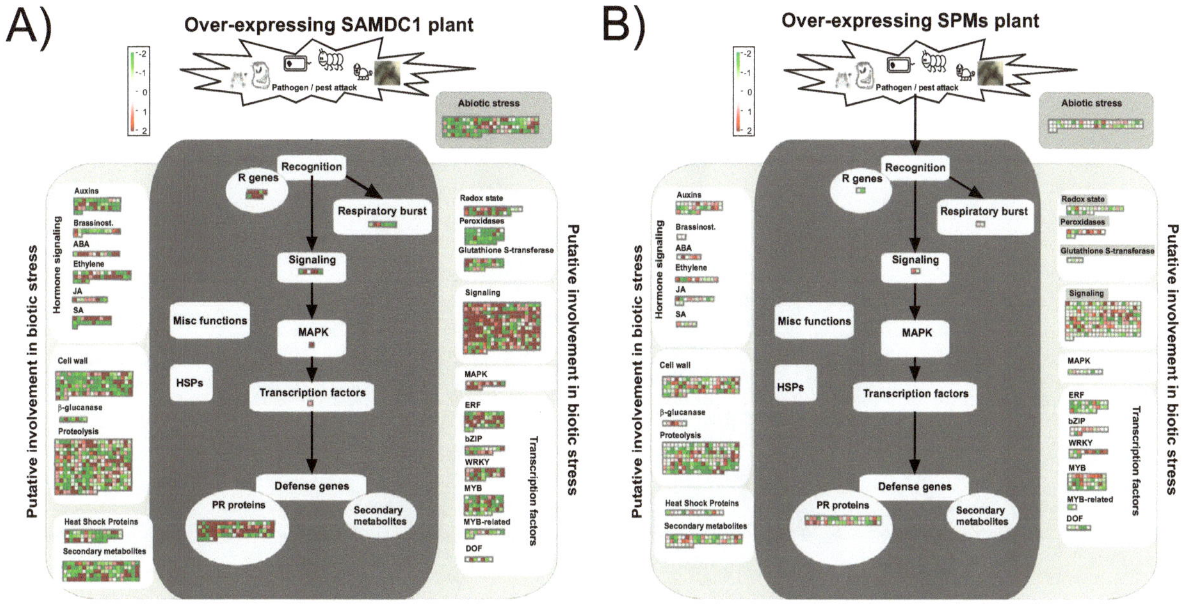

Figure 3: MAPMAN analysis of differentially expressed genes in *Arabidopsis* plants over-expressing the *SAMDC1* genes (A) or the *SPMS* gene (B). Only genes involved in biotic and abiotic stress are shown. The signal from the transgenic plants is expressed as a ratio relative to the signal in wild type Col0 plants converted to a log2 scale, and displayed. The scale is shown in the figure.

FINAL REMARKS AND FUTURE PERSPECTIVES

According to the transcriptomic data, modification of endogenous PA levels alters the expression of an important number of genes, with a significant group of genes involved in stress responses. A "PA-modulon" [35] has been proposed to explain how PAs could affect the coordinated expression of a number of genes in *E. Coli*. In this microorganism, PAs stimulate the synthesis,at translational level, of several key factors including OppA (a periplasmic substrate-binding protein of the oligopeptide uptake system), adenylate cyclase, RNA polymerase sigma subunit, and several transcription factors through their interaction with specific regions of mRNA, [35]. Therein, about one third of the genes whose expression was enhanced more than 2-fold by PAs is controlled by seven transcription factors from this PA-modulon [36]. A possible candidate of PA-modulon in *S. cerevisiae* is COX4, whose biosynthesis is enhanced about 2.5-fold by PAs, at the translational level [37]. The existence of a similar PA-modulon in plants could be one of the possible explanations for some of the gene changes observed in PA accumulating plants, although the identification of members of this putative plant PA modulon is still an unsolved issue. In this sense, future identification of transcription factors controlling gene expression in plants with modified PA metabolism could be an approach to identify target genes of the hypothetical "PA modulon" in plants.

When transcriptomic profiles of *Put* and *Spm* over producer plants are compared, only a set of 150 genes with significant expression changes appeared in common between the *Put* and *Spm* accumulating plants. From them, 71 genes appeared always up-regulated, and enriched in stress-related genes (Fig. **5**). Expression profiles of all 150 common genes analyzed by MAPMAN analysis, are similar for the majority of common genes represented focusing again in biotic and abiotic stress responses (Fig. **4**). Among these genes, some of them code for signalling-related proteins. Interestingly, 7 genes involved in calcium-signalling showed similar expression profiles in *Spm* and *Put* overproducer plants. Regulation by *Spm* of Ca^{2+} allocation through regulating Ca^{2+} permeable channels, including CAXs, has been described as a possible mechanism for the protective role of *Spm* against high salt and drought stress [33, 34]. Moreover, changes of free Ca^{2+} in the cytoplasm of guard cells are involved in stomatal movement that may explain drought tolerance induced by *Spm*. Furthermore, Ca^{2+} signalling genes are one of the gene categories mainly up-regulated in *Spm*-overproducer plants (Fig. **4**). In addition, a great part of this set of genes is also up-regulated in *Put* overproducer plants (Fig. **5**), suggesting that *Put* could also participate in a similar mechanism. All these results point to a possible link between PAs, Ca^{2+} homeostasis and stress responses, which should be further explored [10].

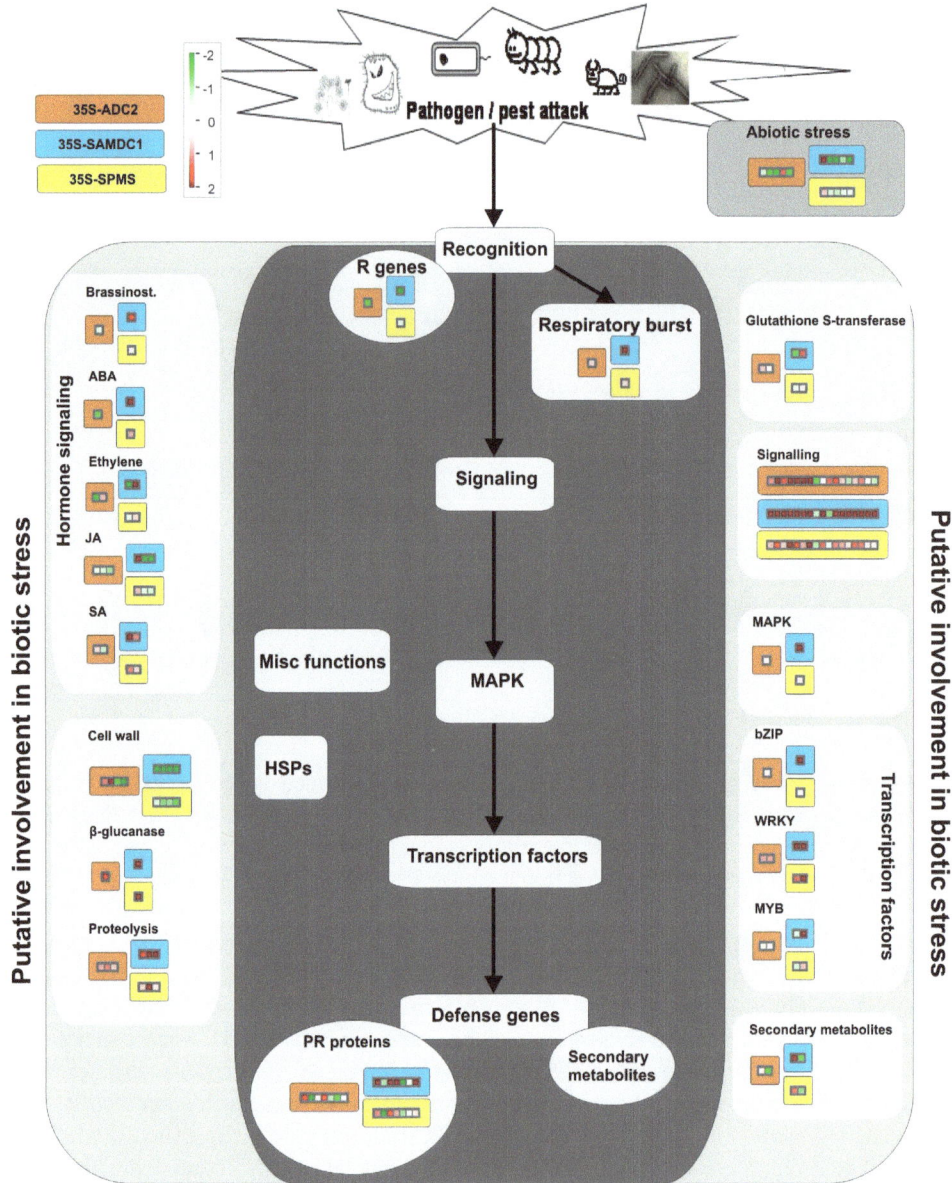

Figure 4: MAPMAN analysis of common differentially expressed genes from *ADC2*, *SAMDC1* and *SPMS* over-expressing plants. Only genes involved in biotic and abiotic stress are shown. Boxes: The signal from the transgenic plants is expressed as a ratio relative to the signal in wild type Col0 plants converted to a log2 scale, and displayed. The scale is shown in the figures. Orange Boxes: *ADC2* over-expressing plants; Blue boxes: *SAMDC1* over-expressing plants; Yellow boxes: *SPMS* over-expressing plants.

Finally, there are some evidences of cross talk between PAs and ABA [10]. Water stress induces the expression of *ADC2*, SPDS1 and *SPMS* [20], in a similar way to ABA treatment [21]. However, this upregulation is not observed in ABA-deficient (*aba2-3*) and ABA-insensitive (*abi1-1*) Arabidopsis mutants [20]. In addition, *Put* accumulation in wild type plants in response to drought was also impaired in *aba2-3* and *abi1-1* mutants. Those results suggest that up-regulation of PA-biosynthetic genes and accumulation of *Put* under water stress are mainly ABA-dependent responses [20]. The existence of cross talk between *Spm* and ABA is also re-inforced by the fact that the expression of *NCED3* is enhanced in plants with elevated *Spm* levels (see Section 4). Homologous over-expression of *SAMDC1* in Arabidopsis leads to elevated *Spm* levels and enhanced tolerance to various abiotic stress conditions. The *SAMDC1* over-expressing plants showed elevated levels of ABA due to the induction of NCED3, a key enzyme involved in ABA biosynthesis [32].

Up-regulated genes

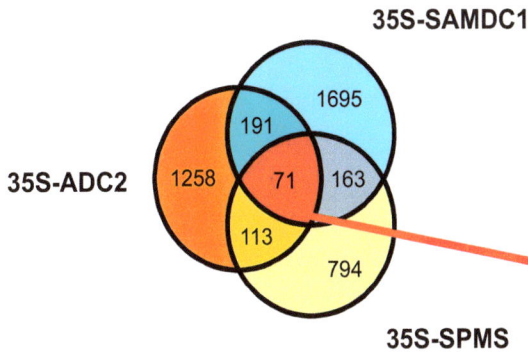

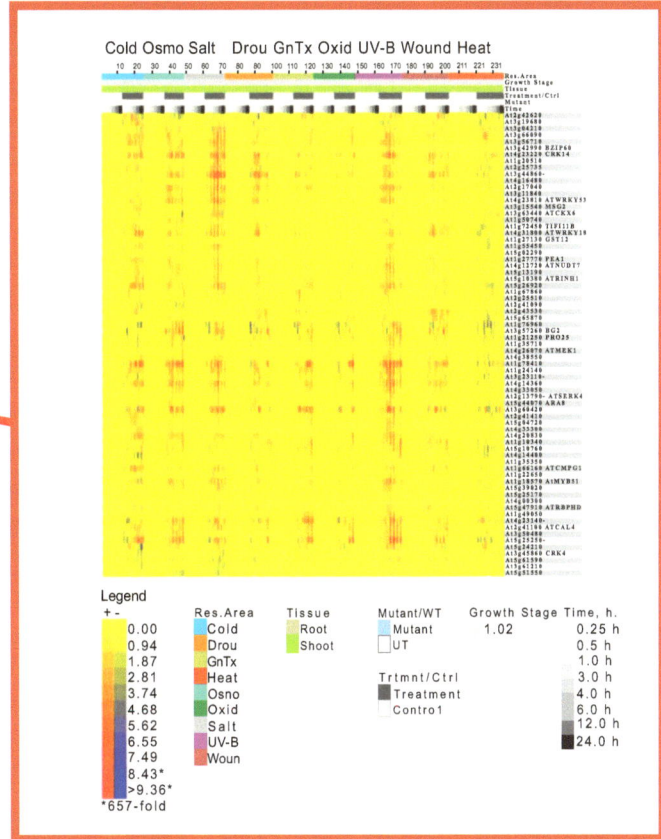

Figure 5: Stress expression profiles of common up-regulated genes in polyamine overproducer plants. Venn diagram: analysis of common up-regulated genes from *ADC2*, *SAMDC1* and *SPMS* over-expressing plants. Values represent the number of transcripts over expressed in transgenic plants relative to wild type Col0 plants. Red Box: E-northern heat map analysis of the 71 common up-regulated genes in the three types of plants using abiotic stress series from BAR [19]. Rows represent genes and columns represent experimental conditions. The color at a point represents the log2 of the ratio of the average of replicate treatments relative to the average of corresponding controls. The scale is shown in the figures.

Regarding to biotic stress, enhanced PA accumulation has been reported during tobacco mosaic virus (TMV) hypersensitive response (HR) [38, 39]. Additionally, *Spm* accumulation has been observed in the intercellular spaces of the necrotic lesions, being an endogenous inducer of the expression of pathogenesis-related (PR) protein-genes [40]. Moreover, exogenous application of *Spm* in tobacco leaves induces a pathway involving mitochondrial dysfunction, activation of MAPK, increased expression of HR marker genes, and caused defense responses and HR-like cell death [41,42,43,44]. It has been proposed [41] that *Spm* is able to activate defense pathways against pathogens by means of a "*Spm* signalling pathway", functioning via the merged signal of *Spm*-activated Ca^{2+}-influx and H_2O_2 produced from *Spm*-degradation by PA oxidases [45]. Both events are able to trigger mitochondrial disfunction, and activation of the cell death programs [41]. The involvement of the *Spm*-signalling pathway has also been confirmed in Arabidopsis plants in response to cucumber mosaic virus (CMV) [46]. These authors used superSAGE analysis to identify a set of genes that were both *Spm* and CMV responsive. Seven *Spm*-up-regulated genes identified in this study are also over-expressed in the *Spm*-overproducer plants including transcription factor AtbZIP60 and MAPK3. AtbZIP60 may control the expression of genes participating in protein folding and secretion which may be required during CMV-elicited HR [47]. MAPK3 is an ortholog of a wound-inducible protein kinase (WIPK) that is *Spm*-induced in the tobacco *Spm*-signalling pathway [41].

All these data indicate that PAs could act as key regulatory molecules in both abiotic and biotic stress processes, by cross-talk with other stress related hormones such as ABA as well as by affecting other secondary messengers such as Ca^{2+}. As discussed in this chapter, it is possible that the nature of these interactions is likely to be dependent on the type of PA involved.

ACKNOWLEDGEMENTS

F Marco and Fundación CEAM are partly supported by Generalitat Valenciana, Bancaixa and programme CONSOLIDER-INGENIO 2010 (GRACCIE, CSD2007-00067). AF Tiburcio has been supported by grants BIO2005-09252-C02-01 and BIO2008-05493-C02-01 (MICINN); P Carrasco has been funded by grant BIO2005-09252-C02-02 (MICINN). AF Tiburcio, P Carrasco and R, Alcázar acknowledge aids in grant from ACTION COST FA0605. OA Ruiz and ME González were supported by: Consejo Nacional de Investigaciones Científicas y Técnicas (CONICET, Argentina), Agencia Nacional de Promoción Científica y Tecnológica (ANPCYT, Argentina), and Comisión de Investigación Científica (CIC, Buenos Aires, Argentina).We also thank to Drs M. Blazquez and J Carbonell for their critical reading of this chapter.

REFERENCES

[1]   Alcázar R *et al.* Involvement of polyamines in plant response to abiotic stress. Biotechnol Lett 2006; 28: 1867-1876.
[2]   Bouchereau A, Aziz A, Larher F, Martin-Tanguy J. Polyamines and environmental challenges: recent development. Plant Sci 1999; 140: 103-125.
[3]   Flores HE. In: Slocum R, Flores HE Eds. The biochemistry and physiology of polyamines in plants. Changes in polyamine metabolism in response to abiotic stress. Boca de Raton FL, CRC Press, 1991.
[4]   Borrell A, Carbonell L, Farrás R, Puig-Parellada P, Tiburcio AF. Polyamines inhibit lipid peroxidation in senescing oat leaves. Physiol Plant 1997; 99: 385-390.
[5]   Marco F, Carrasco, P. Expresion of the pea S-adenosylmethionine decarboxylase gene is involved in developmental and environmental responses. Planta 2002; 214: 641-647.
[6]   Groppa MD, Benavides MP, Tomaro ML. Polyamine metabolism in sunflower and wheat leaf discs under cadmium or copper stress. Plant Science 2003; 164: 293-299.
[7]   Ferrando A, Carrasco P, Cuevas JC, Altabella T, Tiburcio AF. In: Amâncio S, Stulen I Eds. Nitrogen acquisition and assimilation in higher plants. Integrated molecular analysis of the polyamine metabolic pathway in abiotic stress signalling. Kluwer Academic Publishers, The Netherlands, 2004; pp. 207-230.
[8]   Kusano T, Berberich T, Tateda C, Takahashi Y. Polyamines: essential factors for growth and survival. Planta 2008; 228: 367-381.
[9]   Gill SS, Tuteja N. Polyamines and abiotic stress tolerance in plants. Plant Signal Behav 2010; 7: 5(1). PMID: 20023386.
[10]   Alcázar R *et al.* Polyamines: molecules with regulatory functions in plant abiotic stress tolerance. Planta 2010; doi: 10.1007/s00425-010-1130-0.
[11]   The Arabidopsis Genome Initiative. Analysis of the genome sequence of the flowering plant *Arabidopsis thaliana*. Nature 2000; 408: 796-815.
[12]   Watson MW, Malmberg RL. Regulation of Arabidopsis thaliana (L.) Heynh Arginine Decarboxylase by Potassium Deficiency Stress Plant Physiol 1996; 111: 1077-1083.
[13]   Watson MW, Yu W, Galloway GL, Malmberg RL. Isolation and Characterization of a Second Arginine Decarboxylase cDNA from Arabidopsis (Accession No AF009647) Plant Physiol 1997; 114: 1569.
[14]   Janowitz T, Kneifel H, Piotrowski M. Identification and characterization of plant agmatine iminohydrolase, the last missing link in polyamine biosynthesis of plants. FEBS Lett 2003; 544: 258-261.
[15]   Piotrowski M, Janowitz T, Kneifel H. Plant C-N hydrolase and the identification of plant N-carbamoylputrescine amidohydrolase involved in polyamine biosynthesis. J Biol Chem 2003; 278: 1708-1712.
[16]   Hanzawa Y, Imai A, Michael AJ, Komeda Y, Takahashi T. Characterization of the spermidine synthase-related gene family in Arabidopsis thaliana. FEBS Lett 2002; 527: 176-180.
[17]   Panicot M *et al.* A polyamine metabolon involving aminopropyl transferases complexes in Arabidopsis. Plant Cell 2002; 14: 2539-2551.
[18]   Urano K *et al.* Characterization of Arabidopsis genes involved in biosynthesis of polyamines in abiotic stress responses and developmental stages. Plant Cell Environ 2003; 26: 1917-1926.
[19]   Toufighi K, Brady SM, Austin R, Ly E, Provart NJ. The Botany Array Resource: e-Northerns, Expression Angling, and promoter analyses. Plant J 2005; 43: 153-163.

[20]   Alcázar R, Cuevas JC, Patrón M, Altabella T, Tiburcio AF. Abscisic acid modulates polyamine metabolism under water stress in *Arabidopsis thaliana*. Physiol Plant 2006; 128: 448-455.

[21]   Urano K *et al.* Characterization of Arabidopsis genes involved in biosynthesis of polyamines in abiotic stress responses and developmental stages. Plant Cell Environ 2003; 26: 1917-1926.

[22]   Cuevas JC *et al.* Putrescine is involved in Arabidopsis freezing tolerance and cold acclimation by regulating abscisic acid levels in response to low temperature. Plant Physiol 2008; 148: 1094-1105.

[23]   Cuevas JC *et al.* Putrescine as a signal to modulate the indispensable ABA increase under cold stress. Plant Signal Behav 2009; 4: 219-220.

[24]   Alcázar R, García-Martínez JL, Cuevas JC, Tiburcio AF, Altabella T. Overexpression of *ADC2* in Arabidopsis induces dwarfism and late-flowering through GA deficiency. Plant J 2005; 43: 425-436.

[25]   Altabella T, Tiburcio AF, Ferrando. Plant with resistance to low temperature and method of production thereof. Spanish patent application 2009; WO2010/004070

[26]   Alcázar R *et al.* Putrescine accumulation confers drought tolerance in transgenic Arabidopsis plants overexpressing the homologous Arginine decarboxylase 2 gene. Plant Physiol Biochem 2010; doi: 10.1016/j.plphy.2010.02.002

[27]   Urano K, Yoshiba Y, Nanjo T, Ito T, Yamaguchi-Shinozaki K, Shinozaki K. Arabidopsis stress-inducible gene for arginine decarboxylase AtADC2 is required for accumulation of putrescine in salt tolerance. Biochem Biophys Res Commun 2004; 313: 369-375.

[28]   Carmona-Saez P, Chagoyen M, Tirado F, Carazo JM, Pascual-Montano A. GENECODIS: A web-based tool for finding significant concurrent annotations in gene lists. Genome Biol 2007; 8(1): R3.

[29]   Nogales-Cadenas R *et al.* GeneCodis: interpreting gene lists through enrichment analysis and integration of diverse biological information. Nucleic Acids Res (2009); doi: 10.1093/nar/gkp416

[30]   Thimm O *et al.* MAPMAN: a user-driven tool to display genomics data sets onto diagrams of metabolic pathways and other biological processes. Plant J 2004; 37(6): 914-39.

[31]   Usadel B *et al.* Extension of the visualization tool MapMan to allow statistical analysis of arrays, display of corresponding genes, and comparison with known responses. Plant Physiol 2005; 138: 1195-204.

[32]   Yamaguchi K *et al.* The polyamine spermine protects against high salt stress in Arabidopsis thaliana. FEBS Lett 2006; 580: 783-6788.

[33]   Yamaguchi K *et al.* A protective role for the polyamine spermine against drought stress in Arabidopsis. Biochem Biophys Res Commun 2007; 352: 86-490.

[34]   Tusher V, Tibshirani R, and Chu G. Significance analysis of microarrays applied to transcriptional responses to ionizing radiation. Proc Natl Acad Sci USA 2001; 98: 5116-5121.

[35]   Igarashi K, Kashiwagi K. Polyamine Modulon in Escherichia coli: genes involved in the stimulation of cell growth by polyamines. J Biochem 2006; 139(1): 11-6.

[36]   Terui Y *et al.* Enhancement of the synthesis of RpoN, Cra, and H-NS by polyamines at the level of translation in Escherichia coli cultured with glucose and glutamate. J Bacteriol 2007; 189: 2359-2368.

[37]   Uemura T, Higashi K, Takigawa M, Toida T, Kashiwagi K, Igarashi K. Polyamine modulon in yeast-Stimulation of COX4 synthesis by spermidine at the level of translation. Int J Biochem Cell Biol 2009; 41(12): 2538-45.

[38]   Torrigiani P *et al.* Polyamine synthesis and accumulation in the hypersensitive response to TMV in Nicotiana tabacum. New Phytol 1997; 135: 467-473.

[39]   Marini F, Betti L, Scaramagli S, Biodi S, Torrigiani P. Polyamine metabolism is up-regulated in response to tobacco mosaic virus in hypersensitive, but not in susceptible, tobacco. New Phytol 2001;149: 301-309.

[40]   Yamakawa H, Kamada H, Satoh M, Ohashi Y. Spermine is a salicylate-independent endogenous inducer for both tobacco acidic pathogenesis-related proteins and resistance against Tobacco mosaic virus infection. Plant Physiol 1998; 118: 213-1222.

[41]   Takahashi Y, Berberich T, Miyazaki A, Seo S, Ohashi Y, Kusano T. Spermine signalling in tobacco: activation of mitogen-activated protein kinases by spermine is mediated through mitochondrial dysfunction. Plant J 2003; 36: 820-829.

[42]   Takahashi Y *et al.* A subset of the hypersensitive response marker genes including HSR203J is downstream target of a spermine-signal transduction pathway in tobacco. Plant J 2004; 40: 586-595.

[43]   Uehara Y *et al.* Tobacco ZFT1, a transcriptional repressor with a Cys2/His2 type zinc-finger motif that functions in spermine-signaling pathway. Plant Mol Biol 2005; 59: 435-448.

[44]   Mitsuya Y *et al.* Identification of a novel Cys2/His2-type zinc-finger protein as a component of a spermine-signaling pathway in tobacco. J Plant Physiol 2007; 164: 785-793.

[45]   Marina M *et al.* Apoplastic polyamine oxidation plays different roles in local responses of tobacco to infection by the necrotrophic fungus Sclerotinia sclerotiorum and the biotrophic bacterium Pseudomonas viridiflava. Plant Physiol 2008, 147: 2164-2178.

[46]   Mitsuya Y *et al.* Spermine signaling plays a significant role in the defense response of Arabidopsis thaliana to cucumber mosaic virus. J Plant Physiol 2009; 166(6): 626-43

[47]   Iwata Y, Koizumi N. An Arabidopsis transcription factor, AtbZIP60, regulates the endoplasmic reticulum stress response in a manner unique to plants. Proc Natl Acad Sci USA 2005; 102: 5280-5.

CHAPTER 7

Abiotic stress in plants: From Genomics to Metabolomics

A. Roychoudhury, Karabi Datta and S. K. Datta*

Department of Botany, Plant Molecular Biology and Biotechnology Laboratory, University of Calcutta, 35, Ballygunge Circular Road, Kolkata-700019

Abstract: Plants are exposed to various abiotic stresses such as water deficit or ion excess, elevated temperature, high light intensity, salinity, freezing, cold, etc. under field conditions, potentially reducing the yield of crop plants by more than 50%. Investigations of the physiological, biochemical and molecular aspects of stress tolerance have been conducted to unravel the intrinsic mechanisms for mitigation against stress. The new molecular "omic" tools, comprising of genomics, proteomics and metabolomics, have opened up new perspectives in stress biology. Before the advent of genomics era, a gene-by-gene approach was used to decipher the function of genes involved in abiotic stress response. The availability of genome sequences of certain important plant species has enabled the use of strategies like genome-wide expression profiling to identify the genes associated with stress response, followed by the verification of gene function by the analysis of mutants and transgenics. The genomics based approaches provide access to agronomically desirable alleles present at quantitative trait loci (QTLs), thus enabling the improvement of abiotic stress tolerant plants. Marker assisted selection (MAS) is already helping breeders improve drought related traits. The recent upsurge in structural genomics determines the DNA sequence by manual or robotic methods. Further elucidation of the complex networks interacting during stress defenses has been achieved by multi-parallel analysis of transcript levels, microarray analysis, RT-qPCR, Serial analysis of gene expression (SAGE), Massive parallel signature sequencing (MPSS) and more recently oligoarray using the transcriptome of any specie to evaluate abiotic stress response. Analysis of sequence data and gene products would facilitate the identification and cloning of genes at target QTLs and their direct manipulation via genetic engineering. Furthermore, several bioinformatic tools, ESTs and subtractive cDNA libraries have added new dimensions for deciphering the genetic basis of stress tolerance. The current initiatives in functional genomics or proteomic research for the analysis of plant stress tolerance is based on two-dimensional gel electrophoresis (2-DE) and identification of differentially displayed spots by MALDI-TOF, QTOF MS/MS, liquid chromatography coupled with tandem mass spectrometry (LC-MS/MS), characterization of separated proteins by mass spectrometer and database searching. In addition, many proteins are modified by post-translational modifications such as phosphorylation, glucosylation, ubiquitinylation, sumoylation and many others. New techniques in gel-based approaches such as difference gel electrophoresis (DIGE) can provide both qualitative and quantitative data about the differential expression of proteins. Proteomics nowadays is also used to study the relationship between gene expression (transcriptomics) and metabolism (metabolomics). Our limited knowledge of stress-associated metabolism remains a major gap in our understanding of the stress response. Therefore comprehensive profiling of stress-associated metabolites is most relevant to the successful molecular breeding of stress-tolerant crop plants. Metabolomic studies, thus along with transcriptomics and proteomics, and their integration with systems biology, will lead to strategies to alter cellular metabolism for adaptation to abiotic stress conditions. The present review will focus on the current development, progress and applications of genomics, proteomics and metabolomic studies, and implementation of systems biology to meet the challenges of abiotic stress and crop improvement program from a practical standpoint. The accumulating information will provide plant researchers to explore new paradigms to address fundamental and practical questions in a multidisciplinary manner.

INTRODUCTION

Plants frequently encounter abiotic stresses such as extreme temperatures, low water availability, high salt levels, and mineral deficiency/toxicity in both natural and agricultural systems. In the face of a global scarcity of water resources and the increased salinization of soil and water, abiotic stress is already a major limiting factor in plant growth and will soon become even more severe as desertification covers more and more of the world's terrestrial area. Drought and salinity are already widespread in many regions, and has been predicted to cause serious salinization of more than 50% of arable lands by the year 2050 [1]. In many cases, several classes of abiotic stress challenge the plants in combination. For example, high temperature and scarcity of water are commonly encountered in periods of drought, and can be exacerbated by mineral toxicities that constrain root growth. As environmental

*Address correspondence to: **Dr. Swapn K. Datta,** Department of Botany, Plant Molecular Biology and Biotechnology Laboratory, University of Calcutta, 35, Ballygunge Circular Road, Kolkata-700019; E-mail: swpndatta@yahoo.com

conditions begin to change, plants will sense this. After this initial perception, a signal is relayed via several signal transduction cascades that amplify the signal and notify parallel pathways. Higher plants actually have evolved multiple, interconnected strategies that enable them to survive abiotic stress. However, these strategies are not well developed in most agricultural crops. Across a range of cropping systems around the world, abiotic stresses are estimated to reduce yields to less than half of that possible under ideal growing conditions. Understanding the basis of mechanisms that govern the tolerance explains the ecological diversity and species distribution during stress response and provides important information about the performance of field crops. In a world where population growth exceeds food supply, agricultural and plant biotechnologies aimed at overcoming severe environmental stresses need to be fully implemented. Various strategies have been employed to isolate the genes that are involved in the stress response. Several genes that are known to be responsive to stress or are involved in imparting tolerance can be related to cellular functions. Information about such stress-responsive genes has been obtained largely using conventional approaches. However, the challenge still remains to integrate the function of these genes logically to generate a global understanding of the stress response process [2]. The complete genome sequence of rice and *Arabidopsis thaliana* and emerging sequence information for several other plant genomes, such as *Populus*, *Medicago*, lotus, tomato and maize, have given rise to the use of tools which can aid in the determination of the function of many genes simultaneously. The plant engineering strategies for abiotic stress tolerance rely on the expression of genes that are involved in signaling and regulatory pathways or genes that encode proteins conferring stress tolerance or enzymes present in pathways leading to the synthesis of functional and structural metabolites. In contrast to plant resistance to biotic stresses, which is mostly dependent on monogenic traits, the genetically complex responses to abiotic stresses are multigenic, and thus more difficult to control and to develop an engineered product. Traditional approaches to breeding crop plants with improved stress tolerance have also thus far met with limited success-in part because of the difficulty of breeding for multigenic tolerance traits in traditional breeding programs. Desired traits can be introgressed into desirable crop species from wild relatives and, for the cereals, extensive abiotic stress tolerance has been identified in screens of land races and related wild species. It is estimated that only 10-20% of the wild variation has been used in modern wheat varieties. A similar situation appears to apply to rice and barley, although not as extreme as in wheat. Current efforts to improve plant stress tolerance by genetic transformation of specific genes have resulted in several important achievements; however, the genetically complex mechanisms of abiotic stress tolerance make the task extremely difficult. For this reason, biotechnology should be fully integrated with classical physiology and breeding. Acquired plant stress tolerance can be enhanced by manipulating stress-associated genes and proteins and by overexpression of stress-associated metabolites (genetic engineering approach) and by conventional plant breeding combined with the use of molecular markers (Fig. **1**).

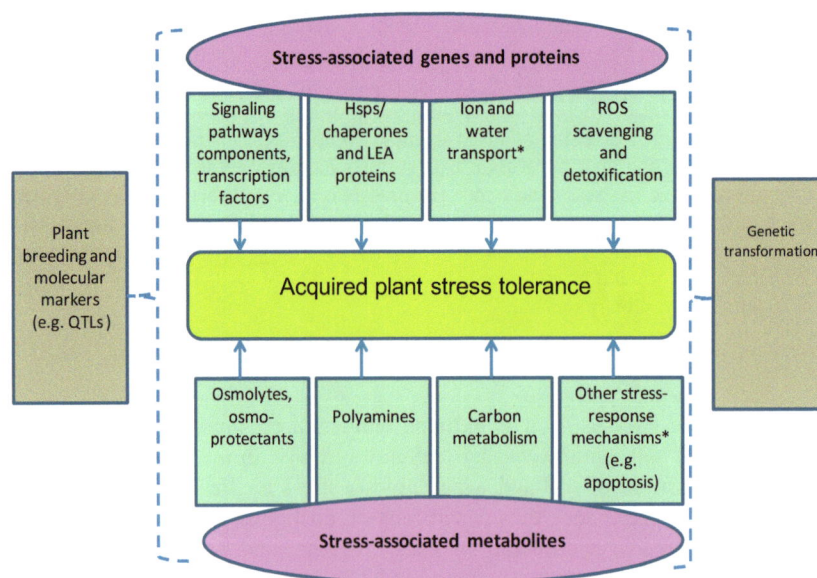

Current Opinion Biotechnol 16:123-132 (2005)

Figure 1: Acquired plant stress tolerance can be enhanced by manipulating stress-associated genes and proteins and by overexpression of stress-associated metabolites. Plant resistance to abiotic stress is a multigenic trait, depending on the combination of many genes, proteins and metabolic pathways all playing in concert. Stress-associated mechanisms that are not discussed in the present review are marked by an asterisk. Acquired plant tolerance to abiotic stress can be achieved both by genetic engineering and by conventional plant breeding combined with the use of molecular markers and quantitative trait loci (QTLs). [Hsp: heat shock protein; LEA: late embryogenesis abundant; ROS: reactive oxygen species]

With the availability of genome sequences, the new molecular 'omic' tools play an increasingly important role to quantitatively study the expression levels of thousands of genes and their products, i.e. proteins and metabolites in parallel, over a time course or across a series of defined conditions. The data obtained using these integrative methods provide valuable information regarding the function and interaction of genes and their products in cellular processes. The 'omic' tools created major interest among researchers and opened up new perspectives in stress biology. Indeed, the last few years saw considerable progress in the analysis of the transcriptome to either salt stress alone or in combination with other abiotic stresses. The real power of the 'omic' approach is the ability to look at the studied response on a number of different levels, including transcripts, proteins or metabolites. The combination of different 'omics' platforms, including genomics, proteomics, metabolomics and mathematical modeling, will enable us to achieve a holistic view of the abiotic stress response in plants and conduct detailed pathway analysis.

GENOMICS BASED APPROACHES TOWARDS STRESS RESPONSE

The application of genomics-type technologies is beginning to have an impact, enhancing our understanding of plant responses to abiotic stresses. The term genomics could be understood as any technology that, preferably in a high-throughput genome-focused fashion, promises insights and answers on how plant genes, proteins, protein activities, and metabolite type and flux respond to external factors. The multidisciplinary genomic approaches for abiotic stress tolerance are schematically represented in Fig. (2A). There is considerable interest at present in using the emerging technologies of genomics as a means to identify key loci controlling stress tolerance and as a tool to screen for allelic variation in the wild and land race gene pools. Functional genomics employs multiple parallel approaches, including global transcript profiling coupled with the use of mutants and transgenics, to study gene function in a high throughput mode and improve crop performance under abiotic stress conditions (Fig. 2B). The aim of these genome-wide efforts is to finally link the genome to the phenome. Delivery of the outcomes of genomics can be through conventional breeding, although genetic transformation offers a more rapid option in some circumstances. The cereals are our dominant source of food, with maize, rice and wheat vying for the number one position. However, the four closely related Triticeae crops namely wheat, barley, rye and triticale occupy the lower yielding environments and cover almost twice the area sown with maize or rice. Although wheat, rye and barley show levels of tolerance to many abiotic stresses well above maize or rice, little information is available about the molecular basis for abiotic stress tolerance in these species and there is still ample scope for improvement.

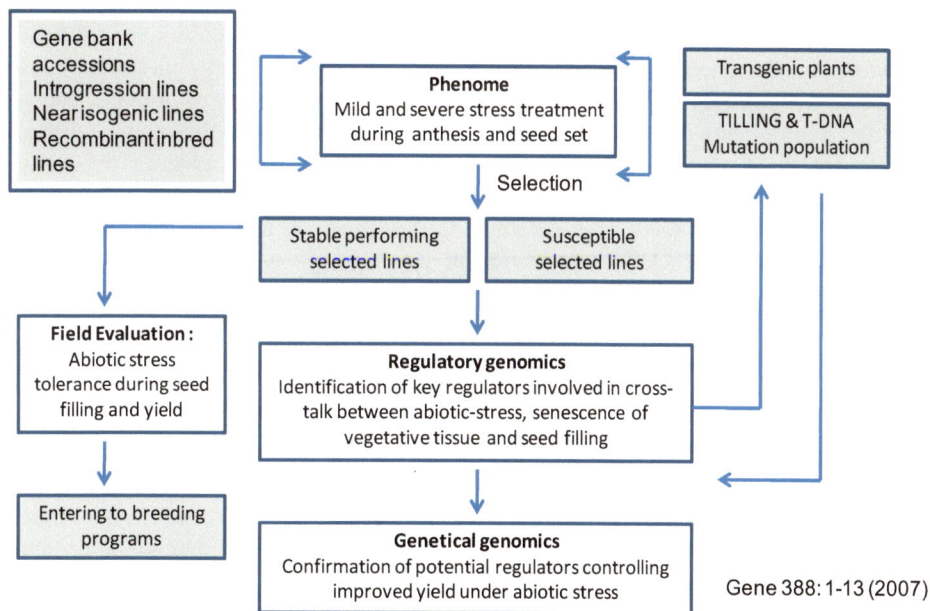

Figure 2A: Schematic representation of the multidisciplinary genomic approaches for abiotic stress tolerance in plants

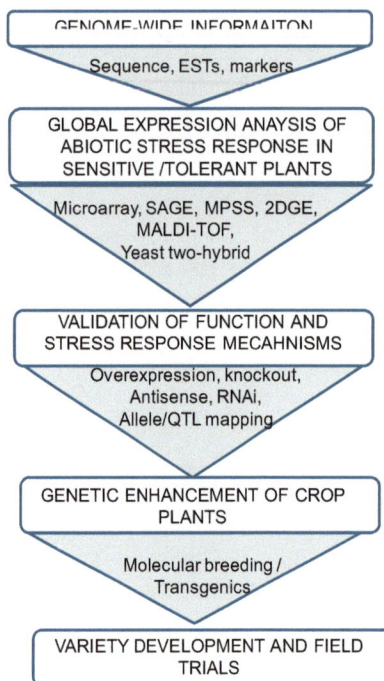

Plant Biotech J 5:361-380 (2007)

Figure 2B: A typical functional genomics approach to improve crop performance under abiotic stress conditions

Until recently, the prevalent strategy of the molecular genetic approach could be termed the 'candidate gene by-gene' approach. This approach aims to dissect single genes in many pathways in order to determine the position of a particular gene in the stress-response cascade and to determine its contribution to tolerance acquisition. The functional genomics strategy begins to tie together physiological and phenotypic observations with information on gene complement, transcription and transcript regulation, the behavior of proteins, protein complexes and pathways, evolutionary adaptive diversification and mutated or engineered phenotypic characters.

The advent of genomics has supplanted the necessity to focus abiotic stress biology on species that show natural tolerance. Increasing numbers of tools and resources in the traditional models has facilitated the identification of stress-relevant genes and pathways. These encompass genome sequences, including bacterial artificial chromosome (BAC) libraries and BAC-end sequences; expressed sequence tag (EST) and cDNA collections; transcript profiling platforms; sophisticated technologies for the real-time non-destructive localization of proteins; and sensors for and profiles of metabolites. Evolutionary and comparative functional studies will probe the knowledge gained from work with *A. thaliana* or rice, their ecotypes, breeding lines and close relatives, and even unrelated species. The availability of the complete *Arabidopsis* and rice genome sequences, together with the collection of several plant ESTs, has understandably shifted the focus from determining the sequences to understanding their function. These sequences will greatly aid the process of determination of the function of the majority of plant genes, whose functions remain to be demonstrated experimentally [3]. The comparative genome sequence analysis between *A. thaliana* and the closely related crucifer *Thellungiella halophila* (*Thellungiella salsuginea*), a species with extreme cold and freezing, drought and salinity tolerance has provided already a lot of information [4]. A comparison of transcriptional regulation in five *A. thaliana* accessions revealed genes that have distinct expression patterns, which identified differences in a number of characters, including stress responses [5].

In general categories, the existing anthology of data has been obtained from reverse genetics, cDNA libraries, ESTs and serial analysis of gene expression (SAGE)-derived transcript populations [6,7,8,9,10,11] and profiling of stress-responsive transcripts, including coverage of nutritional stresses [12]. A vital mission of computational stress genomics is identifying genes that respond to diverse environmental stimuli. Regulation of stress-inducible gene expression, which is determined by chromatin structure, the binding of transcription factors and cis-regulatory DNA sequences, can be inferred computationally by mining transcript profiles and the regional structure and distribution of short response elements in corresponding promoter regions. Several bioinformatic databases for the analysis of plant promoters are available. Reconciling stress-regulated genes with promoter structures and then extracting rules about regulation in a cell and tissue context can, in principle, rely on the increasing number of microarray datasets but genome-wide analyses are still few. These analyses have focused on regulatory networks for which response elements, such as dehydration-responsive element binding protein (DREB) or abscisic-acid-responsive element (ABRE) in plants, or cAMP-responsive element binding protein (CREB) in yeast, are well known [13,14]. The relative paucity of results suggests that additional techniques for the identification of promoter-binding proteins will have to be applied: chromatin immunoprecipitation, protein binding to synthetically generated promoters, and *in vivo* recordings of promoter activity with fluorescent markers [15]. Various profiling platforms can assemble information on abiotic-stress-dependent mechanisms that must then be incorporated by bioinformatics into a plant wide stress database to generate new phenotypes, and to allow the information to be exported to crop species in an iterative fashion.

Transcript Populations and Transcript Profiling

EST sequencing, as important as it continues to be for providing a fast overview of abundant transcripts in a species, is being superseded by more sophisticated approaches. Among these are libraries that are enriched for full-length cDNAs and normalized and subtracted cDNA libraries [8,11]. Such libraries can be made up of transcripts from multiple experimental conditions, in which individual RNA is converted into cDNA, with an oligonucleotide tag attached, that identifies a transcript population. Especially with fully sequenced genomes, SAGE provides, with little effort, hundreds of thousands of tags, with each tag identifying the nature and abundance of a transcript [6, 9]. Conceivably, further advances in SAGE profiling could make this technique a viable alternative to other transcript profiling platforms. Transcript profiling, using GeneChips or long-oligonucleotide array slides, provides important insights into the dynamics of the transcriptional changes that accompany abiotic stress treatments. While both array platforms provide comparable results, they are distinguished by their ease of handling, dynamic range, reproducibility, and sensitivity to nucleotide polymorphisms in the targets. Several contributions have reported on changes to transcript profiles under cold, drought, and high-salinity conditions, mainly in *A. thaliana* and rice, and the variety of species, in which analyses focusing on stress responses have been undertaken, is gradually increasing. In addition, profiling studies have been directed towards understanding transcriptional regulation during nutrient stresses or on probing the effects of stresses on the transcriptome by eliminating regulatory genes that are involved in stress perception or signaling. The precision of global expression profiling is now such that it can report with high confidence stress-dependent changes in transcript abundance that had previously been observed in isolated experiments. The consolidation and integration of individual transcript profiling results (for which the establishment of data standards will be necessary) into a species-wide and, later, a plant wide database would provide control and a high level of confidence. Nonetheless, we have insufficient information about stress-dependent, cell-specific expression patterns (e.g., in guard cells, cells in the vasculature or shoot and root meristems) and current technologies do not readily distinguish the presence of different splicing isoforms or tell us how resident transcripts might be utilized in translation [16].

Candidate Genes/Transcription Factors Up Regulated During Abiotic Stress

Several components of regulatory systems are activated on the perception of stress. This leads to the production of effector molecules, which are directly involved in mitigating stress. The complexity of plant response to the primary abiotic stressors like drought, salinity, cold, heat and chemical pollution, which are often interconnected and cause cellular damage and secondary stresses, such as osmotic and oxidative stress, together with the up regulation of functional proteins and regulatory factors are depicted in Fig. **3**. The regulatory systems are represented by transcription factors and signal transduction components such as kinases and phosphatases [17]. The ABA-independent regulatory cascade is mainly represented by the DREB1 (cold) and DREB2 (salt, dehydration) families of proteins [18, 19]. Another major abiotic stress signal transduction pathway is the ABA-dependent pathway. This pathway is mainly associated with the bZIP class of transcription factors called ABRE-binding factors, ABFs [20, 21, 22]. Several other transcription factors involved in the abiotic stress response have been identified and are being characterized [17]. Expression profiling of mutants and transgenics overexpressing transcription factors can aid in an understanding of their mechanism of action. The analysis of the CBF regulon in *Arabidopsis* revealed that, of 306 cold-responsive genes, only 70% were part of the CBF regulon which included 15 transcription factors [23]. ZAT12 has been identified as a new cold-responsive regulon which functions as a negative regulator of CBF2 [24]. In addition, studies with another cold-responsive regulon, i.e. ICE1, have shown that it acts as a master switch controlling many CBF-dependent and CBF-independent regulons. Microarray analysis has also been performed for transgenic *Arabidopsis* plants constitutively expressing DREB2A, a key transcription factor involved in salt and drought stress regulation. Twenty-one genes were up regulated by DREB2A over-expression. The DRE core motif was found in 14 of these genes, indicating that they may be direct targets for regulation by DREB2A [25].

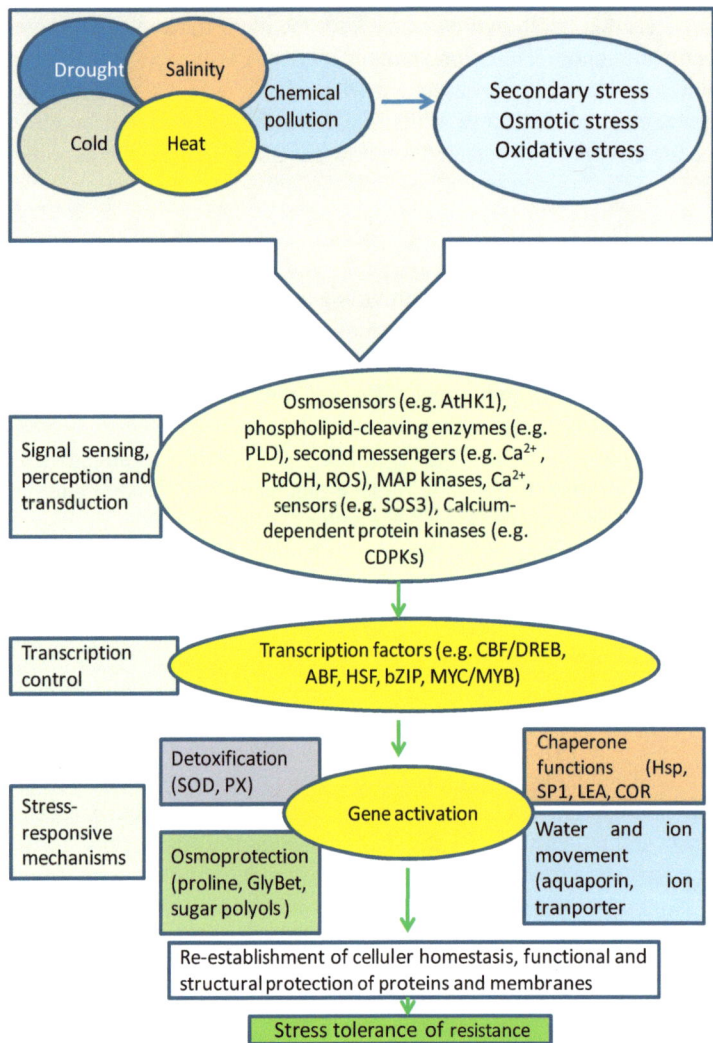

Current Opinion Biotechnol 16:123-132 (2005)

Figure 3: The complexity of the plant response to abiotic stress; the initial stress signals (e.g. osmotic and ionic effects or changes in temperature or membrane fluidity) trigger the downstream signaling process and transcription controls, which activate stress-responsive mechanisms to re-establish homeostasis and to protect and repair damaged proteins and membranes. Inadequate responses at one or more steps in the signaling and gene activation process might ultimately result in irreversible changes in cellular homeostasis and in the destruction of functional and structural proteins and membranes, leading to cell death. [ABF: ABRE binding factor; AtHK1: *Arabidopsis thaliana* histidine kinase-1; bZIP: basic leucine zipper transcription factor; CBF/DREB: C-repeat-binding factor/dehydration-responsive binding protein; CDPK: calcium-dependent protein kinase; COR: cold-responsive protein; MAP: mitogen-activated protein; PLD: phospholipase D; PtdOH: phosphatidic acid; PX: peroxidase; SOD: superoxide dismutase; SP1: stable protein 1]

Microarray analyses to decipher abiotic stress regulation have also been performed for signalling components, such as kinases and phosphatases [26, 27]. J. Heard and coworkers from Monsanto have expressed an NF-YB class CCAAT-binding transcription factor from *Arabidopsis* in maize, and have shown improved drought tolerance of transgenic plants in the field [2]. Such studies have shown that a finely tuned system of regulatory components seems to be in place, and further investigations may help in the identification of suitable targets for crop improvement. A good parallel approach may be to search for candidates in naturally stress-tolerant plants [28]. The significance of this approach is highlighted by the availability of several reports on contrasting stress-tolerant relatives of *Arabidopsis*, tomato, barley and maize [29]. Comparison of *Arabidopsis thaliana* with *Thellungiella halophila* has shown that the latter does not seem to have any special set of stress-responsive genes and does not show a major change in its gene expression profile following stress; however, the major difference is in the expression level of stress-responsive genes prior to exposure to stress conditions [30, 31, 32]. Similar studies performed to compare the gene expression patterns in desiccation-tolerant (ice plant) and non-tolerant plants have shown that the difference is in the expression pattern, rather than in the presence or absence of particular genes [33]. The comparison of gene expression profiling between contrasting genotypes with respect to stress tolerance can be extended to transcription profiling at the QTL level, and the genes identified at such QTLs may potentially be better candidates for mediating stress tolerance [34].

Map Based Cloning of Abiotic Stress-Related Gene Loci

A positional cloning strategy allows the use of a phenotype to determine the position of the mutated gene or natural allele by examining the linkage to markers whose physical location in the genome is already known. The availability of the *Arabidopsis* and rice genome sequences, together with the improvements in the methods used to detect DNA polymorphisms has made map-based cloning a viable option for global functional genomics [35]. A large number of genes in the abiotic stress response pathway have been identified using the map-based cloning approach, including *SOS1*, *SOS2*, *SOS3*, *SOS4*, *SOS5*, *HOS1*, *Sp17*, *STT3*, *FRO1*, *LOS5/ABA3* and *AtCesA8/IRX1* [36,18,37,38,39,40,41,42,43,44]. Efforts are being made to use sequence-indexed knockout mutant resources developed by rice researchers worldwide to identify insertions in stress-associated genes (SAGs). The SAG mutant lines will be screened for response to drought stress, and this information, together with expression information (from a SAG microarray), will be made available in the near future. Targeting induced local lesions in genomes (TILLING) is another high-throughput technology, currently being undertaken for plant species like *Arabidopsis*, lotus, maize, wheat, rice and *Brassica* [45], which will be helpful in determining mutations in a selected stress gene or a variant allele [46], and hence identifying useful alleles for abiotic stress tolerance.

Many of the genes that have been selected for transformation in order to enhance stress responses have functions at the end-points of putative tolerance pathways in a variety of functional categories. In a large number of experiments, enhanced tolerance has been documented, at least under controlled conditions, but the transgenic plants often had slower growth rates. The examples of single-gene transfer chosen here, involving genes that have roles in trehalose and spermidine biosynthesis and the overexpression of alkali ion transporter, have to some degree been captured by high-throughput profiling studies. The engagement of these, and other genes, can credibly contribute to tolerance acquisition. It seems equally comprehensible, however, that the engineering of biochemical end-points in the highly regulated plant metabolic network will result in limited gains, and will result in unintended consequences, such as substrate imbalances or signaling disturbance. The second wave of engineering attempts presently targets a selected group of genes that encode well-studied stress-responsive transcription factors. The most advanced work is on C repeat-binding factors (CBFs) and ABRE-binding factors (ABFs) that control the expression of a set of downstream functions in stress acclimation [47,48,49,50]. The overexpression of these regulatory genes usually has a negative effect on plant growth, but the existence of several DREB/CBF/ABF gene family members, which are distinguished by domains that provide functional redundancy, might indicate functional diversity within the family. At least in one case, growth retardation and stress protection could be separated.

The use of transcription factors will require the engineering of both promoter and coding regions to dissect those elements that govern cross talk between different abiotic and developmental programs. Thus, it can be assumed that the next higher level of gene expression control, sensing and signaling, could provide promising targets. Very recent studies begin to realize this next phase. The ectopic expression of genes that function in mRNA export, protein phosphorylation and calcium sensing has been reported to confer stress tolerance in *A. thaliana*.

Quantitative Trait Loci (QTL)s and Marker Assisted Selection (MAS)

Quantitative trait loci (QTLs) are specific genetic loci in the genome associated with a particular trait. Stress tolerance is a complex trait, and dissection of its QTLs would be of immense value in understanding the stress response and would also be useful for plant breeders [51]. A schematic representation of the position and role of

QTL cloning in the context of Marker Assisted Selection (MAS) is represented in Fig. **4**. Several QTLs involved in stress response have been reported recently [52,53,34,54,55]. Most of the plant QTLs cloned to date has been obtained using a map-based cloning strategy. Lin *et al.* [52] mapped eight QTLs responsible for variation in K^+ or Na^+ content from an F_2 population derived from a cross between a salt-tolerant *indica* variety (Nonabokra) and a susceptible *japonica* variety. Of these, *SKC1*, a major QTL for shoot K^+ content, was mapped to chromosome 1. The *SKC1* was obtained by positional cloning and was found to code for an HKT-type transporter. *SKC1* was shown to play a vital role in maintaining the K^+ to Na^+ ratio. This role of *SKC1* shows that it is an important determinant of salt stress tolerance, and hence would be a valuable QTL for engineering salt stress tolerance in rice [53]. A map-based approach was also demonstrated to be useful in the identification of an important submergence tolerance QTL present on chromosome 9 of rice. Three genes belonging to the ethylene response-factor (ERF) family were identified on this locus, designated as *Sub1*, the variation of one of which, *Sub1A*, resulted in tolerance/susceptibility to submergence. The *Sub1* gene from the submergence-tolerant variety was introgressed in flooding susceptible local rice varieties. The new varieties showed submergence tolerance without compromising on yield or other agronomic traits, demonstrating the efficacy of this locus [56]. The expression profile of genes in a QTL interval associated with the abiotic stress response is also being used to identify target genes. Gorantla *et al.* [51] used information from ESTs sequenced from drought-stressed cDNA libraries to generate a transcript map of rice. Several known stress-responsive genes coding for mitogen-activated protein kinase, OSMYB1, EREBP-like protein, helicase-like transcription factor and 14-3-3-protein homologue could be identified at drought QTL locations. Several introgression lines (ILs) developed in elite rice genetic backgrounds could provide a rich source for the identification of candidate genes and the cloning of QTLs involved in the stress response [57]. Another possible resource for QTL analysis is the *Oryza* Map Alignment Project (OMAP). This project aims to develop bacterial artificial chromosome (BAC)/sequencing tag connector (STC) physical maps of 11 wild and one cultivated rice species, which will ultimately be aligned to the finished rice (*Oryza sativa* ssp. *japonica* cv. Nipponbare) genome sequence [58,59], and will prove to be useful for the identification of stress related QTLs. The analysis of those plants whose genome sequences have been entirely or near completely sequenced provides another, superior source for candidate genes. Given the genetic diversity in ecotypes of *A. thaliana* and the many breeding populations and lines in rice and other crops, this approach has flourished a lot [60].

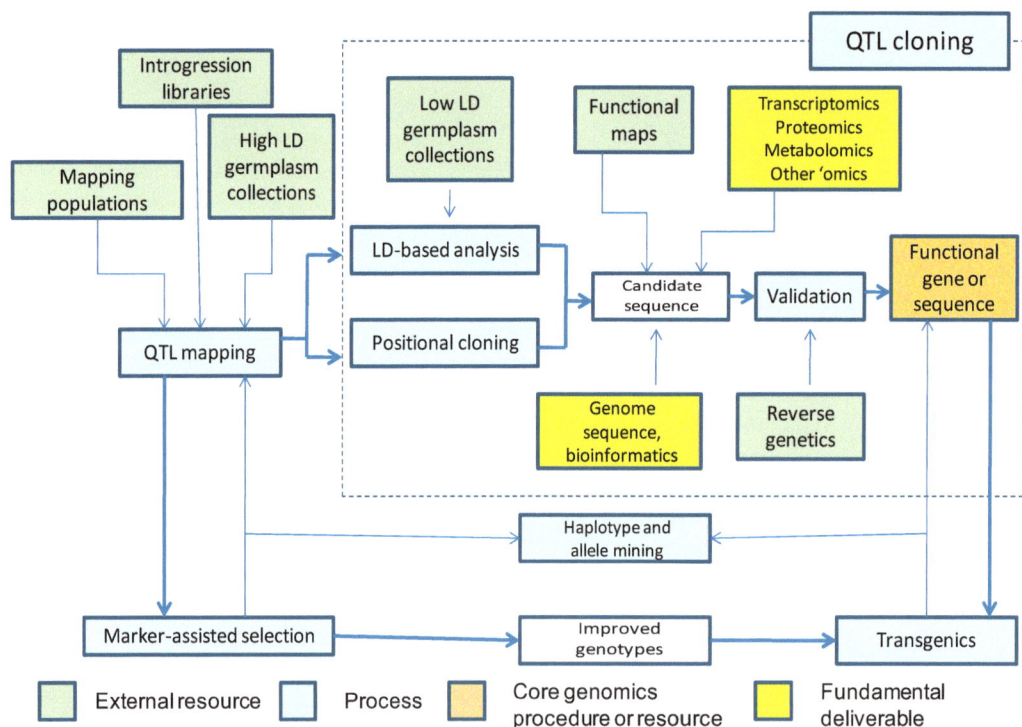

Trends Plant Sc 11 :405-412 (2006)

Figure 4: A schematic representation of the position and role of QTL cloning in the current framework of marker-assisted breeding activities aimed at crop improvement for quantitative traits such as drought tolerance

Once a marker trait association has been established unequivocally, MAS reduces or eliminates the reliance on specific environmental conditions during the selection phase, a major hindrance to the conventional breeding of traits influenced by drought. MAS was used to transfer four QTL alleles for deeper roots from "Azucena", a japonica upland cultivar well adapted to rainfed conditions into IR-64, a rice cultivar having a shallow root system. A marker-assisted backcross (MAB) program was implemented to improve root morphological characters and hence drought tolerance of the upland rice variety "Kalinga III" [55].

Transgenic Approaches

Transgenic approaches have been widely employed to understand abiotic stress response in plants [13, 61]. Interestingly, the majority of such studies have aimed to decipher the function of genes encoding downstream components (effectors), such as those coding for antiporters, heat-shock proteins, superoxide dismutases (SODs) and LEA proteins, rather than upstream components (regulators), such as those coding for transcription factors and kinases. Recently, the functions of genes representing QTLs for abiotic stress have also been confirmed by employing transgenics [56]. Thus, the use of the transgenic system to study a range of genes involved in diverse aspects of the control of the plant adaptive response to abiotic stress would give a clearer picture. A transgenic approach has also been used to dissect the role of stress-responsive promoters [62,63,64]. The major cis-acting elements present in several abiotic stress-responsive promoters include the ABA responsive element (ABRE) from the promoter of ABA-responsive genes [65,66], dehydration responsive element (DRE) from the promoters of cold- and drought-inducible genes [67,68], anaerobic response element (ARE) from the promoter of genes responsive to low oxygen conditions, such as maize *Adh1*, *LDH1* and *PDC1* [69,70,71,72], and heat-shock element (HSE) from the promoter of heat stress-inducible genes [73]. One major drawback of the transgenic approach is that it is not useful for large-scale functional analysis. A possible way of employing transgenics for functional analysis would be the use of PLACs (plant artificial chromosomes), containing large-sized genomic fragments. This would significantly reduce the number of transgenic plants required for functional analysis [74]. One such large-insert library has been prepared for *Arabidopsis* in BIBAC2, a plant transformation competent binary vector, which represented 11.5 X coverage of the *Arabidopsis* genome [75].

Insertional Mutagenesis to Decipher Functions of Abiotic Stress Response Genes

Insertional mutagenesis in plants usually involves the use of T-DNA or transposable elements. It has the advantage of a genome-wide distribution with preferential insertion in gene rich regions [76]. Insertional mutagenesis has been widely employed to characterize abiotic stress-responsive genes, including those coding for AtHKT1 (a high-affinity potassium transporter), CBL1 (calcineurin B-like protein), OsRLK1 (LRR-type receptor-like protein kinase), CIPK3 (calcium associated protein kinase), OSM1/SYP61 (syntaxin) and HOS10 (R2R3-type MYB transcription factor) [77,78,79,80,81,82]. Large-scale forward genetic screens have been used to identify abiotic stress response determinants in a T-DNA-mutagenized *Arabidopsis* population in the *RD29a-LUC* background. More than 200 mutants with altered stress/ABA response were identified from 250, 000 independent insertion lines. These included mutations in genes coding for transcription factors, syntaxin, ABA biosynthetic enzyme, SUMO E3 ligase and the sodium transporter HKT1 [44]. The alternative means of gene tagging, such as the use of traps and activation tagging is useful in studying gene functions. Activation tagging makes use of enhancer elements in the construct, which can activate the transcription of genes near the site of insertion. A retroelement, *Tos17*, has been used for large-scale mutagenesis in rice. *Tos17* insertion lines have been used to study the role of some stress-responsive rice genes, such as *OsMT2B*, a metallothionein gene involved in the scavenging of reactive oxygen species (ROS) [83], and *OsTPC1*, a putative voltage-gated Ca^{2+}-permeable channel involved in the regulation of elicitor-induced hypersensitive cell death [84]. PCR-based screening or sequencing the region flanking the insertion site can screen the tagged lines. In the PCR-based approach, PCR is performed using gene- and insert-specific primers [85] (Krysan *et al.* 1999).

Small RNAs (siRNA) and Micro RNA (miRNA)

A new avenue has been opened by the detection of regulatory systems that depend on small non-coding RNAs or micro RNAs in plants [86, 87]. Some of these RNAs have been shown to be stress-inducible. In addition to affecting the translation process, these siRNAs might participate in the alternative splicing of mRNAs and represent components of an additional level of regulation. A few reports are available in which their function has been correlated with abiotic stress, such as mechanical stress-responsive miRNAs in *Populus*, phosphate starvation-responsive miRNAs in *Arabidopsis* and dehydration-, cold-, salt- and ABA-responsive miRNAs in *Arabidopsis*. The host of genomics tools has provided a wealth of data and provided a better understanding of the changes in cellular metabolism that is induced by abiotic stresses, but fewer results have been forthcoming with respect to the functioning of the whole plant. The conversion of the many data points into understanding is still incomplete. Integration and filtering of data and confirmation by independent means in combination with advanced bioinformatics tools will alleviate this deficit.

Allele Mining

Another important lesson from comparative, functional genomics studies is the recognition by molecular means of enormous adaptive functional diversity in many characters, including stress tolerance. 'Allele mining', as it may be termed, can focus on close relatives of the established models for which sufficient genomic resources are available. Tremendous genetic diversity that can be exploited exists in *Lycopersicon pennellii*, *Hordeum vulgare* ssp. spontaneum, and *T. halophila* (salsuginea), relatives of domesticated tomato, barley and *A. thaliana* respectively. With the *A. thaliana-Thellungiella* species combination, it is already possible to transfer, for example, BAC libraries or full-length cDNA populations from the stress tolerant relative to the stress-sensitive model organism, and to follow this transfer by screening. Such strategies appear possible with several other species combinations, and should provide a way to harness the existing evolutionary adaptive diversity to develop stress-protected crops in which growth and yield are less compromised by abiotic stresses. The search for stress-tolerance alleles that retain growth and yield and the provision of the knowledge to breeding programs present the real challenge for plant genomics.

RNA Interference (RNAi) Technology

Another new tool depends on RNA interference (RNAi)-based reduction or elimination of specific transcripts or functions [88]. The RNAi technology, already widely used in plant biology, can relate the transcript levels of stress-responsive genes to the degree of stress tolerance; hence it is especially potent in quantitative abiotic stress studies. In an attempt to deploy RNAi-mediated gene silencing for large-scale gene inactivation, a collection of gene-specific sequence tags (GST) had been generated for at least 21,500 *Arabidopsis* genes [88] and will be used to create RNAi vectors for functional genomics studies. As an application, hairpin RNA expressing lines have been constructed for 8136 different gene-specific sequence tags. When combined with appropriate screening strategies, these resources will greatly improve our knowledge on plant stress tolerance. However, gene inactivation using anti-sense, co-suppression or RNAi strategies is largely based on a single gene approach and, at present, it is difficult to use these methods at the genome-wide level [89]. GST hairpin RNA (hpRNA) expression clones have been confirmed for 8136 different GSTs to date and, side-by-side, a medium-scale *Arabidopsis* transformation project is underway wherein randomly selected hpRNA clones will be tested for efficacy. For example, effective silencing has been reported for three genes coding for vacuolar-type H^+-ATPase subunit B3. In case of rice, RNAi vectors have been designed [90] for the functional analysis of rice genes.

PROTEOMICS APPROACHES

The systematic analysis of the entire protein complement or proteome is referred to as 'proteomics'. Proteomics particularly focuses on the systematic and detailed analysis of the protein population in a cellular compartment, tissue and whole organism for specific properties such as their identity, quantity, activity and molecular interaction. Analysis of the proteome provides a direct link of genome sequence with biological activity [91, 92]. Analysis of the proteome includes knowledge of the entire protein repertoire as well as studies on other aspects, such as expression levels, post-translational modifications and interactions, to understand the cellular processes at the protein level [93]. On account of its enormous potential, proteomics can be further divided into 'expression proteomics' (study of global changes in protein expression) and 'cell-map proteomics' (systemic study of protein-protein interactions). Proteomics is proving an indispensable tool for examining alteration in the protein profiles caused due to gene mutations, introduction or silencing of genes in response to various stress stimuli in a relatively fast, sensitive and reproducible way. This science is becoming important for generation of information on physiological (e.g. regulatory behaviour and function), biochemical (e.g. metabolic and structural data), genetic (e.g. gene mapping and assigning of the structural genes to the 2D gel map) and architectural (e.g. location of the proteins in the cell) aspects.

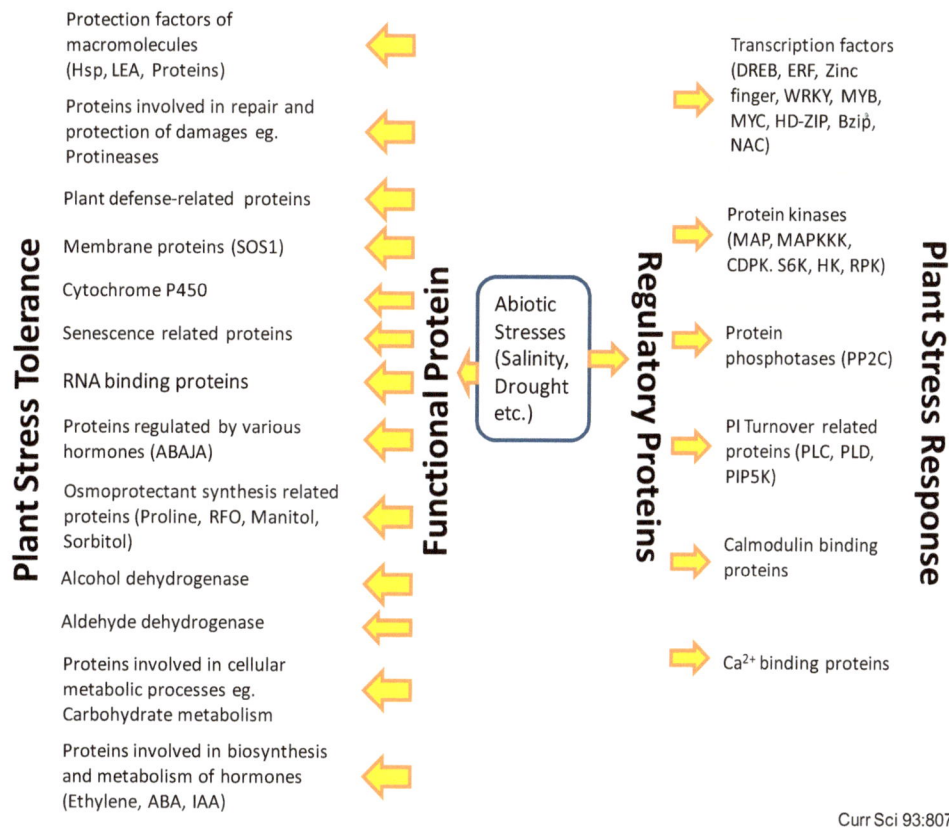

Plant Stress Tolerance

- Protection factors of macromolecules (Hsp, LEA, Proteins)
- Proteins involved in repair and protection of damages eg. Protineases
- Plant defense-related proteins
- Membrane proteins (SOS1)
- Cytochrome P450
- Senescence related proteins
- RNA binding proteins
- Proteins regulated by various hormones (ABAJA)
- Osmoprotectant synthesis related proteins (Proline, RFO, Manitol, Sorbitol)
- Alcohol dehydrogenase
- Aldehyde dehydrogenase
- Proteins involved in cellular metabolic processes eg. Carbohydrate metabolism
- Proteins involved in biosynthesis and metabolism of hormones (Ethylene, ABA, IAA)

Functional Protein

Abiotic Stresses (Salinity, Drought etc.)

Regulatory Proteins

Plant Stress Response

- Transcription factors (DREB, ERF, Zinc finger, WRKY, MYB, MYC, HD-ZIP, Bzip, NAC)
- Protein kinases (MAP, MAPKKK, CDPK. S6K, HK, RPK)
- Protein phosphotases (PP2C)
- PI Turnover related proteins (PLC, PLD, PIP5K)
- Calmodulin binding proteins
- Ca^{2+} binding proteins

Curr Sci 93:807-817 (2007)

Figure 5: High salinity stress-inducible proteins and their possible functions in stress tolerance and response

Majority of plant proteomic studies can be divided into two basic categories. The first category involves protein profiling of biological material with the aim of separating, sequencing and cataloguing as many proteins as possible. Here the objective is to establish the protein framework of a biological system, much as expressed sequence tags (ESTs) that provide a snapshot of the transcript complement. A related and potentially complementary strategy is to target sub-cellular proteomes, thereby dramatically reducing the protein complexity of a particular extract and revealing important information regarding sub-cellular localization. The second basic category of proteomic analyses can be termed comparative proteomics, where the objective is not to identify the entire suite of proteins in a particular sample, but rather to characterize differences between different protein populations. This approach is thus somewhat analogous to comparative DNA microarray profiling. Examples might include proteins from wild type versus mutant plants, stressed versus unstressed or tissues at different developmental stages following responses to external stimuli or environmental stresses. The proteomics of various plants was studied in response to different environmental stresses such as drought, heat, cold and salt. Fig. **5** shows the various groups of abiotic stress inducible proteins and their possible functions in stress tolerance and response.

Gel Staining, Imaging and Analysis

For many years, two basic options were typically followed for protein staining. Colloidal Coomassie Brilliant Blue (CBB) staining is relatively easy, cost-effective and compatible with subsequent protein identification by mass spectrometry (MS), but it is only moderately sensitive, with a detection limit of approximately 10 ng proteins. The other alternative is silver staining, which is more sensitive, detecting as little as 0.5 ng proteins, but not particularly quantitative and less suitable for MS identification. A number of sensitive fluorescent stains, such as SYPRO Ruby and SYPRO Orange were developed that combine the advantage of the other stains, a similar sensitivity to silver stain, but the ease of use and excellent MS compatibility. However, in some cases, the high costs of dyes such as SYPRO Ruby may be prohibitive, particularly with large projects involving many gels. Thus, preparative gels were stained with CBB and analytical gels were visualized by silver staining in most of the studies of salt stress.

Imaging of Protein Location and Metabolites

One way to advance stress genomics is provided by imaging tools. Random fusions of cDNAs with green fluorescent protein (GFP) are one approach [94], which has revealed unexpected complexities of expression and dynamic changes in cells. Meanwhile, a classical approach, involving a calcium (aequorin) reporter combined with a yellow fluorescent protein (YFP), has been used to monitor cell-type-specific responses to drought, salt and cold in roots. Different imaging applications, such as fluorescence imaging to report ion flux or metabolite pool dynamics, have become available. The more widespread, preferably high-throughput, use of imaging would provide a new dimension to abiotic stress analyses because it would allow us to examine the tissue- and cell-specific action of adaptive protective mechanisms that are not revealed by whole-plant genomics.

Profiling of Abiotic Stress-Related Proteins

The importance of protein profiling has long been acknowledged in plant abiotic stress studies. Previous studies have provided useful information on individual enzymes or transporters, measuring their stress-dependent changes in quantity and activity. The results of such studies have formed the current hypotheses on stress responsive networks, in which protein modifications, protein-protein interactions, stress-dependent protein movements, de novo synthesis, and controlled degradation play significant roles, literally spanning the time and space of the stress response. The need for protein studies is also underscored, for example, by the recognition that rapid calcium spikes (which may be modulated in space, amplitude and frequency) lead to protein modifications that precede transcript changes [95] (Kiegle *et al.* 2000). Consequently, large scale high-throughput proteome analyses must be integrated with transcriptome and metabolome analyses if we are to obtain a comprehensive understanding of the stress response.

As a shot-gun approach to unravel biochemical and molecular changes elicited in stressed cells, electrophoresis of proteins isolated from non-induced (control) and induced (stressed) cells has been practised in a number of studies. The bulk of the early work on stress proteins was carried out employing 1-dimensional (1D) protein gel electrophoresis. Thus, the initial detection of pathogenesis-related (PR) proteins and heat shock proteins (HSPs) in plants was made by this approach. In reality, large-scale proteome studies are still limited. However, more recently, 2D is being routinely employed for the objective of analysing stress proteins in plants. At present, there are two major approaches to proteome profiling. The first and traditional approach uses two-dimensional gel separation (2D-PAGE) combined with mass spectrometry (MS) to measure changes in protein quantity [96, 97]. In one of the earlier studies using a proteomic approach, it was shown that salt treatment causes the accumulation of a lectin-like protein (SalT) [98]. Moons *et al.* [99,100] used 2D gel electrophoresis to study the role of ABA and salt on salt tolerance donors of Indica rice (*Oryza sativa* L.) cv. Pokkali and Nona Bokra and salt-sensitive cultivar Taichung. A novel histidine-rich protein and two types of LEA proteins were identified in higher levels in roots of tolerant rice varieties compared to those of sensitive varieties. It was also found that ABA and jasmonates antagonistically regulated the expression of salt-inducible proteins associated with water deficit or defence responses. Thus, six salt-inducible proteins (peroxidase, SalT, pathogenesis-related (PR) protein-10, PR-1, [101]; and two unknown proteins) also accumulated after ABA treatment [102]. Salt-responsive proteins in roots of the salt-tolerant rice variety, Pokkali and the salt-sensitive variety IR-64, IR-29 were studied by a proteomic approach. Among the salt responsive proteins identified were an ABA and stress responsive protein (ASR1), ascorbate peroxidase and caffeoyl CoA O-methyltransferase (CCOMT). CCOMT was markedly upregulated by salt stress in Pokkali, but changed little in IR-29. CCOMT is involved in suberin and lignin biosynthesis and increased lignification may help to reduce the bypass water flow that allows Na^+ ions to enter rice roots via an apoplastic route [103,104].

The proteomes of the salt-tolerant mutant (RH8706-49) and salt-sensitive mutant of wheat (H8706-34) were analysed using 2-DE and MALDI-TOF-MS [105]. Five candidate proteins belonging to the chloroplast were identified: H^+-transporting two-sector ATPase, glutamine synthetase 2 precursor, putative 33 kDa oxygen evolving protein of photosystem II and ribulose-1, 5 bisphosphate carboxylase/oxygenase small subunit. They are likely to play a crucial role in keeping the function of the chloroplast and the whole cells when the plant is under stress.

Proteins, whose abundance increased under salt stress in leaf sheath, root and leaf blade of three indica rice cultivars Nipponbare, IR-36 and Pokkali were analysed using 2-DE, and N terminal and internal amino acid sequence analysis [106]. Eight proteins showed one-to-threefold up regulation in leaf sheath, in response to 50 mM NaCl for 24 h. Among these, three proteins were unidentified (LSY081, LSY262 and LSY363), while five proteins were identified as fructose bisphosphate aldolases, photosystem II (PSII) oxygen evolving complex protein, oxygen evolving enhancer protein 2 (OEE2) and superoxide dismutase (SOD). The expression of SOD was a common response to cold, drought, salt and ABA stresses, while the expression of LSY081, LSY363 and OEE2 was enhanced by salt and ABA stress. LSY262 was expressed in leaf sheath and root, while fructose biphosphate aldolase, PSII oxygen evolving complex protein and OEE2 were expressed in leaf sheath and blade. LSY363 was expressed in leaf sheath, but was below the level of detection in the leaf blade and root. These results indicate that specific proteins are expressed in specific organs of rice, indicating a coordinated response to salt stress.

Soluble proteins extracted from leaves of two wheat species (*Triticum durum* sensitive cv. Ben Bachir and *Triticum aestivum* tolerant cv. Tanit) differing in their sensitivity were analysed by 2-DE in order to detect NaCl-induced changes. The greatest alterations in the polypeptide profiles following salt stress were found in the most sensitive cultivar: among the 12 spots (molecular mass 15-31 kDa) specifically considered in the acidic region of the gel, 11 declined and even disappeared in the leaf proteins of the NaCl-sensitive variety, while in the tolerant species two more new polypeptides were induced by NaCl [107]. The effect of salt stress on the polypeptide levels was examined using 2-DE in roots of two contrasting wheat (*T. durum*) cultivars, i.e. sensitive cv. Ben Bachir and tolerant cv. Chilli. The net synthesis of a 26 kDa polypeptide was significantly changed in the tolerant cultivar [108].

Arabidopsis thaliana cell-suspension cultures were used to investigate the effects of salinity (NaCl) and hyper osmotic stress (sorbitol) on plant cellular proteins (Ndimba *et al.* 2005). Two-dimensional electrophoreses revealed a total of 2949 protein spots. Two hundred and sixty-six protein spots showed significant changes in abundance across five independent experiments. MALDI-TOF-MS identified 75 salt and sorbitol-responsive spots of ten functional categories, including H^+-transporting ATPases, signal transduction related proteins, transcription/translation-related proteins, detoxifying enzymes, amino acid and purine biosynthesis related proteins, proteolytic enzymes, heat-shock proteins, carbohydrate metabolism-associated proteins and proteins with no known biological functions.

The role played by ABA in regulating salt-induced protein synthesis was investigated in roots of *Solanum lycopersicon* Mill cv. Alisa Craig (AC) and the near isogenic ABA-deficient mutant, *flacca* (*flc*), using 2-DE analysis. The polypeptide profiles of salt-treated AC and *flc* roots were similar, suggesting that the synthesis of most novel polypeptides in salt-treated roots is not dependent on an elevated level of endogenous ABA [109]. Matrix-assisted laser desorption/ionization-time of flight (MALDI-TOF) and electrospray ionization (ESI) are the two most commonly used MS techniques [110,111]. The 2D + micro sequencing have been employed in order to identify the drought-responsive proteins that accumulate during the phase of water deprivation in *Pinus pinaster* seedlings. Of a total of 1000 protein spots resolved on the gel in this study, 38 responded to stress. When internal micro-sequences obtained for 11 proteins were analysed, 10 could be identified through sequence homology-based search. Importantly, the identified proteins were found to be associated with diverse processes such as photosynthesis, cell elongation, antioxidant metabolism and lignification. The detailed characterization of stress proteins and their corresponding genes has proven to be of immense practical value.

Proteomic studies for the analysis of salt tolerance were carried out in plant species such as *Arabidopsis*, rice, barley, wheat, tomato, tobacco, halophytes like *Breguiera gymnorrhiza* and *Suaeda aegyptiaca*, using leaves, roots or cell suspensions. A proteomics study was conducted in leaves of the halophyte *Suaeda aegyptica* using 2-DE and LC/MS/MS to identify the mechanisms of salt responsiveness [112]. Among the 700 protein spots, 102 showed significant response to salt treatment. Forty proteins of 12 different expression groups were analysed using LC/MS/MS. Among these, 27 spots were identified, including proteins involved in oxidative stress tolerance, glyceine betaine synthesis, cytoskeleton remodelling, photosynthesis, ATP production, protein degradation, cyanide detoxification and chaperone activities. Ramanjulu *et al.* [113] studied the effects of salt and drought on extracellular proteins in barley and identified several salt-induced proteins. 2D-gel electrophoresis revealed in the leaf extract of the mangrove plant, *Bruguiera gymnorhiza*, a 33 kDa protein with pI 5.2, whose quantity increased as a result of NaCl treatment. The N-terminal amino acid sequence of this protein displayed significant homology with the mature region of the oxygen evolving enhancer protein 1 (OEE1) precursor [114]. The proteins induced in response to anaerobic stress, called anaerobic polypeptides (ANPs) accounts for more than 70% of total protein synthesis after 5 h of anaerobic stress. The anoxically treated cells up-regulate ethanolic fermentation and glycolytic pathways as a strategy to survive under such stress conditions. These observations provided the basis for the recent work in which transgenic rice plants have been produced that overexpressed pyruvate decarboxylase (PDC), an enzyme that rate-limits the ethanolic fermentation process. Transgenic plants overexpressing PDC were found to possess relatively higher flooding tolerance [70,72]. There are several other examples of a similar nature, wherein useful research strategies have emerged based on proteome information. Roychoudhury *et al.* [115] have reported that late embryogenesis abundant (LEA) proteins are present in higher amounts in the salt-tolerant rice cultivars compared to the salt-sensitive or aromatic cultivars. The overexpression of *lea* gene *Rab16A* from the salt-tolerant rice Pokkali generated enhanced salt-tolerance in transgenic tobacco, which could resist up to 300 mM NaCl [116]. Similarly, when another *lea* gene from barley, namely *HVA1*, was transformed into rice cells, these plants maintained higher growth rates than the non-transformed plants (control) under water and salt stress conditions [117]. Earlier, it was reported that 100 kDa HSPs are accumulated to a significant level in diverse plant species in response to high temperature stress [118]. When recently the *hsp100* gene was over-expressed in the transgenic *Arabidopsis* plants, it caused a significant increase in high temperature tolerance. The future avenues for further increasing stress tolerance warrant that several stress tolerance-related genes must be pyramided. The realization of this goal can only be achieved if major breakthroughs are made in further identification of the stress-related proteins and isolation and cloning of the requisite genes. Genomics and proteomics research will be of great help in constantly expanding the information on newer stress responsive genes and proteins. For instance, Moons *et al.* [119] have reported that complete submergence of rice seedlings for 60 h increased the accumulation of a 97 kDa protein in roots. When peptides generated by *in situ* tryptic digestion of this 2D protein spot were analysed, significant homology to plant pyruvate orthophosphate dikinase (PPDK) protein was revealed. This study thus associated PPDK protein to flooding stress response in rice. Recently, the patterns of protein synthesis during hypoxic and anoxia conditions in maize by employing 2D method has also been done. In this study, expression of as many as 262 individual proteins was shown to alter with changes in O_2 tension regime. Further, of 48 protein spots analysed by MS, 46 were identifiable on the basis of database search. Thus, applications based on results of protein analysis are enormous in production of abiotic stress-tolerant transgenic plants.

Although in excess of 1000 proteins can be readily identified, few stress-dependent changes in the quantities of these proteins have be identified, possibly due to problems in isolation or quantification or the inability to recognize protein modification. A second approach focuses on specific modifications of the proteome, such as membrane protein phosphorylation [120] or populations of nitrosylated proteins [121]. Unexpected, yet illuminating results have emerged from this approach, such as very divergent phosphorylation sites of receptor like- kinases from the same subfamily or the

discovery of candidates for NO signaling pathways. Unlike transcriptome analyses, for which mature platforms exist, proteome analyses require novel, specific toolboxes. A specific protein chip has been developed for large-scale kinase assays to study the potential substrates of AtMPK3 and AtMPK6, the two mitogen associated protein kinases (MAPKs) known for their involvement in various stress responses [120]. The identified substrates contained a large number of ribosomal proteins. One interpretation and hypothesis would assume that, when under stress, the MAPKs might directly affect mRNA loading by changing the phosphorylation state of ribosomal proteins. Bae *et al.* (2003) used 2D-PAGE and MALDI-TOF-MS to study the *Arabidopsis* nuclear proteome and changes in the nuclear proteome in response to cold stress. One hundred and eighty-four protein spots were identified, with the expression of almost 30% of these proteins altered in response to cold stress. These included several proteins previously reported to be involved in stress, including heat-shock proteins, transcription factors (AtMYB2 and OBF4), DNA-binding proteins (DRT102 and Dr1), catalytic enzymes (phosphoglycerate kinase, serine acetyltransferase and glyceraldehyde-3-phosphate dehydrogenase), syntaxin, calmodulin and germinlike proteins. Proteome analysis of rice was performed by Salekdeh *et al.* [103] to study changes in response to drought stress. More than 1000 protein spots were identified on 2D-PAGE, with the expression of 42 proteins altered by stress. MS established the identities of 16 of these drought-responsive proteins. A systematic proteomic study was conducted to investigate the salt stress-responsive proteins of the roots of indica rice cv. Nipponbore using 2-DE and MALDI-TOFTOF [97]. More than 1100 protein spots were reproducibly detected, including 34 that were up regulated and 20 that were down regulated. MS analysis and database searching revealed the identification of 12 spots representing ten different proteins. Three spots were identified as the same protein, enolase. While four of them were previously confirmed as salt stress-responsive proteins, six were novel, i.e. UDP glucose pyrophosphorylase, cytochrome-*c* oxidase subunit 6b-1, glutamine synthetase root isozyme, putative nascent polypeptide associate complex alpha chain, putative splicing factor-like protein and putative actin-binding protein. These proteins are involved in regulation of carbohydrate, nitrogen and energy metabolism, ROS scavenging, mRNA and protein processing and cytoskeletal stability. This study gave new insights into salt stress response in rice roots and demonstrated the power of the proteomic approach in plant biology studies. The rice root proteome was also studied to identify salt stress-responsive proteins. 2D-PAGE identified 54 proteins whose expression changed in response to salt stress. Dani *et al.* [122] used *Nicotiana tabacum* plants as a model to investigate changes in soluble apoplast composition induced in response to salt stress, using a vacuum infiltration procedure, 2-DE and LC/MS/MS. The study of proteome in response to salt stress identified 20 proteins whose expression changed in response to stress. These included several well-known stress associated proteins, together with chitinases, germin-like protein and lipid transfer proteins. 2-DE analyses revealed about 150 polypeptide spots in the pH range of 3.0-10.0 independent protein extracts, with a high level of reproducibility between the two sample sets. Quantitative evaluation in treated and untreated samples revealed 20 polypeptides whose abundance changed in response to salt stress. An enhanced accumulation of protein species commonly induced by biotic and abiotic stresses was observed. In particular, two chitinases and a germin-like protein (GLP) increased significantly and two-lipid transfer proteins (LTPs) were expressed entirely *de novo*. Some apoplastic polypeptides, involved in cell-wall modifications during plant development, remained largely unchanged. Proteome analysis was performed to study the effect of cold stress on rice anthers at the young microspore stage. More than 3000 proteins were resolved on 2D-PAGE, 70 of which showed differential expression in response to cold stress. Seven of the 18 proteins identified by MALDI-TOF-MS were partially degraded, reflecting the effect of cold stress at the young microspore stage (Imin *et al.* 2004). Similarly, in an analysis of the rice cold stress proteome, proteins from unstressed seedlings were compared with those from seedlings exposed to temperatures of 15, 10 and 5°C. Of a total of 1700 protein spots separated by 2D-PAGE, 60 proteins were up regulated with a decrease in temperature. MALDI TOF-MS or ESI/MS/MS established the identities of 41 of these proteins, and these mainly included chaperones, proteases, detoxifying enzymes, and enzymes linked to cell wall biosynthesis, energy pathways and signal transduction. These results emphasize the importance of maintaining protein quality control via chaperones and proteases, together with an increase in cell wall components, during the cold stress response [123].

Another methodology to label proteins from pools to be compared with different dyes was originally introduced by Unlu *et al.* [124] and has been further developed and is now termed as differential in gel electrophoresis (DIGE) [125]. This has actually been represented in Fig. **6**. The separate pools of proteins are covalently labelled with N-hydroxysuccinimidyl derivatives of the fluorescent cyanine dyes (Cy2, Cy3, Cy5). These fluorescent dyes are designed to modify the ε-amino group of lysine residues in proteins. By this means, approximately 3% of the available protein and one single lysine per protein molecule are labeled. The technique allows the analysis of up to three pools of protein samples simultaneously on a single 2-D gel, thereby minimizing the problem of gel-to-gel variability. In a standard protocol, two of the dyes (Cy3, Cy5) are used to label two protein samples to be compared, and the third dye (Cy2) is used to label an internal standard that consists of equal amounts of all the samples to be analyzed within the overall experiment. The 2-DE DIGE technique, for example, was used to monitor the alterations in the proteome of cold-stress-treated *Arabidopsis* plants [126].

Taken together, the above-mentioned studies suggested the usefulness of proteomic approaches in identifying functional proteins responsive to salt stress alone or in combination with other abiotic stresses in plants. The proteins identified in response to salt stress by proteomic approach are implicated in diverse physiological and defense processes. While some are probably part of the general stress response to help the plant to survive in stress conditions, others may contribute to the negative physiological effects of salt. However, our knowledge of salt stress-responsive proteins is still far from complete. Clearly, there is a need to examine and analyse a much larger number of salt stress-responsive regulatory proteins in various other plant species, including woody species in order to understand more about the molecular complexity of salt stress. The following emerging technologies of proteomic research need to be intensively employed in this venture.

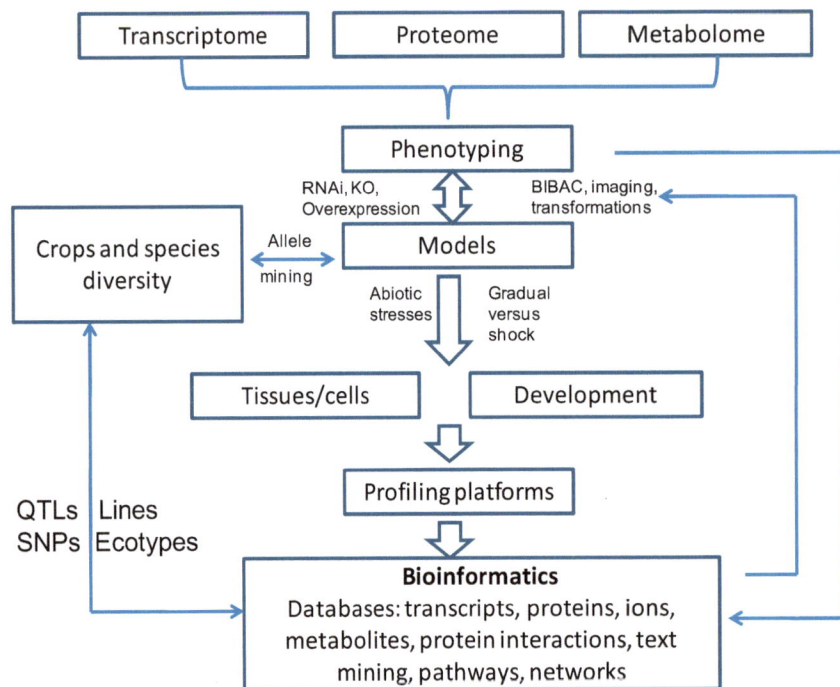

Trend Plant Sci 13:165-171 (2008)

Figure 6: Flow chart of stress systems biology; the chart connects the systems approach to the analysis of plant stress response pathways with gene mining and the transfer of knowledge from models to crops.

Computer-Aided Technologies

Image-editing and comparative analyses are crucial steps in any comparative proteomics project. These steps require substantial manual intervention and time investment. Several commercially available software packages are available to assist in image acquisition, spot-editing, quantification, annotation, comparisons and generation of web-formatted datasets. Commonly used software packages for the analysis of 2-DE gels include image-master 2D Elite, PDQuest, Melanie 4, Phoretix and Progenesis, and MasterScanTM. No image analysis software is completely accurate because the variability in 2D gel images prevents 100% correspondence between analogous spots in images. 2-DE gels of salt-stress responsive proteins were analysed with image Master 2Delite software in rice leaf sheath, tobacco leaf apoplast, Melanie software in rice roots and *Suaeda aegyptiaca* leaves. Further support for proteomics has been lent by identifying proteins by mass that requires access to a protein sequence database. The most commonly used databases are SWISSPROT, TrEMBL, NCBInr, pdEST and non-reduntant collection of protein sequences at the US National Centre for Biotechnology Information (NCBI). SWISS-PROT is an annotated collection of protein sequences. The NCBI database contains translated protein sequences from the entire collection of DNA sequences kept at GeneBank. Some of the important databases are as follows: (a) SWISS-2D-PAGE for protein identification; (b) NCBI/BLAST and SWISS-PROT which are sequence databases; (c) SWISS-MODEL for three-dimensional structure; (d) PROSITE (e.g. PIR, SWISS-PROT) for domain structure, and finally (e) GenBank and EMBL which are DNA data banks. The science, of bioinformatics is a cardinal part of the present-day proteomics science, as development of sophisticated software for an efficient analysis and storage of data with partially automated comparison of multiple 2D gels is needed in scaling proteomics to meet challenges put by the genomics research. Some of the bioinformatic tools, available for plant proteome analyses, includes the Proteins of *Arabidopsis thaliana* Database (PAT), MIPS *Arabidopsis thaliana* Database (MAtDB) and Rice Proteome Database (RPD).

Studying Protein-Protein Interactions

The protein-protein interactions can be analysed directly through proteomics science by performing co-precipitation studies with a 'bait' protein followed by mass spectrometric 'read-out' of the bound proteins. The yeast two-hybrid method is the most commonly used technique to study protein-protein interactions. A large-scale yeast two-hybrid analysis was performed by Cooper *et al.* [127] to identify rice proteins involved in stress and development. One of the proteins identified in the yeast two hybrid analyses was protein phosphatase type 2A regulatory B subunit. This protein not only interacted with itself, but also with a carboxypeptidase, an orthologue of a wheat translation initiation factor, an inositol phosphatase-like protein (IPP) and a stress-regulated 14-3-3 protein. All of these proteins seem to be involved in the stress response. The IPP is known to negatively regulate the response to both ABA and stress. Further, IPP also interacts with the dehydration repressible zinc-finger protein.

Post-Translational Modification of Proteins

The post-translational modification of stress-regulated proteins and transcription factors through phosphorylation, for instance, is a key regulatory event in many cellular processes, including signalling, targeting and metabolism [128, 22]. Other modifications include glucosylation, ubiquitinylation, sumoylation etc. Current developments in proteomics enable the global analysis of post-translationally modified proteins. The MS has been employed to characterize function-critical posttranslational modifications, including phosphorylation and glycosylation. The current scope of proteomics is much broader than was indicated by its classical definition that included only the 2D-based analysis. It includes protein identification, study of post-translational modifications and protein-protein interactions and the determination of function. Several newer methods of protein analysis such as affinity purification, antibody usage, phage display, etc. have been combined with 2D to effectively address to these objectives.

Limitations of Current Proteomic Approaches

The 2D currently has an insufficient dynamic range for complete proteome analysis owing to its limited loading capacity and the detection limits of staining. Further, specific classes of proteins are excluded or under-represented in 2D gel patterns. These include very acidic or basic proteins, excessively large or small proteins and membrane proteins. Clearly, the detection and quantification of low-abundance proteins such as transcription factors, protein kinases and other regulatory proteins is incompatible with the standard 2D + MS approach. These arguments call for further technical developments in the 2D methods. It is suggested that development of 'protein chips', analogous to microarray chips for nucleic acids, could provide convenient high throughput solution to proteome analysis. However, as proteins are more complex and more diverse compared with nucleic acids; development of such chips for proteomics has proved difficult so far. Finally, it is possible that other methods of protein analysis such as HPLC and capillary IEF (CIEF) may be combined with MS in the years to come, for meeting the deficiencies that are encountered in current 2D protocols.

METABOLOMICS FOR PLANT STRESS RESPONSE

Environmental stress could be defined in plants as any change in growth condition(s), within the plant's natural habitat, that alters or disrupts its metabolic homeostasis. Such change(s) in growth condition requires an adjustment of metabolic pathways, aimed at achieving a new state of homeostasis, in a process that is usually referred to as acclimation. From the standpoint of metabolomics, at least three different types of compounds are important for these processes, (1) compounds involved in the acclimation process such as antioxidants or osmoprotectants; (2) byproducts of stress that appear in cells because of the disruption of normal homeostasis by the alteration(s) in growth conditions; and (3) signal transduction molecules involved in mediating the acclimation response. The signal transduction molecules could be newly synthesized compounds or compounds that are released from their conjugated form(s), such as the plant hormone salicylic acid, or they could be different by-products of stress metabolism. The second type of signaling molecules could include by-products of membrane degradation, different ROS or various oxidized compounds such as phenolic compounds or even some antioxidants [129,130]. Despite being a relatively new approach in plant biology, metabolomics is becoming one of the major tools in studying plant stress responses. Significant new discoveries have already been made in the field by using metabolomics alone or in combination with other omics disciplines. Metabolomics requires automated raw data processing software that can handle data from various instruments, extensive mass spectral libraries and powerful database management systems that can store both raw and metadata.

Stress- Up Regulated Metabolites

The various plant metabolites, involved in abiotic stress responses, commonly known as compatible solutes or osmolytes, include compounds such as polyols like mannitol, myo-inositol and sorbitol; dimethylsulfonium compounds such as dimethylsulfoniopropionate, glycine betaine; sugars such as sucrose, trehalose and fructan; or amino acids such as proline and ectoine that serve as osmoprotectant to protect plants under extreme salt, drought and desiccation stresses [116,131]. The saturation level of membrane fatty acids can significantly alter chilling tolerance. Many small molecules protect plants from oxidative damage associated with a variety of stresses. Ascorbic acid, glutathione, tocopherols, anthocyanins and carotenoids protect plant tissues by scavenging ROS generated during oxidative stress. The plant defense response is also associated with the activation of the general phenylpropanoid pathway and induction of lignin biosynthesis [132,133]. Salicylic acid, methyl salicylate, jasmonic acid, methyl jasmonate and other small molecules, produced as a result of stress, can also serve as signaling molecules activating systemic defense and acclimation responses. In view of the above, it is clear that the field of detailed time-course metabolic profiling analysis of plants subjected to stress could lead to the identification of many of the compounds mentioned above. These could be further tested by direct measurements, correlated with changes in transcriptome and proteome expression, and confirmed by mutant analysis. With respect to the studies described above, metabolomic research could in fact be the most important tool in identifying the early compounds that signal the perception of stress because these would act even before any change(s) in the transcriptome or proteome could be detected. Major approaches currently used in plant metabolomics research include metabolic fingerprinting, metabolite profiling and targeted analysis [134,135,136].

Metabolic Fingerprinting

Metabolic fingerprinting is largely used to identify metabolic signatures or patterns associated with a particular stress response without identification or precise quantification of all the different metabolites in the sample. Fingerprinting can be performed with a variety of analytical techniques, including NMR [137], MS [138], Fourier transform ion cyclotron resonance mass spectrometry, or Fourier transform infrared (FT-IR) spectroscopy [139]. One of the limitations of NMR spectroscopy is its low sensitivity, which makes it difficult to detect low abundance cellular metabolites. MS has an advantage over NMR in terms of resolving power, providing higher sensitivity and lower limit of detection. However, MS generates more complex spectrum because of the formation of product ions and adducts, and its results comes in a form of discriminant ions. This can provide a significant challenge for data validation. Using MS with different classification tools, a larger subset of metabolites associated with the phenotype can be identified. Metabolic fingerprints can be analyzed with a variety of pattern recognition and multivariate statistic techniques [140]. Most metabolomics data sets are underdetermined, meaning they contain many more variables than samples [141], and for proper statistical analysis, it is important to reduce the number of variables to obtain uncorrelated features in the data. For instance, to minimize the cosuppression effect, samples can be separated using very rapid gradients with a short chromatographic column and the HPLC-MS data can then be analyzed using multivariate analysis to identify the discriminant ions. To confirm the fingerprinting results, samples are then re-analyzed with long HPLC gradient. This two-step fingerprinting/validating strategy was used to characterize the wound response in *Arabidopsis* [142]. Focusing only on those molecules that demonstrate to be discriminant between groups rather than peak identification significantly reduces the cosuppression effect.

Metabolite Profiling

Metabolite profiling is aimed at a simultaneous measurement of all or a set of metabolites in a sample. Multiple analytical techniques can be used for metabolite profiling [140,136]. These techniques include NMR, GC-MS, liquid chromatography-mass spectrometry (LC-MS), capillary electrophoresis-mass spectrometry (CE-MS) and FT-IR spectroscopy. To date, GC-MS is the most developed analytical platform for plant metabolite profiling. Using GC-MS, it is possible to profile several hundred compounds belonging to diverse chemical classes including sugars, organic acids, amino acids, sugar alcohols, aromatic amines and fatty acids. The major advantage of GC-MS for metabolomics is the availability of both commercially and publicly available EI spectral libraries [135]. The limitation of the GC-MS profiling is that it can only analyze volatile compounds or compounds that can be volatilized following chemical derivatization. For non-volatile compounds, LC-MS and CE-MS provide a better alternative. LC-MS application in Metabolomics is steadily increasing especially after the recent adoption of the ultra performance liquid chromatography technology that can dramatically increase separation efficiency and decrease analysis time [143,144]. CE-MS provides a viable alternative for metabolite profiling due to its high resolving power, low sample volume requirements and the ability to separate cations, anions and uncharged molecules simultaneously [145].

Targeted Analysis

Targeted profiling is used when it is necessary to determine the precise concentration of a limited number of known metabolites and provides a very low limit of detection. Targeted analysis has been widely used to follow the dynamics of a limited number of metabolites known to be involved in a particular stress. Targeted analysis can also be used for comparative metabolite profiling of a large number of known metabolites. For truly quantitative measurement, targeted compounds should be available in a pure form and preferably labeled with stable isotope, which provides a significant challenge for plant stress research because many plant metabolites involved in stress response and their intermediates are not available in a pure form. Uniform metabolic labeling combined with MS has been successfully used for quantitative metabolic profiling in microorganisms [146,147,148].

Metabolomic Research and Abiotc Stress

Metabolomics was used to study temperature [48,149,150], water and salinity [139,151,4,152,153], sulfur [154,155,156], phosphorus [157], oxidative [158] and heavy metal [159] stress as well as a combination of multiple stresses [160] in plants.

Temperature Stress

Metabolite profiling was used to understand the dynamics of the *Arabidopsis* response to temperature stress [149]. The authors performed GC-MS profiling of *Arabidopsis* plants subjected to heat and cold stress and identified a set of known metabolites as well as unknown mass spectral tags that specifically respond to heat or cold stress or to both. Cold stress appeared to cause a more dramatic metabolic response. In a subsequent study, these data were coanalyzed with transcript profiling data to uncover mechanism underlying cold acclimation in *Arabidopsis* [150]. Global GC-TOF-MS metabolite profiling of cold stressed *Arabidopsis* plants that differ in freezing tolerance in comparison with plants overexpressing the C-repeat/dehydration responsive element-binding factor (CBF) 3 revealed that *Arabidopsis* metabolome is extensively reconfigured in response to low temperature and suggested a prominent role for the CBF cold response pathway in this process [48]. In an example of targeted profiling, Morsy *et al.* [161] studied the carbohydrate metabolism of rice under chilling, salt and osmotic stress in different genotypes differing in chilling tolerance. Using a quantitative HPLC assay, the authors measured the levels of soluble carbohydrates in the chilling-tolerant and chilling-sensitive genotypes under chilling stress and identified differences in carbohydrate accumulation. The chilling-tolerant genotype accumulated galactose and raffinose under stress, while these sugars declined in the chilling-sensitive genotype.

Water Deficit and Salinity Stress

Metabolic fingerprinting of salt stress in tomato was used to identify metabolic changes in fruits under salinity stress [139]. The authors studied two tomato varieties subjected to salinity stress through metabolic fingerprinting using FT-IR spectroscopy. The compounds identified corresponded to saturated and unsaturated nitrile compounds, cyanide-containing compounds and a strong broad peak of NH_2 (an amino radical) and other nitrogen containing compounds. More detailed metabolic analysis of salt stress response was performed in a time-course experiment on salt stressed *Arabidopsis* cell cultures [153]. GC-MS and LC-MS profiling was performed at 0.5, 1, 2, 4, 12, 24, 48 and 72 h after a salt treatment at 100 mM NaCl. The short-term responses to salt stress included the induction of the methylation cycle for the supply of methyl groups, the phenylpropanoid pathway for lignin production and glycine betaine production. The long-term effects were the coinduction of glycolysis and sucrose metabolism and coreduction of the methylation cycle. GC-MS profiling, in combination with microarray analysis, was also used to compare salinity stress competence in the extremophile *Thellungiella halophila* with *Arabidopsis* [4]. The authors found drastic differences in metabolic profiles of the two species. Generally, *Thellungiella* had a higher metabolite levels both without and with salt stress when compared with *Arabidopsis*. In *Arabidopsis*, 150 mM salt stress caused increase in sucrose, proline and an unknown metabolite (putative complex sugar). In *Thellungiella*, the response was more complex. In addition to having higher levels of many metabolites before stress, changes in other sugars, sugar alcohols, organic acids and phosphate were detected [4]. An integrated study of the early and late changes in transcript and metabolite profiles revealed difference in the dynamics of grapevine response to water and salinity stress [152] and showed the differences in molecular response to water deficit and salinity. GC-MS profiling and anion-exchange chromatography with UV detection revealed that concentration of glucose, malate and proline is higher in water-deficit-treated plants, than in salt-stressed plants. These differences in metabolite levels were correlated with differences in transcript levels of many genes involved in energy metabolism and nitrogen assimilation, suggesting a higher demand in water-deficit-treated plants to adjust osmotically, detoxify ROS and cope with photoinhibition than in salt-stressed plants [152]. In another interesting study, transcript expression and metabolite profiling were used to study the salt-tolerant *Populus euphratica* plant grown within its native habitat, the Negev desert, in order to understand the mechanism underlying stress acclimation [151].

Sulphur and Phosphorus Stress

In a metabolomics study of sulfur deficiency in *Arabidopsis*, Nikiforova *et al.* [156] used untargeted GC-MS and LC-MS profiling to monitor the response of 134 known metabolites and 6023 unknown non redundant ion traces to sulfur starvation. Based on the profiling data, the coordinated network of metabolic regulation induced by sulfur stress was successfully reconstructed [156]. These data were subsequently analyzed together with transcriptomics data to reconstruct gene-metabolite correlation networks involved in *Arabidopsis* response to sulfur deprivation [155]. Combination of transcriptomics and metabolomics approaches was used to investigate transcriptional and metabolic responses of bean plants growing under P deficient and P-sufficient conditions [157]. GS-TOF-MS profiling of bean roots under phosphorus stress conditions identified a set of metabolites significantly changing in P-deficient roots. Most metabolites, including amino acids, polyols and sugars, were increased in P-stressed plants [157].

Heavy metal stress

Metabolic consequences of stress induced by heavy metals in plants were studied using NMR-based metabolic fingerprinting [162] and metabolite profiling [159]. Metabolic fingerprinting using NMR spectroscopy combined with multivariate statistics analysis was used to discriminate between control and cadmium-treated *Silene cucubalus* cell cultures [162]. Compounds that showed an increase in cadmium-treated cells were identified as malic acid and acetate, while glutamate and branched chain amino acids decreased. Metabolite profiling of *Arabidopsis* cells exposed to cesium stress using NMR showed that metabolite changes because of a Cs stress, include mainly sugar metabolism and glycolytic fluxes, and depended on potassium levels in the cell [159].

Oxidative Stress

Despite the well-established role of several metabolic systems, including the ascorbate-glutathione system, in oxidative stress response, only few reports exist on using metabolomics to study oxidative stress response in plants. Baxter *et al.* [158] studied the dynamics of metabolic change in response to oxidative stress caused by menadione in heterotrophic *Arabidopsis* cells. The authors used GC-MS profiling to measure the levels of 50 polar metabolites following stress treatment and correlated metabolic changes to changes in mRNA levels measured in the same sample. In this study, oxidative stress initially caused dramatic inhibition of the TCA cycle and large sectors of amino acid metabolism followed by backing up of glycolysis and diversion of carbon into the oxidative pentose phosphate pathway [158]. Transcriptomics analysis of the same samples also revealed a coordinated transcriptional response to cope with the stress with a major switch from anabolic to catabolic metabolism.

Combined Stress

Traditionally, abiotic stress conditions are studied in plants by applying a single stress condition such as drought, salinity or heat, and analyzing the different molecular aspects of plant acclimation. This type of analysis is, however, in sharp contrast to the conditions that occur in nature, in which plants are routinely subjected to a combination of different abiotic stresses. Drought and heat stress represent an excellent example of two different abiotic stress conditions that occur in the field simultaneously. Metabolite profiling of plants subjected to drought, heat stress or a combination of drought and heat stress revealed that plants subject to a combination of drought and heat stress accumulated sucrose and other sugars such as maltose and glucose [160]. In contrast, Pro that accumulated in plants subjected to drought did not accumulate in plants during a combination of drought and heat stress. Heat stress was found to exaggerate the toxicity of Pro to cells, suggesting that during a combination of drought and heat stress, sucrose replaces Pro in plants as the major osmoprotectant. These findings of different metabolic responses to stress combination in comparison with each individual stress highlight the need for further studies of different stress combinations at the metabolic level [163].

In the future, we envision more studies that include metabolomics as an integral part of the systems biology approach to study plant response to a variety of stress conditions.

COMBINATION OF OMICS PLATFORMS

Integrated analysis of metabolite and transcript or metabolite and protein levels in several plant systems already identified several important features of plant metabolic regulation. Currently, most plant stress response studies use largely either one or a combination of two approaches, whereas integrated studies of the plant stress response, using a combination of all three approaches, are just appearing. Integration of the transcriptomics and metabolomics data, to elucidate gene-to-gene and metabolite-to-gene networks in *Arabidopsis*, under sulfur deficiency was described by Hirai *et al.* [164], while combined metabolomics and proteomics approach to study the *Arabidopsis* response to a cesium stress was described by Le Lay *et al.* [159]. It is important to mention that combined analysis of the metabolomics data with other omics data is quite challenging because of the data integration problem [165]. This hampers the wide use of combined data sets and requires further development of data integration and data fusion approaches. Another issue with combined 'omics' studies is related to sample collection and processing. Most of the combined studies use different samples for transcripts, proteins and metabolite measurements. This can introduce a significant error in the subsequent analysis of the combined data and cause lack of correlation in RNA, protein and metabolite levels because of the time difference in quenching metabolism between different samples. Ideally, all three types of molecules should be analyzed from the same biological sample, and the proper quenching and sampling technique that allows for preservation of RNA, proteins and metabolites should be used [166,167]. Furthermore, omics data should be combined with mathematical modeling of the biological systems [168,169,170,171]. Mathematical modeling of plant stress response using 'omics' data is quite limited. It is partially because of the lack of proper time-course data sets and insufficient 'topdown' modeling approaches that can utilize large transcriptomics, proteomics and metabolomics datasets.

TOWARDS SYSTEMS BIOLOGY

The practice of integrating physiological, morphological, molecular, biochemical and genetic information has long been applied to biological research, and in diverse fields such as plant breeding and ecology. 'Omics' research approaches have produced copious data for living systems, which have necessitated the development of systems biology to integrate multidimensional biological information into networks and models. Systems biology essentially is the study of interactions among biological components using models and/or networks to integrate genes, metabolites, proteins, regulatory elements and other biological components. Applications of systems biology to plant science have been rapid, and have increased our knowledge about circadian rhythms, multigenic traits, stress responses and plant defenses, and have advanced the virtual plant project. There is an immediate need for systems integration in plant biology, considering the large datasets generated from different omics technologies such as genomics, proteomics, transcriptomics, interactomics and metabolomics.

Systems biology approach allows not only to analyze the topology of the biochemical and signaling networks involved in stress response but also to capture the dynamics of the response. Systems biology research requires close interaction of biologists and mathematicians in all steps of experimental design, data collection and data analysis and mining. One of the most critical aspects for successful systems biology study is the type of high throughput data available for mathematical modeling. Time-course experiments can provide information on system's dynamics, but the exact time points for sample collection following initial perturbation should be properly selected, based on the systems behavior, to capture both fast and slow responses. Data providing absolute quantities of metabolites are more suited for mathematical modeling than semi-quantitative data currently provided by many metabolomic studies. Additionally, data on enzyme activities rather than protein levels are required by most dynamic metabolic modeling approaches.

Various Networks in Systems Biology

The network construction and analysis is one of the most common approaches in Systems Biology, which include components like genes, proteins, cis-elements, metabolites and other molecules. The various networks include:

Gene to Metabolite Networks

This defines the interactions among genes and metabolites. The study of gene-to-metabolite networks is more complex in plants, particularly in comparison with mammals, because of the greater diversity and larger numbers of metabolites produced by plants as an adaptation to their sessile life style and during adverse stress conditions. Gene-to-metabolite networks have been recently characterized for stress responses, plant defense and hormone- induced responses [172,173,174].

Plant Interactome and Protein Interaction Networks

An interaction network based on the yeast two-hybrid system for essentially all *Arabidopsis* MADS box domain DNA-binding proteins revealed both specific heterodimeric and homodimeric interactions [175]. In a similar study, the interactions between myeloblastosis (MYB) protein and R/B-like basic-helix-loop-helix (BHLH) were characterized, which helped to distinguish the functions of different MYB proteins with similar sequences [176]. The research helped to characterize the functions of MYB DNA-binding proteins and BHLH transcriptional factors, both of which are involved in a variety of processes, including cell cycle, cell proliferation and cell lineage establishment. Recently, a comprehensive study has helped to identify a protein-protein interaction network of over 70 proteins in wheat for abiotic stress response and development [177]. Besides the yeast two-hybrid system and anti-tag co-immunoprecipitation, protein microarrays were also used to probe the interaction between proteins globally in *Arabidopsis* [178].

Transcriptional regulatory networks

The transcription regulatory network describes the interaction between transcriptional regulatory genes and downstream genes. The data are often based on interactions between regulatory genes and downstream genes, as defined by mutant studies, global gene expression profiling, computational prediction of cis-elements and protein-DNA interaction studies using gel-shift assays. Examples of such networks, which also can incorporate information from the existing literature, have been constructed to infer signaling events involved in several abiotic stress responses.

Gene Regulatory Networks

A gene regulatory network describes how genes interact with one another during the biological process to perform a function [179]. The relationships among the genes in the network can be defined in terms of activation, repression and other types of functional interactions.

Systems Biology and Abiotic stress

All the above four types of networks have been used to study plant responses to abiotic stress. The gene-to-metabolite network has been used to study Arabidopsis thaliana responses to nutritional deficiency [180,155]. A protein-protein interaction network was used to identify the key groups of genes involved in abiotic stress responses and flowering control in wheat [177]. A transcriptional regulatory network was built to elucidate the molecular mechanisms of dehydration and cold stress responses in Arabidopsis [17]. A gene regulatory network was successfully applied to explain the dynamics of aperture opening and closure in Arabidopsis in response to environmental stimuli. In addition, a combined study of transcriptomics, QTLs, mutant studies and yeast two-hybrid assays has helped to characterize genes involved in stress response and seed germination in rice. Using systems biology platforms, plant stress response networks and dynamics were generated, which enabled the discovery of several important plant stress response genes [127]. The flow chart of stress systems biology is schematically shown in Fig. **6**. Morioka et al. [181] used, for example, the linear dynamical system to model gene expression and metabolite time series data from Arabidopsis grown under sulfur starvation conditions. Using this variation of Markov model, the authors not only were able to detect known changes in gene expression and metabolite accumulation but also identified novel changes involved in the stress response [181]. This study shows the power of the systems biology approach in understanding and predicting the behavior of the plant system under stress. These examples illustrate the power of using different types of networks to understand complex responses to stress, especially indicating how stresses induce the same and related genes. Thus, plant systems biology has come of age, given its potential to provide crucial understanding of the regulatory networks controlling abiotic stress responses in plants. Advances in plant systems biology are needed to take full advantage of the ever-increasing numbers of omics technologies.

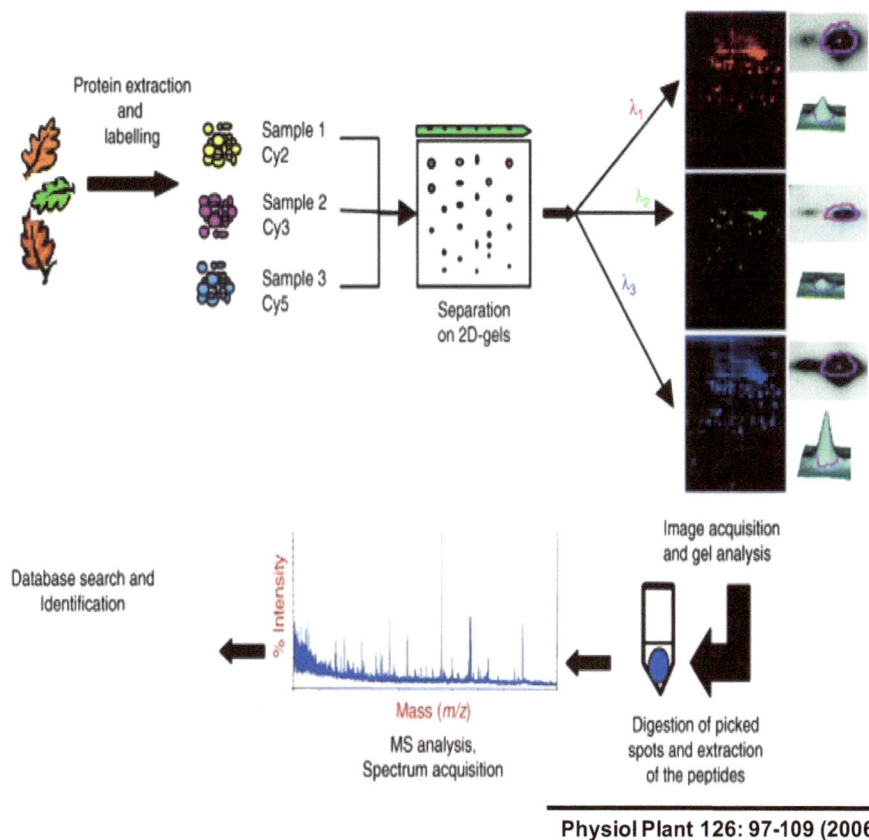

Physiol Plant 126: 97-109 (2006)

Figure 7: Illustration of the difference gel electrophoresis method applied to plant tissues and followed by mass spectrometry analysis. After extraction of proteins from different samples, they are labelled with Cy dyes, then mixed and separated on bidimensional gels. After acquisition of images using a multiple-laser scanner, proteomic patterns are analysed and proteins of interest are picked and digested using trypsin. Determination of the mass of individual peptides and, if necessary, sequencing are performed on a mass spectrometry (MS) analyser. The results obtained are then entered into various database search engines to obtain identification of proteins.

CONCLUSION

The mining and exploitation of the data obtained from genomics and the related research areas of genome wide transcriptomics, proteomics, and metabolomics will bring us into a new era of understanding of biological systems. Advances in various areas of omics and its integration into systems biology research are being made possible by combining expertise from biology, chemistry, instrumentation, computer science, physics, and mathematics. Given that the era of such true interdisciplinary cooperation is only starting, many exciting discoveries are to be expected in the coming years. Understanding of Abiotic stress biology will allow us to develop water use efficient crops enabling them grow in adverse soils in response to unpredictable climate change.

ACKNOWLEDGEMENTS

We gratefully acknowledge the financial assistance provided by the DBT Program Support, Government of India, New Delhi and TATA Innovation Fellowship to SKD.

REFERENCES

[1]   Wang W, Vinocur B, Altman A. Plant responses to drought, salinity and extreme temperatures: towards genetic engineering for stress tolerance. Planta 2003; 218: 1-14.

[2]   Valliyodan B, Nguyen HT. Understanding regulatory networks and engineering for enhanced drought tolerance in plants. Curr Opin Plant Biol 2006; 9: 189-195.

[3]   Datta SK, Chakraborty S, Datta K. In: Kharwal MC Ed. Food Legumes for Nutritional Security and Sustainable Agriculture. Impact of functional genomics of food crops: Rice to legumes, The Indian Society of Genetics and Plant Breeding publication. 2008; pp. 291-314.

[4]   Gong Q, Li P, Ma S, Indu Rupassara S, Bohnert HJ. Salinity stress adaptation competence in the extremophile *Thellungiella halophila* in comparison with its relative *Arabidopsis thaliana*. Plant J 2005; 44: 826-839.

[5]   Chen WJ *et al*. Contribution of transcriptional regulation to natural variations in *Arabidopsis*. Genome Biol 2005; 6: R32.

[6]   Jung SH, Lee JY, Lee DH. Use of SAGE technology to reveal changes in gene expression in *Arabidopsis* leaves undergoing cold stress. Plant Mol Biol 2003; 52: 553-567.

[7]   Mahalingam R *et al*. Characterizing the stress/defense transcriptome of *Arabidopsis*. Genome Biol 2003; 4: R20

[8]   Seki M *et al*. RIKEN *Arabidopsis* full-length (RAFL) cDNA and its applications for expression profiling under abiotic stress conditions. J Exp Bot 2004; 55: 213-223

[9]   Sharp RE *et al*. Root growth maintenance during water deficits: physiology to functional genomics. J Exp Bot 2004; 55: 2343-2351

[10]   Poroyko V *et al*. The maize root transcriptome by serial analysis of gene expression. Plant Physiol 2005; 138: 1700-1710

[11]   Wong CE *et al*. Expressed sequence tags from the Yukon ecotype of *Thellungiella* reveal that gene expression in response to cold, drought and salinity shows little overlap. Plant Mol Biol 2005; 58: 561-574

[12]   Rabbani MA *et al*. Monitoring expression profiles of rice genes under cold, drought, and high-salinity stresses and abscisic acid application using cDNA microarray and RNA gel-blot analyses. Plant Physiol 2003; 133: 1755-1767

[13]   Datta SK. Recent development in transgenics for abiotic stress tolerance in rice. JIRCAS working report, Tsukuba, Japan, 2002; pp. 43-53.

[14]   Zhang W, Ruan J, Ho TH, You Y, Yu T, Quatrano RS. Cis-regulatory element based targeted gene finding: genome-wide identification of abscisic acid- and abiotic stress-responsive genes in *Arabidopsis thaliana*. Bioinformatics 2005; 21: 3074-3081.

[15]   Gibbons FD, Proft M, Struhl K, Roth FP. Chipper: discovering transcription-factor targets from chromatin immunoprecipitation microarrays using variance stabilization. Genome Biol 2005; 6: R96.

[16]   Kawaguchi R, Girke T, Bray EA, Bailey-Serres J Differential mRNA translation contributes to gene regulation under non-stress and dehydration stress conditions in *Arabidopsis thaliana*. Plant J 2004; 38: 823-839.

[17]   Yamaguchi-Shinozaki K, Shinozaki K. Transcriptional regulatory networks in cellular responses and tolerance to dehydration and cold stresses. Annu Rev Plant Biol 2006; 57: 781-803.

[18]   Liu J, Ishitani M, Halfter U, Kim CS, Zhu JK. The *Arabidopsis thaliana SOS2* gene encodes a protein kinase that is required for salt tolerance. Proc Natl Acad Sci USA 2000; 97: 3730-3734.

[19]   Nakashima K *et al*. Organization and expression of two *Arabidopsis* DREB2 genes encoding DRE binding proteins involved in dehydration- and high-salinity responsive gene expression. Plant Mol Biol 2000; 42: 657-665.

[20]   Choi H, Hong J, Ha J, Kang J, Kim SY. ABFs, a family of ABA-responsive element binding factors. J Biol Chem 2000; 275: 1723-1730.

[21]   Uno Y *et al*. *Arabidopsis* basic leucine zipper transcription factors involved in an abscisic acid-dependent signal transduction pathway under drought and high-salinity conditions. Proc Natl Acad Sci USA 2000; 97: 11632-11637.

[22]   Roychoudhury A, Gupta B, Sengupta DN. Trans-acting factor designated OSBZ8 interacts with both typical abscisic acid responsive elements as well as abscisic acid responsive element like sequences in the vegetative tissues of indica rice cultivars. Plant Cell Rep 2008; 27: 779-794.

[23]   Fowler S, Thomashow MF. *Arabidopsis* transcriptome profiling indicates that multiple regulatory pathways are activated during cold acclimation in addition to the CBF cold response pathway. Plant Cell 2002; 14: 1675-1690.

[24]   Vogel JT, Zarka DG, Van Buskirk HA, Fowler SG, Thomashow MF. Roles of the CBF2 and ZAT12 transcription factors in configuring the low temperature transcriptome of *Arabidopsis*. Plant J 2005; 41: 195-211.

[25]   Sakuma Y *et al*. Functional analysis of an *Arabidopsis* transcription factor, DREB2A, involved in drought responsive gene expression. Plant Cell 2006; 18: 1292-1309.

[26]   Umezawa T, Yoshida R, Maruyama K, Yamaguchi-Shinozaki K, Shinozaki K. SRK2C, a SNF1-related protein kinase 2, improves drought tolerance by controlling stress-responsive gene expression in *Arabidopsis thaliana*. PNAS USA 2004; 101: 17306-17311.

[27]   Osakabe Y, Maruyama K, Seki M, Satou M, Shinozaki K, Yamaguchi-Shinozaki K. Leucine-rich repeat receptor-like kinase 1 is a key membrane-bound regulator of abscisic acid early signaling in Arabidopsis. Plant Cell 2005; 17: 1105-1119.

[28]   Xiong L, Zhu J-K. Molecular and genetic aspects of plant responses to osmotic stress. Plant Cell Environ 2002; 25: 131-139.

[29]   Bohnert HJ, Gong Q, Li P, Ma S. Unraveling abiotic stress tolerance mechanisms-getting genomics going. Curr Opin Plant Biol 2006; 9: 180-188.

[30]   Taji T *et al*. Comparative genomics in salt tolerance between *Arabidopsis* and *Arabidopsis*-related halophyte salt cress using *Arabidopsis* microarray. Plant Physiol 2004; 135: 1697-1709.

[31]   Kant S, Kant P, Raveh E, Barak S. Evidence that differential gene expression between the halophyte, *Thellungiella halophila* and *Arabidopsis thaliana* is responsible for higher levels of the compatible osmolyte proline and tight control of Na$^+$ uptake in *T. halophila*. Plant Cell Environ 2006; 29: 1220-1234.

[32]   Wong CE *et al.* Transcriptional profiling implicates novel interactions between abiotic stress and hormonal responses in *Thellungiella*, a close relative of *Arabidopsis*. Plant Physiol 2006; 140: 1437-1450.

[33]   Vinocur B, Altman A. Recent advances in engineering plant tolerance to abiotic stress: achievements and limitations. Curr Opin Biotechnol 2005; 16: 123-132.

[34]   Salvi S, Tuberosa R. To clone or not to clone plant QTLs: present and future challenges. Trends Plant Sci 2005; 10: 297-304.

[35]   Jander G, Norris SR, Rounsley SD, Bush DF, Levin IM, Last RL. *Arabidopsis* map-based cloning in the post genome era. Plant Physiol 2002; 129: 440-450.

[36]   Liu J, Zhu JK. A calcium sensor homolog required for plant salt tolerance. Science 1998; 280: 1943-1945.

[37]   Shi H, Ishitani M, Kim C, Zhu JK. The *Arabidopsis thaliana* salt tolerance gene *SOS1* encodes a putative Na$^+$/H$^+$ antiporter. Proc Natl Acad Sci USA 2000; 97: 6896-6901.

[38]   Shi H, Kim Y, Guo Y, Stevenson B, Zhu JK. The *Arabidopsis SOS5* locus encodes a putative cell surface adhesion protein and is required for normal cell expansion. Plant Cell 2003; 15: 19-32.

[39]   Lee H, Xiong L, Gong Z, Ishitani M, Stevenson B, Zhu JK. The *Arabidopsis HOS1* gene negatively regulates cold signal transduction and encodes a RING finger protein that displays cold-regulated nucleo-cytoplasmic partitioning. Genes Dev 2001; 15: 912-924.

[40]   Xiong L, Ishitani M, Lee H, Zhu JK. The *Arabidopsis LOS5/ABA3* locus encodes a molybdenum cofactor sulfurase and modulates cold stress- and osmotic stress-responsive gene expression. Plant Cell 2001; 13: 2063-2083.

[41]   Shi H, Xiong L, Stevenson B, Lu T, Zhu JK. The *Arabidopsis* salt overly sensitive 4 mutants uncover a critical role for vitamin B6 in plant salt tolerance. Plant Cell 2002; 14: 575-588.

[42]   Yamanouchi U, Yano M, Lin H, Ashikari M, Yamada K. A rice spotted leaf gene, *Spl7*, encodes a heat stress transcription factor protein. Proc Natl Acad Sci USA 2002; 99: 7530-7535

[43]   Chen Z *et al.* Disruption of the cellulose synthase gene, *AtCesA8/IRX1*, enhances drought and osmotic stress tolerance. Plant J 2005; 43: 273-283.

[44]   Koiwa H, Bressan RA, Hasegawa PM. Identification of plant stress-responsive determinants in *Arabidopsis* by large scale forward genetic screens. J Exp Bot 2006; 57: 1119-1128.

[45]   Gilchrist EJ, Haughn GW. TILLING without a plough: a new method with applications for reverse genetics. Curr Opin Plant Biol 2005; 8: 211-215.

[46]   Henikoff S, Comai L. Single-nucleotide mutations for plant functional genomics. Annu Rev Plant Biol 2003; 54: 375-401.

[47]   Haake V, Cook D, Riechmann JL, Pineda O, Thomashow MF, Zhang JZ. Transcription factor CBF4 is a regulator of drought adaptation in *Arabidopsis*. Plant Physiol 2002; 130: 639-648.

[48]   Cook D, Fowler S, Fiehn O, Thomashow MF. A prominent role for the CBF cold response pathway in configuring the low-temperature metabolome of *Arabidopsis*. Proc Natl Acad Sci USA 2004; 101: 15243-15248.

[49]   Maruyama K *et al.* Identification of cold-inducible downstream genes of the *Arabidopsis* DREB1A/CBF3 transcriptional factor using two microarray systems. Plant J 2004; 38: 982-993.

[50]   Oh SJ *et al.* *Arabidopsis* CBF3/DREB1A and ABF3 in transgenic rice increased tolerance to abiotic stress without stunting growth. Plant Physiol 2005; 138: 341-351.

[51]   Gorantla M, Babu PR, Lachagari VBR, Feltus FA, Paterson A, Reddy AR. Functional genomics of drought stress response in rice: transcript mapping of annotated unigenes of an *indica* rice (*Oryza sativa* L. cv. Nagina 22). Curr Sci 2005; 89: 496-514.

[52]   Lin HX *et al.* QTLs for Na$^+$ and K$^+$ uptake of the shoots and roots controlling rice salt tolerance. Theor Appl Genet 2004; 108: 253-260.

[53]   Ren ZH *et al.* A rice quantitative trait locus for salt tolerance encodes a sodium transporter. Nat Genet 2005; 37: 1141-1146.

[54]   Xu Y, McCouch SR, Zhang Q. How can we use genomics to improve cereals with rice as a reference genome? Plant Mol Biol 2005; 59: 7-26.

[55]   Tuberosa R, Salvi S. Genomics-based approaches to improve drought tolerance of crops. Trends Plant Sci 2006; 11: 405-412.

[56]   Xu K *et al.* *Sub1A* is an ethylene-response-factor-like gene that confers submergence tolerance to rice. Nature 2006; 442: 705-708.

[57]   Li L *et al.* Tiling microarray analysis of rice chromosome 10 to identify the transcriptome and relate its expression to chromosomal architecture. Genome Biol 2005; 6: R52.

[58]   Wing RA *et al.* The *Oryza* map alignment project: the golden path to unlocking the genetic potential of wild rice species. Plant Mol Biol 2005; 59: 53-62.

[59]   Ammiraju JS *et al.* The *Oryza* bacterial artificial chromosome library resource: construction and analysis of 12 deep-coverage large-insert BAC libraries that represent the 10 genome types of the genus *Oryza*. Genome Res 2006; 16: 140-147.

[60]   Sanchez AC, Subudhi PK, Rosenow DT, Nguyen HT. Mapping QTLs associated with drought resistance in sorghum (*Sorghum bicolor* L Moench). Plant Mol Biol 2002; 48: 713-726.

[61]   Bajaj S, Mohanty A. Recent advances in rice biotechnology-towards genetically superior transgenic rice. Plant Biotechnol J 2005; 3: 275-307.

[62]   Haralampidis K, Milioni D, Rigas S, Hatzopoulos P. Combinatorial interaction of cis elements specifies the expression of the *Arabidopsis AtHsp90-1* gene. Plant Physiol 2002; 129: 1138-1149.

[63]   Trindade LM, Horvath BM, Bergervoet MJ, Visser RG. Isolation of a gene encoding a copper chaperone for the copper/zinc superoxide dismutase and characterization of its promoter in potato. Plant Physiol 2003; 133: 618-629.

[64]   Yamaguchi-Shinozaki K, Shinozaki K. Organization of cis-acting regulatory elements in osmotic- and cold-stress-responsive promoters. Trends Plant Sci 2005; 10: 88-94.

[65]   Ingram J, Bartels D. The molecular basis of dehydration tolerance in plants. Annu Rev Plant Biol 1996; 47: 377-403.

[66]   Grover A *et al.* Understanding molecular alphabets of the plant abiotic stress response. Curr Sci 2001; 80: 206-216.

[67]   Yamaguchi-Shinozaki K, Shinozaki K. A novel cis-acting element in an *Arabidopsis* gene is involved in responsiveness to drought, low-temperature, or high-salt stress. Plant Cell 1994; 6: 251-264.

[68]   Thomashow MF. Plant cold acclimation, freezing tolerance genes and regulatory mechanisms. Annu Rev Plant Biol 1999; 50: 571-599.

[69]   Dolferus R, Jacobs M, Peacock WJ, Dennis ES. Differential interactions of promoter elements in stress responses of the *Arabidopsis Adh* gene. Plant Physiol 1994; 105: 1075-1087.

[70]   Setter TL *et al.* Physiology and Genetics of Submergence Tolerance in Rice. Ann Bot 1997; 79: 67-77.

[71]   Dennis ES *et al.* Molecular strategies for improving waterlogging tolerance in plants. J Exp Bot 2000; 51: 89-97.

[72]   Quimio CA *et al.* Enhancement of submergence tolerance in transgenic rice overproducing pyruvate decarboxylase. J Plant Physiol 2000; 156: 516-521.

[73]   Schoffl F, Prandl R, Reindl A. Regulation of the heat shock response. Plant Physiol 1998; 17: 1135-1141.

[74]   Somerville C, Somerville S. Plant functional genomics. Science 1999; 285: 380-383.

[75]   Chang YL, Henriquez X, Preuss D, Copenhaver GP, Zhang HB. A plant-transformation-competent BIBAC library from the *Arabidopsis thaliana* Landsberg ecotype for functional and comparative genomics. Theor Appl Genet 2003; 106: 269-276.

[76]   An G, Lee S, Kim SH, Kim SR. Molecular genetics using T-DNA in rice. Plant Cell Physiol 2005; 46: 14-22.

[77]   Rus A *et al. AtHKT1* is a salt tolerance determinant that controls Na$^+$ entry into plant roots. PNAS USA 2001; 98: 14150-14155.

[78]   Zhu J *et al. OSM1/SYP61*, a syntaxin protein in *Arabidopsis* controls abscisic acid-mediated and non-abscisic acid-mediated responses to abiotic stress. Plant Cell 2002; 14: 3009-3028.

[79]   Zhu J *et al. HOS10* encodes an R2R3-type MYB transcription factor essential for cold acclimation in plants. Proc Natl Acad Sci USA 2005; 102: 9966-9971.

[80]   Cheong YH, Kim KN, Pandey GK, Gupta R, Grant JJ, Luan S. CBL1, a calcium sensor that differentially regulates salt, drought, and cold responses in *Arabidopsis*. Plant Cell 2003; 15: 1833-1845.

[81]   Kim KN, Cheong YH, Grant JJ, Pandey GK, Luan S. CIPK3, a calcium sensor-associated protein kinase that regulates abscisic acid and cold signal transduction in *Arabidopsis*. Plant Cell 2003; 15: 411-423.

[82]   Lee S, Kim S-H, Kim S-J, Lee K, Han S-K. Trapping and characterization of cold-responsive genes from T-DNA tagging lines in rice. Plant Sci 2004; 166: 69-79.

[83]   Wong HL, Sakamoto T, Kawasaki T, Umemura K, Shimamoto K. Down-regulation of metallothionein, a reactive oxygen scavenger, by the small GTPase *OsRac1* in rice. Plant Physiol 2004; 135: 1447-1456.

[84]   Kurusu T, Yagala T, Miyao A, Hirochika H, Kuchitsu K. Identification of a putative voltage-gated Ca^{2+} channel as a key regulator of elicitor-induced hypersensitive cell death and mitogen-activated protein kinase activation in rice. Plant J 2005; 42: 798-809.

[85]   Krysan PJ, Young JC, Sussman MR. T-DNA as an insertional mutagen in *Arabidopsis*. Plant Cell 1999; 11: 2283-2290.

[86]   Jones-Rhoades MW, Bartel DP. Computational identification of plant microRNAs and their targets, including a stress-induced miRNA. Mol Cell 2004; 14: 787-799.

[87]   Sunkar R, Zhu JK. Novel and stress-regulated microRNAs and other small RNAs from *Arabidopsis*. Plant Cell 2004; 16: 2001-2019.

[88] Hilson P *et al.* Versatile gene-specific sequence tags for *Arabidopsis* functional genomics: transcript profiling and reverse genetics applications. Genome Res 2004; 14: 2176-2189.

[89] Parinov S, Sundaresan V. Functional genomics in *Arabidopsis*: large-scale insertional mutagenesis complements the genome sequencing project. Curr Opin Biotechnol 2000; 11: 157-161.

[90] Miki D, Shimamoto K. Simple RNAi vectors for stable and transient suppression of gene function in rice. Plant Cell Physiol 2004; 45: 490-495.

[91] Pandey A, Mann M. Proteomics to study genes and genomes. Nature 2000; 405: 837-846.

[92] Agrawal GK, Rakwal R. Rice proteomics, a cornerstone for cereal food crop proteomes. Mass Spectrom Rev 2006; 25: 1-53.

[93] Peck SC. Update on proteomics in *Arabidopsis*. Where do we go from here? Plant Physiol 2005; 138: 591-599.

[94] Cutler SR, Ehrhardt DW, Griffitts JS, Somerville CR. Random GFP::cDNA fusions enable visualization of subcellular structures in cells of *Arabidopsis* at a high frequency. PNAS USA 2000; 97: 3718-3723.

[95] Kiegle E, Moore CA, Haseloff J, Tester MA, Knight MR. Cell-type specific calcium responses to drought, salt and cold in the *Arabidopsis* root. Plant J 2000; 23: 267-278.

[96] Pinheiro C, Kehr J, Ricardo CP. Effect of water stress on lupin stem protein analyzed by two-dimensional gel electrophoresis. Planta 2005; 221: 716-728.

[97] Yan S, Tang Z, Su W, Sun W. Proteomic analysis of salt stress-responsive proteins in rice roots. Proteomics 2005; 5: 235-244.

[98] Claes B, Dekeyser R, Villarroel R, Bulcke VM, Bauw G, Montagu MV. Characterization of rice gene showing organ specific expression in response to salt stress and drought. Plant Cell 1990; 2: 19-27.

[99] Moons A, Bauw G, Prinsen E, Montagu MV, Van der Straeten D. Molecular and physiological responses to abscisic acid and salts in roots of salt-sensitive and salt-tolerant indica rice varieties. Plant Physiol 1995; 107: 177-186.

[100] Moons A, Prinsen E, Bauw G, Montagu MV. Antagonistic effects of abscisic acid and jasmonates on salt stress-inducible transcripts in rice roots. Plant Cell 1997; 9: 2243-2259.

[101] Datta K *et al.* Over expression of the cloned rice thaumatin-like protein (PR-5) gene in transgenic rice plants enhances environmental friendly resistance to *Rhizoctonia solani* causing sheath blight disease Theor Appl Genet 1999; 98:1138-1145.

[102] Datta SK, Muthukrishnan S (Eds.) Pathogenesis-related proteins in plants. CRC Press, USA, 1999.

[103] Salekdeh GH, Siopongco J, Wade LJ, Ghareyazie B, Bennett J. Proteomic analysis of rice leaves during drought stress and recovery. Proteomics 2002; 2: 1131-1145.

[104] Salekdeh GH, Siopongco J, Wade LJ, Ghareyazie B, Bennet J. A proteomic approach to analyzing drought and salt responsive in rice. Field Crops Res 2002; 76: 199-219.

[105] Huo CM, Zhao BC, Ge RC, Shen YZ, Huang ZJ. Proteomic analysis of the salt tolerance mutant of wheat under salt stress. Yi Chuan Xue Bao 2004; 31: 1408-1414.

[106] Abbasi F, Komatsu S. A proteomic approach to analyze salt responsive proteins in rice leaf sheath. Proteomics 2004; 4: 2072-2081.

[107] Ouerghi Z, Remy R, Ouelhazi L, Ayadi A, Brulfert J. Two-dimensional electrophoresis of soluble leaf proteins isolated from two wheat species (*Triticum durum* and *Triticum aestivum*) differing in sensitivity towards NaCl. Electrophoresis 2000; 21: 2487-2491.

[108] Majoul T, Chahed K, Zamiti E, Ouelhazi L, Ghrir R. Analysis by two-dimensional electrophoresis of the effect of salt stress on the polypeptide patterns in roots of a salt-tolerant and salt-sensitive cultivar of wheat. Electrophoresis 2000; 21: 2562-2565.

[109] Chen CSS, Plant AL. Salt-induced protein synthesis in tomato roots: The role of ABA. J Exp Bot 1999; 50: 677-687.

[110] Mann M, Hendrickson RC, Pandey A. Analysis of proteins and proteomes by mass spectrometry. Annu Rev Biochem 2001; 70: 437-473.

[111] Mann M, Pandey A. Use of mass spectrometry-derived data to annotate nucleotide and protein sequence databases. Trends Biochem Sci 2001; 26: 54-61.

[112] Askari H, Edqvist J, Hajheidari M, Kafi M, Salekdeh GH. Effects of salinity levels on proteome of *Suaeda aegyptiaca* leaves. Proteomics 2005; 6: 2542-2554.

[113] Ramanjulu S, Kaiser W, Dietz KJ. Salt and drought stress differentially affect accumulation of extracellular proteins in barley. Z Naturforsch 1999; 54: 337-347.

[114] Sugihara K, Hanagata N, Dubinsky Z, Baba S, Karube I. Molecular characterization of cDNA encoding oxygen evolving enhancer protein 1 increased by salt treatment in the mangrove *Bruguiera gymnorrhiza*. Plant Cell Physiol 2000; 41: 1279-1285.

[115] Roychoudhury A, Basu S, Sarkar SN, Sengupta DN. Comparative physiological and molecular responses of a common aromatic indica rice cultivar to high salinity with non-aromatic indica rice cultivars. Plant Cell Rep 2008; 27: 1395-1410.

[116] RoyChoudhury A, Roy C, Sengupta DN. Transgenic tobacco plants over expressing the heterologous *lea* gene *Rab16A* from rice during high salt and water deficit display enhanced tolerance to salinity stress. Plant Cell Rep 2007; 26: 1839-1859.

[117] Xu D, Duan X, Wang B, Hong B, Ho T-HD, Wu R. Expression of a late embryogenesis abundant protein gene, *HVA1*, from barley confers tolerance to water deficit and salt stress in transgenic rice. Plant Physiol 1996; 110: 249-257.

[118] Singla SL, Pareek A, Grover A. Plant Hsp100 family with special reference to rice. J Biosci 1998; 23: 337-345.

[119] Moons A, Valcke R, Van Montagu M. Low-oxygen stress and water deficit induce cytosolic pyruvate orthophosphate dikinase (PPDK) expression in roots of rice, a C$_3$ plant. Plant J 1998; 15: 89-98.

[120] Nuhse TS, Stensballe A, Jensen ON, Peck SC. Phosphoproteomics of the *Arabidopsis* plasma membrane and a new phosphorylation site database. Plant Cell 2004; 16: 2394-2405.

[121] Lindermayr C, Saalbach G, Durner J. Proteomic identification of S-nitrosylated proteins in *Arabidopsis*. Plant Physiol 2005; 137: 921-930.

[122] Dani V, Simon WJ, Duranti M, Croy RR. Changes in the tobacco leaf apoplast proteome in response to salt stress. Proteomics 2005; 5: 737-745.

[123] Cui S, Huang F, Wang J, Ma X, Cheng Y, Liu J. A proteomic analysis of cold stress responses in rice seedlings. Proteomics 2005; 5: 3162-3172.

[124] Unlu M, Morgan ME, Minden JS. Difference gel electrophoresis: a single gel method for detecting changes in protein extracts. Electrophoresis 1997; 18: 2071-2077.

[125] Tonge R *et al*. Validation and development of fluorescence two-dimensional differential gel electrophoresis proteomics technology. Proteomics 2001; 1: 377-396.

[126] Amme S, Matros A, Schlesier B, Mock H-P. Proteome analysis of cold stress response in *Arabidopsis thaliana* using DIGE-technology. J Exp Bot 2006; 57: 1537-1546.

[127] Cooper B *et al*. A network of rice genes associated with stress response and seed development. Proc Natl Acad Sci USA 2003; 100: 4945-4950.

[128] Mukherjee K, Choudhury AR, Gupta B, Gupta S, Sengupta DN. An ABRE-binding factor, OSBZ8, is highly expressed in salt tolerant cultivars than in salt sensitive cultivars of indica rice. BMC Plant Biol 2006; 6:18.

[129] Mittler R. Oxidative stress, antioxidants and stress tolerance. Trends Plant Sci 202; 7: 405-410.

[130] Mittler R, Vanderauwera S, Gollery M, Van Breusegem F. Reactive oxygen gene network of plants. Trends Plant Sci 2004; 9: 490-498.

[131] Roychoudhury A, Basu S, Sengupta DN. Effects of exogenous abscisic acid on some physiological responses in a popular aromatic indica rice compared with those from two traditional non-aromatic indica rice cultivars. Acta Physiol Plant 2009; 31: 915-926.

[132] Basu S, Roychoudhury A, Saha PP, Sengupta DN. Comparative analysis of some biochemical responses of three indica rice varieties during polyethylene glycol-mediated water stress exhibits distinct varietal differences. Acta Physiol Plant 2010; DOI: 10.1007/s11738-009-0432-y

[133] Basu S, Roychoudhury A, Saha PP, Sengupta DN. Differential antioxidative responses of indica rice cultivars to drought stress. Plant Growth Regul 2010; 60: 51-59.

[134] Fiehn O. Metabolomics- the link between genotypes and phenotypes. Plant Mol Biol 2002; 48: 155-171.

[135] Halket JM, Waterman D, Przyborowska AM, Patel RK, Fraser PD, Bramley PM. Chemical derivatization and mass spectral libraries in metabolic profiling by GC/MS and LC/MS/MS. J Exp Bot 2005; 56: 219-243.

[136] Shulaev V. Metabolomics technology and bioinformatics. Brief Bioinform 2006; 7: 128-139.

[137] Krishnan P, Kruger NJ, Ratcliffe RG. Metabolite fingerprinting and profiling in plants using NMR. J Exp Bot 2005; 56: 255-265.

[138] Goodacre R, York EV, Heald JK, Scott IM. Chemometric discrimination of unfractionated plant extracts analyzed by electrospray mass spectrometry. Phytochemistry 2003; 62: 859-863.

[139] Johnson HE, Broadhurst D, Goodacre R, Smith AR. Metabolic fingerprinting of salt-stressed tomatoes. Phytochemistry 2003; 62: 919-928.

[140] Sumner LW, Mendes P, Dixon RA. Plant metabolomics: large-scale phytochemistry in the functional genomics era. Phytochemistry 2003; 62: 817-836.

[141] Kohane IS, Kho AT, Butte AJ. Microarrays for Integrative Genomics. The MIT Press, Cambridge, 2003.

[142] Grata E *et al*. Development of a two-step screening ESI-TOF-MS method for rapid determination of significant stress-induced metabolome modifications in plant leaf extracts: the wound response in *Arabidopsis thaliana* as a case study. J Sep Sci 2007; 30: 2268-2278.

[143] Giri S, Krausz KW, Idle JR, Gonzalez FJ. The metabolomics of (1/2)-arecoline 1-oxide in the mouse and its formation by human flavin-containing monooxygenases. Biochem Pharmacol 2007; 73: 561-573.

[144] Granger JH, Williams R, Lenz EM, Plumb RS, Stumpf CL, Wilson ID. A metabolomic study of strain- and age-related differences in the Zucker rat. Rapid Commun Mass Spectrom 2007; 21: 2039-2045.

[145] Soga T, Ohashi Y, Ueno Y, Naraoka H, Tomita M, Nishioka T. Quantitative metabolome analysis using capillary electrophoresis mass spectrometry. J Proteome Res 2003; 2: 488-494.

[146] Mashego MR *et al.* MIRACLE: mass isotopomer ratio analysis of U-^{13}C-labeled extracts. A new method for accurate quantification of changes in concentrations of intracellular metabolites. Biotechnol Bioeng 2004; 85: 620-628.

[147] Lafaye A, Labarre J, Tabet JC, Ezan E, Junot C. Liquid chromatography-mass spectrometry and ^{15}N metabolic labeling for quantitative metabolic profiling. Anal Chem 2005; 77: 2026-2033.

[148] Wu L *et al.* Quantitative analysis of the microbial metabolome by isotope dilution mass spectrometry using uniformly ^{13}C-labeled cell extracts as internal standards. Anal Biochem 2005; 336: 164-171.

[149] Kaplan F *et al.* Exploring the temperature stress metabolome of *Arabidopsis.* Plant Physiol 2004; 136: 4159-4168.

[150] Kaplan F *et al.* Transcript and metabolite profiling during cold acclimation of *Arabidopsis* reveals an intricate relationship of cold-regulated gene expression with modifications in metabolite content. Plant J 2007; 50: 967-981.

[151] Brosche M *et al.* Gene expression and metabolite profiling of *Populus euphratica* growing in the Negev desert. Genome Biol 2005; 6: R101.

[152] Cramer GR *et al.* Water and salinity stress in grapevines: early and late changes in transcript and metabolite profiles. Funct Integr Genomics 2007; 7: 111-134.

[153] Kim JK, Bamba T, Harada K, Fukusaki E, Kobayashi A. Time-course metabolic profiling in *Arabidopsis thaliana* cell cultures after salt stress treatment. J Exp Bot 2007; 58: 415-424.

[154] Nikiforova VJ *et al.* Towards dissecting nutrient metabolism in plants: a systems biology case study on sulphur metabolism. J Exp Bot 2004; 55: 1861-1870.

[155] Nikiforova VJ, Daub CO, Hesse H, Willmitzer L, Hoefgen R. Integrative gene-metabolite network with implemented causality deciphers informational fluxes of sulphur stress response. J Exp Bot 2005; 56: 1887-1896.

[156] Nikiforova VJ *et al.* Systems rebalancing of metabolism in response to sulfur deprivation, as revealed by metabolome analysis of *Arabidopsis* plants. Plant Physiol 2005; 138: 304-318.

[157] Hernandez G *et al.* Phosphorus stress in common bean: root transcript and metabolic responses. Plant Physiol 2007; 144: 752-767.

[158] Baxter CJ *et al.* The metabolic response of heterotrophic *Arabidopsis* cells to oxidative stress. Plant Physiol 2007; 143: 312-325.

[159] Le Lay P *et al.* Metabolomic, proteomic and biophysical analyses of *Arabidopsis thaliana* cells exposed to a caesium stress. Influence of potassium supply. Biochimie 2006; 88: 1533-1547.

[160] Rizhsky L, Liang H, Shuman J, Shulaev V, Davletova S, Mittler R. When defense pathways collide. The response of *Arabidopsis* to a combination of drought and heat stress. Plant Physiol 2004; 134: 1683-1696.

[161] Morsy MR, Jouve L, Hausman JF, Hoffmann L, Stewart JM. Alteration of oxidative and carbohydrate metabolism under abiotic stress in two rice (*Oryza sativa* L.) genotypes contrasting in chilling tolerance. J Plant Physiol 164: 157-167.

[162] Bailey NJ, Oven M, Holmes E, Nicholson JK, Zenk MH. Metabolomic analysis of the consequences of cadmium exposure in *Silene cucubalus* cell cultures via ^{1}H NMR spectroscopy and chemometrics. Phytochem 2003; 62: 851-858.

[163] Mittler R. Abiotic stress, the field environment and stress combination. Trends Plant Sci 2006; 11: 15-19.

[164] Hirai MY *et al.* Elucidation of gene-to-gene and metabolite-to gene networks in *Arabidopsis* by integration of metabolomics and transcriptomics. J Biol Chem 2005; 280: 25590-25595.

[165] Mehrotra B, Mendes P. In: Saito K, Dixon RA, Willmitzer L Eds. Plant Metabolomics, Vol. 57. Bioinformatics approaches to integrate metabolomics and other systems biology data. Springer-Verlag, Berin, Heidelberg, 2006; pp. 105-115.

[166] Weckwerth W, Wenzel K, Fiehn O. Process for the integrated extraction, identification and quantification of metabolites, proteins and RNA to reveal their co-regulation in biochemical networks. Proteomics 2004; 4: 78-83.

[167] Martins AM, Sha W, Evans C, Martino-Catt S, Mendes P, Shulaev V. Comparison of sampling techniques for parallel analysis of transcript and metabolite levels in *Saccharomyces cerevisiae.* Yeast 2007; 24: 181-188.

[168] Kitano H. Systems biology: a brief overview. Science 2002; 295: 1662-1664.

[169] van der Greef J, Stroobant P, van der Heijden R. The role of analytical sciences in medical systems biology. Curr Opin Chem Biol 2004; 8: 559-565.

[170] Kell DB. Theodor Bucher Lecture. Metabolomics, modelling and machine learning in systems biology - towards an understanding of the languages of cells. Delivered on 3 July 2005 at the 30th FEBS Congress and the 9th IUBMB conference in Budapest. FEBS J 2006; 273: 873-894.

[171] Sims KJ, Alvarez-Vasquez F, Voit EO, Hannun YA. A guide to biochemical systems modeling of sphingolipids for the biochemist. Methods Enzymol 2007; 432: 319-350.

[172] Goossens A *et al.* A functional genomics approach toward the understanding of secondary metabolism in plant cells. Proc Natl Acad Sci USA 2003; 100: 8595-8600.

[173] Carrari F *et al.* Integrated analysis of metabolite and transcript levels reveals the metabolic shifts that underlie tomato fruit development and highlight regulatory aspects of metabolic network behavior. Plant Physiol 2006; 142: 1380-1396.

[174] Zulak KG *et al.* Gene transcript and metabolite profiling of elicitor-induced opium poppy cell cultures reveals the coordinate regulation of primary and secondary metabolism. Planta 2007; 225: 1085-1106.

[175] de Folter S *et al.* Comprehensive interaction map of the *Arabidopsis* MADS box transcription factors. Plant Cell 2005; 17: 1424-1433.

[176] Zimmermann IM, Heim MA, Weisshaar B, Uhrig JF. Comprehensive identification of *Arabidopsis thaliana* MYB transcription factors interacting with R/B-like BHLH proteins. Plant J 2004; 40: 22-34.

[177] Tardif G *et al.* Interaction network of proteins associated with abiotic stress response and development in wheat. Plant Mol Biol 2007; 63: 703-718.

[178] Popescu SC *et al.* Differential binding of calmodulin related proteins to their targets revealed through high-density *Arabidopsis* protein microarrays. Proc Natl Acad Sci USA 2007; 104: 4730-4735.

[179] Alvarez-Buylla ER, Benítez M, Dávila EB, Chaos A, Espinosa-Soto C, Padilla-Longoria P. Gene regulatory network models for plant development. Curr Opin Plant Biol 2007; 10: 83-91.

[180] Hirai MY *et al.* Integration of transcriptomics and metabolomics for understanding of global responses to nutritional stresses in *Arabidopsis thaliana.* Proc Natl Acad Sci USA 2004; 101: 10205-10210.

[181] Morioka R, Kanaya S, Hirai MY, Yano M, Ogasawara N, Saito K. Predicting state transitions in the transcriptome and metabolome using a linear dynamical system model. BMC Bioinformatics 2007; 8: 343.

CHAPTER 8

Epigenome and Abiotic Stress Tolerance in Plants

Sanjay Kapoor[1] and Meenu Kapoor[2]*

[1]Department of Plant Molecular Biology, University of Delhi South Campus Benito Juarez road, New Delhi, India and [2]University School of Biotechnology, Guru Gobind Singh Indraprastha University, Delhi, India

Abstract: Epigenome refers to the genomic content of a cell layered with all the covalent/noncovalent modifications on the DNA and histones. Remodeling of the chromatin is induced by these changes in response to environmental or developmental signals without any change in the underlying nucleotide sequence. These modifications are stably inherited through mitotic and meiotic divisions. Every differentiated cell type possesses a characteristic epigenome that is dynamic and differs from the epigenome of neighboring cell type, while the genotype of the cells remains constant. DNA methylation, histone modifications, inclusion of histone variants in the nucleosomes, small RNAs and their effector molecules all contribute towards changing the landscape of the epigenomes. Valuable insights into genomic distribution of methyl groups has been obtained by employing various high throughput technologies. Orchestered changes in the epigenetic landscape also occurs in response to abiotic stress. Plants respond to these stresses by modulating the expression of stress-regulated genes through covalent/noncovalent modifications of chromatin. This chapter highlights the contribution of epigenetic modifications in shaping the epigenome and mediating gene regulation in response to abiotic stress.

INTRODUCTION

Epigenetics involves the study of stable or heritable changes in the chromatin states without any alteration in the underlying nucleotide sequence of DNA. The term "Epigenetic" was coined by Conrad Waddington (1905-1975) to describe, "the casual interaction of genes with their environment" [1]. All the biological phenomenon known at that time that could not be explained by genetic principles, such as, paramutation in maize, position effect variegation in Drosophila, imprinting of genes in mammals, were all categorized under epigenetics [2]. Since then, and more specifically in the past decade efforts have been made by various research groups independently and through international consortia to unravel the molecular basis of this phenomenon. Biochemical, genetic and molecular studies have provided conclusive evidence for covalent modifications of DNA and its associated milieu of histone proteins and non-covalent alterations of chromatin such as inclusion of histone variants such as H2A.Z and H3.3 in the nucleosomes and effects of non-coding RNAs as the key molecular players in the observed epigenetic phenomenon. Covalent modifications include methylation of DNA and/ or histone proteins and other post-translational modifications such as acetylation, phosphorylation, ubiquitination, ADP ribosylation, proline isomerization and sumoylation of chromatin proteins. The addition of these charge-altering modifications on DNA and histones disturb the DNA protein interactions thereby modulating packaging of DNA and its accessibility to regulatory proteins [3]. Recent studies have provided evidences that these alterations in chromatin structure are mediated through recruitment of effector molecules that recognize and distinguish between the vast repertoire of chromatin marks and thus bring about local changes in gene expression. The sum total of all the epigenetic marks on the chromatin adds a layer of information over and above the underlying DNA sequence and this constitutes the epigenome. While the genotype specifying the array of differentiated cells in an organism remains the same, its epigenome is dynamic and sensitive to environmental and developmental signals. In fact, every differentiated cell type has its own epigenome. Despite these complexities, epigenome sequencing projects have been undertaken to precisely map the position of these covalent marks on the chromatin.

DNA methylation is one of the most important modifications in eukaryotes and it most commonly involves addition of methyl groups to the cytosine residues. Cytosine DNA methyltransferases are the enzymes that catalyze the transfer of methyl group from S-adenosyl L-methionine (SAM) to the 5th carbon of cytosines. These enzymes exhibit preferences for either hemi-methylated or non-methylated DNA resulting in either maintenance of the

Address correspondence to: Dr. Meenu Kapoor, University School of Biotechnology, Guru Gobind Singh Indraprastha University, Delhi, India. Tel: 011-23900237; Fax: 011-23865941; E-mail: kapoorsk@genomeindia.org

Narendra Tuteja, Sarvajeet Singh Gill and Renu Tuteja [Eds.]

previously established DNA patterns during replication or establishment of novel, post-replication, patterns of methylation. DNA methylation plays important biological roles in development of both plants and animals. It contributes structurally in the formation of heterochromatin and in defending the genome from invading transposons by suppressing the activity of these elements. It has also been implicated in regulation of gene expression, silencing of genes, imprinting or parent-of-origin-specific silencing of alleles, seed development, transition of plants from vegetative to reproductive stage, flowering time and in mediating responses to abiotic stress. In order to understand the underlying molecular mechanism of DNA methylation that manifests and regulate these vast arrays of responses in eukaryotic genome, it is imperative to first precisely determine the location and distribution of these epigenetic marks at the whole genome level. Various methodologies have been used for high throughput epigenome analysis, each method having its merits and demerits. Affinity purification using monoclonal antibodies against methylated cytosines (anti-m5C antibody) or the DNA recognizing domain of methyl CpG binding proteins have been the method of choice for enrichment of methylated fractions while for unmethylated fractions, CXXC (CAP) affinity purification method has been recently described [4,5]. These enriched fractions when used as probes in genomic arrays, either tiling or expression arrays, give a fair resolution. The affinity purification methods, however, require a higher density of methylated cytosines in the genome and have been successfully applied in analyzing both plant and mammalian genomes. Besides this, methylation-sensitive enzymes that recognize methylated cytosines in their recognition sequences, such as *Mcr*BC, have been used to fractionate the genome into methylated and non-methylated DNA enriched fractions, which are subsequently used as probes for hybridization with microarrays [6]. Treatment of genomic DNA with sodium bisulphate deaminates the non-methylated cytosines into uracil leaving the methylated cytosines unaltered. This method has been the most popular method to map cytosine methylation at single base resolution [7]. One drawback of this method is that its sensitivity is dependent on complete conversion of all cytosine residues and hence on the completion of sodium bisulphate reaction. More recently, however, whole genome sequencing using the sequencing-by-synthesis technology offered by 454 (Roche), Solexa Genome Analyzer (Illumina) and SOLiD (Applied Biosystems, ABI) have become the methods of choice for identifying the covalent modification tags with the sharpest possible resolution [8,9,10].

One basic difference observed between mammalian and plant epigenomes is that in vertebrates, cytosines present exclusively as CpG are methylated, whereas in plants, cytosines present in all the sequence contexts, i.e. CpG, CpNpG or CpNpN have been observed to be methylated [11]. Accordingly, the repertoire of cytosine DNA methyltransferase enzymes in plants is richer than in animals and plant-specific methyltarnsferase that methylate the asymmetric cytosines (CpNpG, CpNpN) are present. Five methyltransferase genes (*Dnmt1, Dnmt3a, Dnmt3b, Dnmt3L* and *Dnmt2*) are present in mammals that encode the enzymes for maintenance and *de novo* methylation of cytosines, while, in plants eleven genes have been identified in *Arabidopsis* and ten in rice that could perform similar functions [12,13,14]. Many of these genes may function redundantly as was observed in the case of *MET1* mutants of *Arabidopsis*. *Met1* mutants displayed a 70% reduction in global methylation patterns and severe developmental abnormalities but still the plants survived and in some cases the transgenics showed limited fertility [15,16]. A global loss of cytosine methylation of this magnitude in animals would prove lethal as was observed in null mutants of *Dnmt1* in mouse [17,18]. These early mutant studies in plants clearly hinted and gave the much needed foresight that plant systems could be used as model systems for DNA methylation studies. However, revealing the role of specific genes in a background of redundant genes has proved to be quite challenging in higher plants due to repeated back crosssings that are required to obtain isogenic lines. Recently, the genome sequence of a simple land plant, *Physcomitrella patens* was released and in the last 3-4 years concerted efforts have generated genetic and genomic resources for high throughput functional genomics studies in this simple land plant [19,20,21]. *Physcomitrella,* by virtue of its dominant haploid phase that alternates with a short sporophytic phase, uncomplicates the forward genetic studies and allows the phenotypes to be analyzed in the same generation without the need for backcrossing as in higher plants.

Studies involving whole genome methylation profiling in vertebrates and plants have also revealed differences in patterns and regions that get methylated in these two systems. In mammals, a global methylation pattern has been observed and all kinds of sequence elements, genic, intergenic, repeats and transposons with the exception of CpG islands in gene promoters, were found to be tagged with methyl groups [22,23]. In plant genomes such as that in *Arabidopsis*, a mosaic pattern of methylation has emerged, which is similar to the pattern seen in invertebrate genomes of sea squirt, *Ciona intestinalis* and in fungi, *Neurospora crasa* [11,24,25]. In these, regions of heavy methylation are interspersed with non-methylated fragments. Methylation of repeat sequences and transposons in plants involve the short interfering RNA (siRNA) and occurs by the process of RNA-directed DNA methylation (RdDM) [26]. In plants with larger genomes, such as maize, though methylation patterns have not been interrogated at whole genome levels, but the presence of large numbers of transposons and their degraded relics indicate a more global pattern unlike that observed in plants with smaller genomes [27,28]. One surprising finding of methylome profiling has been the presence of methylated CpGs in the coding regions (gene body) of genes of both plants and animals. In *Arabidopsis*, while repetitive DNA is specifically targeted for methylation, 33% of all expressed genes, comprising mostly of high percentage of housekeeping genes showed gene body methylation concentrated in the center of the coding regions away from the 5' and 3' ends of genes [11]. This pattern of methylation has been suggested to have role in shutting off the transcriptional noise that may arise due to activation of cryptic initiation sites within the genes in the absence of methylation. This pattern of genic methylation has also been observed in invertebrate and insect genomes. The conservation of this methylation pattern across evolutionary lineages reflects the fact that this was also present in the common ancestors of plants and animals some 1.6 billion years ago.

Histone proteins, two each of H2A, H2B, H3 and H4, form the core of nucleosomes around which 147 bp of DNA is wrapped. The N-terminal regions of H3 and H4 are prominently modified at more than 60 different amino acid residues either before or after their incorporation into nucleosomes [3]. Besides, the addition of dimethyl or trimethyl groups on the same amino acids further contributes to the repertoire of tags and encrypt a coded language on these proteins. Inclusion of histone variants with their own specific modifications along with the combination of various covalent tags on core histones forms the histone code. At any developmental stage the codes on histone proteins and their variants mediate chromatin remodeling, regulate DNA methylation and hence modulate the expression of genes. Incorporation of the histone variant H2A.Z near the 5' ends of the genes has been shown to coordinate transcription in a temperature dependent manner [29]. At ambient temperatures nucleosomes containing H2A.Z at transcription start site (TSS) tend to bind the DNA more tightly than the nucleosome core containing H2A. This prevents the accessibility of DNA to RNA pol II and/or other regulatory proteins from binding to the critical *cis*-elements, thereby, preventing transcription of some genes at this temperature. Alternatively, H2A.Z package the DNA to occlude its accessibility to repressor proteins and DNA methylating complexes thereby allowing transcription of some genes at lower temperatures [24]. At higher temperatures, however, genes showing enhanced transcription tend to accumulate nucleosomes deficient in H2A.Z at TSS while in down regulated genes this loss of H2A.Z exposes the DNA to repressors and DNA methylating complexes thereby preventing gene transcription [30]. Modifications of histone proteins are also sensitive to environmental signals and are altered in response to biotic and abiotic stress conditions.

Non-coding RNAs have emerged as important regulatory molecules that play a major role in shaping the epigenome of a cell. These small RNAs act in concert with components of RNA interference (RNAi) machinery and function as specificity determinants of RNA-induced silencing complexes (RISC) that silence transposons and repeat sequences by recruiting DNA methylation complexes at these sites. RNA directed DNA methylation also targets promoters of genes to invoke Transcriptional Gene Silencing silencing (TGS) [26,31,32]. Post-transcriptional gene silencing (PTGS) mediated by siRNA and microRNAs may not be truly epigenetic as transcript degradation cannot be stably inherited and is more developmental stage/ generation specific.

EPIGENETIC CHANGES DURING ABIOTIC STRESS

In the past decade a series of elegant experiments have implicated abscisic acid (ABA) and its downstream *trans*-activating factors in epigenetic regulation of a number of genes. ABA has been known to play pivotal roles in regulating many plant processes such as seed development, seed dormancy, seed germination, seedling growth, transition of plants from vegetative to reproductive growth, root development and abiotic stress tolerance. The levels of ABA fluctuate during seed development exhibiting peak expression during early embryo maturation and during accumulation of seed proteins, seed dormancy and in response to desiccation. A vast array of transcription factor genes such as plant-specific B-domain protein encoding *ABI3*, *ABSCISIC ACID-INSENSITIVE 3*; *VP1*, *VIVIPAROUS1*; *LEC2*, *LEAFY-COTYLEDON 2*; *FUSCA3* (*FUS3*) and *APETELA2* (*ABI4*), bZIP (*ABI5*) AND *HAP3* subunit of CCAAT binding factor (LEC1) are activated in response to ABA which then coordinate the expression of downstream target genes. This cascade of events has been demonstrated through elegant studies related to orchestrated accumulation of the major seed storage protein phaseolin in the bean plant (*Phaseolis vulgaris*) during seed maturation. This protein is encoded by the *PHAS* gene that expresses only during embryo development while it remains inactive in all the vegetative tissues. The tissue-specific expression of *PHAS* promoter was observed to be stringently regulated by ordered modification of H3 and H4 histone proteins in the region of 3 TATA boxes in its promoter region [33]. This promoter is activated by a seed-specific transcriptional activator PvALF, *Phaseolus vulgaris* ABI3-LIKE FACTOR, which is ABA inducible. A chronological order of chromatin changes has been studied using an estrogen receptor-inducible system that helped to analyze the phased activities of the promoter. This involved a potentiating step, followed by the actual activation of the promoter. In the presence of PvALF alone, remodeling of chromatin was observed at the *PHAS* promoter and this was associated with H3-K9 and H4-K12 acetylation. However, subsequent activation of transcription occurs in the presence of ABA is correlated with increase in trimethylated H3-K4 and a concomitant decrease in acetylation of H3 and H4 at lysine 9 and lysine 12, respectively [34]. Other ABA-responsive genes such as *LEC2*, *FUS3* and *ABAI3* that are involved in accumulation of storage proteins and lipids during seed maturation are known to be down regulated prior to germination by B3 domain containing transcription factors encoded by *VAL* genes. This regulation is mediated by recruitment of CHD3 type chromatin remodeling factors encoded by *PICKLE* [35,25].

Plants subjected to water deficiency stress exhibit adaptive responses by modulating expression of genes encoding histone variants. Though compelling evidence for a structural role for these variants in response to environmental stress is lacking but studies indicate that they modulate physiological and morphological changes in plants. Tomato plants subjected to prolonged drought conditions showed induction of genes encoding H1 histone variant, HIS [36]. *His1-s* antisense plants exhibited higher stomatal conductance, transpiration and photosynthetic rate as compared to wild type plants [37]. Dynamic and reversible modifications of H3 histones have been observed in submerged rice plants. These stressed plants showed specific induction of rice ADH1 and PDC1 genes that correlates with addition of a methyl group to dimethylated H3K4 and an increase in acetylation of H3 at these loci [38]. However, the involvement of chromatin proteins in modulating gene expression of stress-induced genes in response to appearance and disappearance of environmental stress still needs to be elaborated.

Chromatin remodeling mediated by histone modifications such as acetylation/ deacetylation has been shown to play critical roles in regulating expression of many stress regulated genes. In rice, expression of 9 HDACs was observed to be differentially regulated in response to jasmonic acid, salicylic acid and ABA and under abiotic stresses such as salt, osmotic and cold stress [39]. Transcription of many cold-regulated genes in *Arabidopsis* is activated by the transcriptional activator CBF1 that binds to CRT/DRE motifs in the promoter of cold responsive genes. CBF1 has been shown to mediate chromatin regulation by interacting with histone acetyltransferase, GCN5 and transcriptional adaptor proteins ADA2 *in vitro* [40]. In *Arabidopsis,* the plant-specific histone deacetylase, AtHD2C was observed to be involved in ABA, salt, osmotic and drought tolerance [41]. Transgenic plants over-expressing *AtHD2C* showed increased tolerance to salt and drought stress and activation of stress induced *RD29B* and *RAB18* genes in seeds.

Deacetylation of histone H4 by HDAC, complexed with a WD-40 repeat protein, HOS15, represses stress-regulated genes *RAD29A*, *COR15A* and *ADH*. HOS15 specifically interacts with H4 and mediates deacetylation and repression of stress-regulated genes. Though the expression of many stress-regulated genes is enhanced in *hos15* mutants, its tolerance towards salt stress, exogenous ABA, heat and oxidative stress by H_2O_2 remains unchanged while it shows hypersensitivity to freezing temperature [42].

Under environmental stress, plants tend to respond by adapting to stress which could also be manifested by changes in gene expression patterns mediated by differential DNA methylation. Maize seedlings show genome-wide hypomethylation specifically in the root tissues where m5C content of the genome declines from 38.4% to 24.7% when plants are subjected to 5 days of chilling. The m5C content further declines to 22.5% after plants are allowed to recover at 23°C for 7 days. This demethylation occurs in a non-random manner with DNA wrapped around the nucleosome core undergoes hypomethyaltion, while the linker DNA remains hypermethylated. This alternative pattern of methylation, with hypo-methylation of core DNA and hypermethylation of linker DNA, has been shown to affect the expression of specific genomic fragments, such as those encoding a putative reterotransposon and a coding sequence leading to their specific activation under cold stress [43]. Similarly, active demethylation of glycerophosphodiesterase-like genes (*NtGPDC*) has been observed in detached tobacco leaves treated with aluminum or subjected to salt and low temperature stress. CpGs at *NtGPDC* loci showed specific demethylation which correlated well with its specific induction under stress [44].

Small non-coding RNAs (siRNA and microRNAs) have emerged as key regulators of gene expression mediating their effect through target degradation, translational arrest or chromatin remodeling. These small RNA molecules are known to be induced in response to environmental stress and are involved in adaptive responses manifested by plants. Many microRNA genes are differentially expressed in response to cold, dehydration and salt stress in *Arabidopsis* and rice [45,46,47]. These small RNAs have been implicated in gene silencing pathways where they function as specificity determinants of RISC and RITS (RNA-induced transcriptional silencing complex) leading to methylation of repetitive sequences and transposons through RNA-directed DNA methylation process [48,49]. The key components involved in this process include Dicer-like proteins (DCL), Argonautes (AGO), RNA-dependent RNA polymerases (RdRP) and DNA methylating enzymes, cytosine DNA methyltransferases (DNMT). Genes encoding these proteins have been identified and characterized in rice and *Arabidopsis* [50,51,52,53]. Microarray-based gene expression analysis of these genes under three abiotic stress conditions (salt, drought and cold) revealed differential expression profiles in rice. Expression of 1 DCL, 7 Argonautes, 3 RdRP and 6 Cytosine DNA methyltransferases were observed to be affected under stress (Fig. **1**). Interestingly, *OsAGO3*, *OsRDR4* and *OsRDR3* were observed to be specifically activated in response to dehydration stress while, among cytosine DNA methyltransferases, except *OsCMT2*, that shows enhanced transcript accumulation in response to cold and salt stress, all other cytosine methyltransferase exhibited a downward trend in expression under the abiotic stress conditions analyzed [50,51]. Recent studies have shown that small RNA mediated DNA methylation and adaptive responses of plants under abiotic stress can be transmitted through meiotic divisions and maintained transgenerationally [54]. *Arabidopsis* plants subjected to salt, flood, cold and UVC stresses exhibit 2-6 fold increase in homologous recombination frequency as compared to wild-type untreated plants. This trait gets transmitted to the F1 progeny, which also shows enhanced tolerance to stress under control conditions. Both increase in homologous recombination and stress tolerance, however, declines in subsequent generations. These adaptive changes correlate well with differential DNA methylation in parents and in the F1 progeny but are not maintained in F2 and later generations. Both homologous recombination frequency and DNA methylation are impaired in *dcl2 dcl3* double mutants, whereas, *dcl2* mutation alone affects DNA methylation, indicating the involvement of siRNA-mediated DNA methylation and possible chromatin reorganization playing a key role in this transgenerational affect in plants [54].

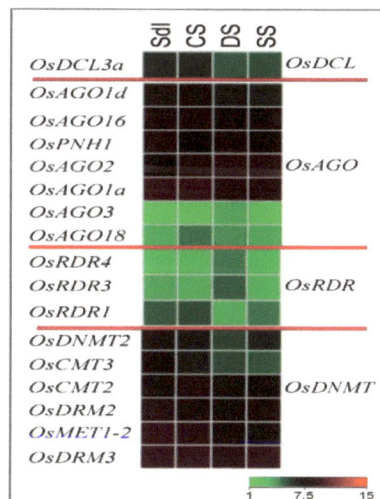

Figure 1: Microarray-based expression profiling of rice Dicer-like (*OsDCL*), argonautes (*OsAGO*), RNA-dependent RNA Polymerase (*OsRDR*) and cytosine DNA methyltransferase (*OsDNMT*) genes under three abiotic stress conditions. The color scale bar below represents the log to the base 2 expression values. Sdl, seedling; CS, cold stress; DS, drought stress; SS, salt stress.

CONCLUSIONS

There is convincing evidence to support some level of association between stress tolerance and epigenetics. Although, short term adaptation to stress may be achieved by genic level regulatory mechanisms, epigenetics seems to be involved in creating relatively long term memory of the stress experience by involving altered DNA methylation and smRNA silencing pathways, which may lead to anticipatory preparedness for a particular stress.

Recent studies have shown the stress induced epigenetic changes to be transgenerational for at least one generation, thereby, proving that these changes are able to cross the meiotic barrier and at the same time be limited to fewer number generations to minimize the penalty on plant vigor. It will be, however, for the future research to elucidate whether epigenomic changes in fact have an adaptive significance in combating abiotic stresses.

REFERENCES

[1]   Waddington CH. The Epigenotype. Endeavour 1942; 1:18–20.
[2]   Goldberg AD, Allis CD, Bernstein E. Epigenetics: A landscape takes shape. Cell 2007; 128: 635-638.
[3]   Kouzarides T. Chromatin modifications and their function. Cell 2007; 128: 693–705
[4]   Cross, SH, Charlton, JA, Nan, X, Bird, AP Purification of CpG islands using a methylated DNA binding column. Nat Genet 1994; 6: 236–244.
[5]   Illingworth, R *et al.* A novel CpG island set identifies tissue-specific methylation at developmental gene loci. PLoS Biol 2008; 6: e22.
[6]   Schumacher A *et al.* Microarray-based DNA methylation profiling: technology and applications. Nucleic Acids Res 2006; 34: 528–542.
[7]   Frommer, M *et al.* A genomic sequencing protocol that yields a positive display of 5-methylcytosine residues in individual DNA strands. Proc Natl Acad Sci USA 1992; 89: 1827–1831.
[8]   Fan, J B *et al.* Illumina universal bead arrays. Methods Enzymol 2006; 410: 57–73.
[9]   Korshunova, Y *et al.* Massively parallel bisulphate pyrosequencing reveals the molecular complexity of breast cancer-associated cytosine-methylation patterns obtained from tissue and serum DNA. Genome Res 2008; 18: 19–29.
[10]   Cokus *et al.* Shotgun bisulphite sequencing of the Arabidopsis genome reveals DNA methylation patterning. Nature 2008; 452: 215-219.
[11]   Zhang *et al.* Genome-wide high-resolution mapping and functional analysis of DNA methylation in arabidopsis. Cell 2006; 126: 1–13.
[12]   Chen T and Li E. Structure and function of eukaryotic DNA methyltransferases. Curr Top Dev Biol 2004; **60**:55–89
[13]   Ponger L and Li WH. Evolutionary diversification of DNA methyltransferases in eukaryotic genomes. Mol Biol Evol 2005; 22:1119-1128.
[14]   Sharma R, Mohan Singh RK, Malik G, Deveshwar P, Tyagi AK, Kapoor S and Kapoor M. Rice Cytosine DNA Methyltransferases: Gene Expression Profiling during Reproductive Development and Abiotic Stress. FEBS J 2009; 276: 6301-6311
[15]   Finnegan EJ, Peacock, WJ, Dennis, ES. Reduced DNA methylation in Arabidopsis thaliana results in abnormal plant development. Proc Natl Acad Sci USA 1996; 93: 8449-54.
[16]   Kankel MW, Ramsey DE, Stokes TL, Flowers SK, Haag JR, Jeddeloh JA, Riddle NC, Verbsky ML, Richards EJ. Arabidopsis MET1 cytosine methyltransferase mutants. Genetics 2003; 163: 1109-1122.
[17]   Li E, Bestor TH, Jaenisch R. Targeted mutation of the DNA methyltransferase gene results in embryonic lethality. Cell 1992; 69: 915–926
[18]   Okano M, Bell DW, Haber DA, Li E. DNA methyltransferases Dnmt3a and Dnmt3b are essential for *de novo* methylation and mammalian development. Cell 1999; 99: 247–257.
[19]   Frank W, Decker DL, Reski R. Tools to study Physcomitrella. Plant Biol 2005; 7: 220-227.
[20]   Reski R, Frank W. Moss (*Physcomitrella patens*) functional genomics - Gene discovery and tool development with implications for crop plants and human health. Briefings in Functional Genomics and Proteomics 2005; 4: 48-57.
[21]   Rensing SA. *et al.* The Physcomitrella genome reveals evolutionary insights into the conquest of land by plants. Science 2008; 319: 64.
[22]   Rabinowicz PD, Palmer LE, May BP, Hemann MT, Lowe SW, McCombie WR, Martienssen RA. Genes and transposons are differentially methylated in plants, but not in mammals. Genome Res 2003; 13: 2658-2664.
[23]   Eckhardt, F. *et al.* DNA methylation profiling of human chromosomes 6, 20 and 22. Nat Genet 2006; 38: 1378–1385.
[24]   Zilberman D, Coleman-Derr D, Ballinger T, Henikoff S. Histone H2A.Z and DNA methylation are mutually antagonistic chromatin marks. Nature 2008; 456: 125-129.
[25]   Suzuki M, Wang HHY, McCarty DR. Repression of the LEAFY COTYLEDON 1/B3 regulatory network in plant embryo development by VP1/ABSCISIC ACID INSENSITIVE 3-LIKE B3 genes. Plant Physiol 2007; 143: 902–911.
[26]   Mette MF, Aufsatz W, van der Winden J, Matzke MA, Matzke AJ Transcriptional silencing and promoter methylation triggered by double-stranded RNA. EMBO J 2000; 19: 5194–5201
[27]   SanMiguel P, Tikhonov A, Jin YK, Motchoulskaia N, Zakharov D, Melake-Berhan A, Springer PS, Edwards KJ, Lee M, Avramova Z, Bennetzen JL. Nested retrotransposons in the intergenic regions of the maize genome. Science 1996; 274: 765-768.
[28]   Palmer LE, Rabinowicz PD, O'Shaughnessy AL, Balija VS, Nascimento LU, Dike S, de la Bastide M, Martienssen RA, McCombie WR. Maize genome sequencing by methylation filtration. Science 2003; 302: 2115-2117.

[49]   Wassenegger M, Heimes S, Riedel L, Sanger HL. RNA-directed de novo methylation of genomic sequences in plants. Cell 1994; 76: 567-576.

[50]   Zhu J, Jeong JC, Zhu Y, Sokolchik I, Miyazaki S, Zhu J-K, Hasegawa PM, Bohnert, HJ, Shi H, Yun D-J, Bressan RA. Involvement of Arabidopsis HOS15 in histone deacetylation and cold tolerance. Proc Natl Acad Sci USA 2008; 105: 4945-4950.

[29]   Kumar SV, Wigge PA. H2A.Z-containing nucleosomes mediate the thermosensory response in *Arabidopsis*. Cell 2010; 140: 136-147.

[30]   Meneghini MD, Wu M, Madhani HD. Conserved histone variant H2A.Z protects euchromatin from the ectopic spread of silent chromatin. Cell 2003; 112: 725-736.

[31]   Chan SW. *et al.* RNA silencing genes control *de novo* DNA methylation. Science 2004; 303: 1336

[32]   Chan SW, Henderson IR, Jacobsen SE. Gardening the genome: DNA methylation in *Arabidopsis thaliana*. Nat Rev Genet 2005; 6: 351–360.

[33]   Li G, Chandler SP, Wolffe AP, Hall, TC. Architectural specificity in chromatin structure at the TATA box in vivo: Nucleosome displacement upon b-phaseolin gene activation. Proc Natl Acad Sci USA 1998; 95: 4772-4777.

[34]   Ng DW-K, Chandrasekharan MB, Hall TC. Ordered Histone Modifications Are Associated with Transcriptional Poising and Activation of the phaseolin Promoter. Plant Cell 2006; 18: 119-132.

[35]   Ogas J, Kaufmann S, Henderson J, Somerville C. PICKLE is a CHD3 chromatin-remodeling factor that regulates the transition from embryonic to vegetative development in *Arabidopsis*. Proc Natl Acad Sci USA 1999; 96: 13839-13844.

[36]   Kahn TL, Fender SE, Bray EA, O'Connell MA. Characterization of Expression of Drought-and Abscisic acid-Regulated Tomato genes in the Drought-resistant Species of *Lycopersicon pennellii*. Plant Physiol 1993; 103: 597-605.

[37]   Scippa GS, Michele MD, Onelli E, Patrignani G, Chiatante D, Bray EA. The histone-like protein H1-S and the response of tomato leaves to water deficit. J Exp Botany 2004; 55: 99-109.

[38]   Tsuji H, Saika H, Tsutsumi N, Hirai A, Nakazono M. Dynamic and reversible changes in histone H3-Lys4 methylation and H3 acetylation occurring at submergence-inducible genes in rice. Plant Cell Physiol 2006; 47:995-1003.

[39]   Fu W, Wu K, Duan J. Sequence and expression analysis of histone deacetylases in rice. Biochem Biophys Res Commun 2007; 356: 843-850.

[40]   Stockinger EJ, Mao Y, Regier MK, Triezenberg SJ, Thomashow MF. Transcriptional adaptor and histone acetlytransferase proteins in *Arabidopsis* and their interactions with CBF1, a transcriptional activator involved in cold-regulated gene expression. Nucleic Acids Res 2001; 29: 1524-1533.

[41]   Sridha S, Wu K. Identification of AtHD2C as a novel regulator of abscisic acid responses in *Arabidopsis*. Plant J 2006; 46: 124-133.

[42]   Zhu *et al* Involvement of Arabidopsis HOS15 in histone deacetylation and cold tolerance . Proc Natl Acad Sci 2007; 105: 4945-50.

[43]   Steward N, Ito M, Yamaguchi Y, Koizumi N, Sano H. Periodic DNA Methylation in Maize Nucleosomes and Demethylation by Environmental Stress. J Biol Chem 2002; 277: 37741-37746.

[44]   Choi C-S, Sano H. Abiotic-stress induces demethylation and transcriptional activation of a gene encoding a glycerophosphodiesterase-like protein in tobacco plants. Mol Gen Genomics 2007; 277: 589-600.

[45]   Sunkar R, Zhu JK. Novel and stress-regulated microRNAs and other small RNAs from Arabidopsis. Plant Cell 2004; 16: 2001–2019.

[46]   Shukla LI, Chinnusamy V, Sunkar R. The role of microRNAs and other endogenous small RNAs in plant stress responses. Biochim Biophy Acta 2008; 1779: 743-748.

[47]   Sanan-Mishra N, Kumar V, Sopory SK, Mukherjee SK. Cloning and Validation of novel miRNA from basmati rice indicates cross-talk between abiotic and biotic stresses. Mol Gen Genomics 2009; 282: 463-467.

[48]   Wassenegger M, Heimes S, Riedel L and Sänger HL. RNA-directed de novo methylation of genomic sequences in plants. Cell 1994; 76: 567-576.

[49]   Matzke MA, Birchler JA. RNAi-mediated pathways in the nucleus. Nat Rev Genet 2005; 6: 24-35.

[50]   Kapoor M *et al.* Genome-wide identification, organization and phylogenetic analysis of Dicer-like, Argonaute and RNA-dependent RNA Polymerase gene families and their expression analysis during reproductive development and stress in rice. BMC Genomics 2008; 9: 451.

[51]   Sharma R *et al.* Rice Cytosine DNA Methyltransferases: Gene Expression Profiling during Reproductive Development and Abiotic Stress. FEBS J 2009; 276: 6301-6311.

[52]   Margis R *et al.* The evolution and diversification of Dicers in plants. FEBS Lett 2006; 580: 2442-2450.

[53]   Nonomura K, Morohoshi A, Nakano M *et al.* A germ cell specific gene of the ARGONAUTE family is essential for the progression of premeiotic mitosis and meiosis during sporogenesis in rice. Plant Cell 2007; 19: 2583-2594.

[54]   Boyko *et al.* Transgenerational adaptation of Arabidopsis to stress requires DNA methylation and the function of Dicer-like proteins. PLoS One 2010; 5: e9514.

CHAPTER 9

Rhizotoxic Ions: 'Omics' Approaches for Studying Abiotic Stress Tolerance in Plants

Cheng-Ri Zhao, Yoshiharu Y Yamamoto and Hiroyuki Koyama*

Applied Plant Science, Faculty of Applied Biological Sciences, Gifu University, Gifu, 501-1193, Gifu, Japan

Abstract: Rhizotoxic ions inhibit root growth, resulting in reduced yield of various crop plants. It is thus improvement of tolerance of roots to rhizotoxic ions, one of the most important targets in plant breeding, to improve the productivity of crops in various soil types. Molecular breeding such as marker assisted selection and transgenic breeding would be promising approaches in current plant breeding, while identification of critical genes is needed to realize these approaches. In this chapter, we introduce recent progress of transcriptomics and other –omics studies for complex response of plant roots to rhizotoxic ions, and those that for identifying critical genes for tolerance to rhizotoxic ions.

INTRODUCTION

Root development is one of the most important factors in plant growth, because roots supply the above-ground parts of the plant with water and nutrients [1]. In the natural environment, however, roots are easily disturbed by various factors in soils, such as rhizotoxic ions [2]. For example, aluminum (Al) in acid soils severely inhibits root growth. This is one of the most serious stress factors in acid soils, and results in decreased yields of various crop plants [1]. Some other ions, such as sodium (Na), easily translocate from the roots to the shoots. Accumulation of such ions in the shoots can also lead to yield losses [3]. However, these translocatable ions are also toxic to roots at low concentrations. Some other heavy metal ions, such as copper (Cu) and cadmium (Cd), also have rhizotoxic effects on plants. Therefore, improving the tolerance of roots to rhizotoxic ions is an important topic in plant breeding, to improve the productivity of crops growing in various problem soils.

Physiological studies indicate that plant roots have various mechanisms of tolerance to rhizotoxic ions. For example, there are two distinct mechanisms of tolerance to Al rhizotoxicity; internal Al resistance and Al-exclusion [1]. Various tolerance mechanisms have been also identified for other rhizotoxic ions, such as maintenance of Na^+/K^+ homeostasis in NaCl tolerance [4] and synthesis of binding molecules in heavy metal tolerance [5]. Because some tolerance mechanisms are linked to induction of gene expression, transcriptome analyses are a useful approach for identifying genes that are critical for tolerance to rhizotoxins. In fact, a number of studies have identified genes that are responsive to rhizotoxic ions, and those that are critical for tolerance. Some of these studies have combined transcriptomic with other '-omics' approaches and bio-informatics. In this chapter, we focus on recent progress in transcriptomic and other '-omics' studies in identifying genes that are responsive to rhizotoxic ions.

RHIZOTOXICITY

Hydroponic cultures are frequently used to determine the effects of rhizotoxins on growing roots. It is notable that the severity of cationic rhizotoxins in hydroponic cultures is affected by co-existing cations and the pH of the medium [2]. Cationic rhizotoxins are concentrated at the plasma membrane (PM) surface because of its negative charge, which is generated by weak acid ligands such as phospholipids of the membrane lipid [6]. The negative charge of the PM surface decreases at acidic pH (pH<5.5) because negatively charged ligands are protonated. In addition, high levels of co-existing cations, such as Ca^{2+}, neutralize the negative charge of the PM, and thus, can alleviate toxicity of cationic rhizotoxins. Al rhizotoxicity is a typical example that fits the above model. Trivalent Al^{3+} is the most toxic species among the monomeric forms of Al (i.e. Al^{3+}, $Al(OH)^{2+}$, $Al(OH)_2^+$, $Al(OH)_3^0$). Activity of bulk-phase Al^{3+}, $\{Al^{3+}\}_{bulk}$, decreases at pH >4.5, but Al toxicity becomes severe around pH 5.0 because $\{Al^{3+}\}_{PM}$

---

*Address correspondence to: Dr. Hiroyuki Koyama, Faculty of Applied, Biological Sciences, Gifu University, 1-1, Yanagido, Gifu, 501-1193, Gifu, Japan; Tel(Fax): 81-58-293-2911; E-mail: koyama@gifu-u.ac.jp

Narendra Tuteja, Sarvajeet Singh Gill and Renu Tuteja [Eds.]

increases with increasing PM negativity due to increasing negativity of the PM surface [7]. On the other hand, a high concentration of $\{Ca^{2+}\}_{PM}$ alleviates Al rhizotoxicity by neutralizing the negative charge of the PM surface [6]. Toxicity of other rhizotoxic cations is regulated by the same model.

For studies on rhizotoxicity, low ionic strength solutions would be preferable to regular solutions, because they would enhance toxicity and could minimize any indirect effects during treatments due to unusual high concentration of toxicants. However, low ionic strength solutions could not be used for long-term treatments because they may induce nutrient deficiency. Therefore, one of the best methods to evaluate the effects of rhizotoxins is the use of short-term treatments in low ionic strength solutions. Transcriptomic responses Arabidopsis thaliana roots obtained using this experimental system will be summarized in the following sections.

TRANSCRIPTOMIC RESPONSE OF THE ROOTS OF ARABIDOPSIS THALIANA TO VARIOUS RHIZOTOXIC IONS

Physiological studies indicate that there are both common and specific responses to rhizotoxic ions in roots. For example, accumulation of reactive oxygen species (ROS) commonly occurs when roots are exposed to various rhizotoxins [8]. On the other hand, some responses are specific to particular stressors, for example, the release of organic acids from roots in response to Al-rhizotoxicity [9,10]. Identification of the molecular basis of these common and specific responses to rhizotoxins is one of the most interesting outcomes of studies using '-omics' approaches. Comparison of plants' transcriptomes among various treatments is one approach to identify specific and common responses induced by various stressors. This approach was used to identify a group of genes that are all induced by drought, cold and high-salinity treatments. These genes are regulated by the same transcription system, which is controlled by DREB (dehydration responsive element binding protein) transcription factors [11]. A similar approach was used to determine the specific and common responses of *Arabidopsis* to rhizotoxic stressors [8]. In addition, genes associated with damage and those associated with resistance could be identified by comparing the transcriptomes between sensitive and resistant cultivars, or between mutant and wild-type plants. This concept has been applied to identify genes associated with resistance to rhizotoxic stressors, and those associated with damage, as summarized in the following subsections.

Comparative Transcriptomics among Different Rhizotoxic Treatments

Recently, we performed a comparative microarray in Arabidopsis to identify specific and common gene expression responses to rhizotoxic ion treatments [8]. The data sets were obtained from root samples exposed to short-term, low ionic strength Al, NaCl, Cd, and Cu toxic solutions. Briefly, rhizotoxic treatments were applied at the same level of severity and for the same duration (90% inhibition of root growth; 24 h). We compared genes that were most reproducibly and most highly up-regulated, *i.e.,* genes showing the uppermost 2.5% of -fold change values in three independent replications. In this experimental design, specific responsive genes are grouped together with other genes that might be associated with resistance or damage responses to the particular stressor (see Fig. **1**). For example,

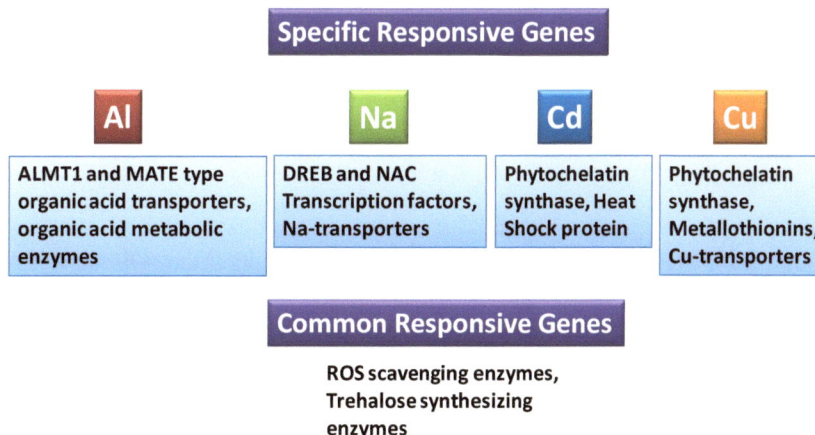

Figure 1: Typical responsive genes to various rhizotoxic ions.

the Al-specific gene group contained many genes encoding transporters, including a known Al-resistance gene of *Arabidopsis*, *AtALMT1* [9,10], which encodes an Al-activated malate transporter. The NaCl-specific gene group contained many transcription factors including *DREB*, while the Cu-specific group contained a gene encoding metallothionein and various genes encoding catalytic enzymes in secondary metabolism [8]. The Cd-specific group contained genes for heat shock proteins (HSPs); however, the roles of HSPs in Cd resistance remain unclear. On the other hand, several genes encoding ROS scavenging enzymes (*e.g.*, peroxidases and superoxide dismutases) were induced by all stressors. Thus, ROS accumulation is a common response to all of these rhizotoxic ions [8,12-14]. Some of these scavenging enzymes might be involved in defense mechanisms to all rhizotoxic stressors. The contributions of these scavenging enzymes to resistance to rhizotoxic ion treatments have been discussed in other transcriptomic studies [15,16].

Comparison of Transcriptome Between Cultivars Contrasting in Resistance

Studies in molecular physiology and genetics indicate that the expression level of critical genes is correlated with phenotypic differences among varieties and ecotypes [17]. For example, the degrees of Cu- and Cd-resistance of *Silene vulgaris* and *Arabidopsis halleri* were correlated with the expression of phytochelatin synthase, metallochioneis and a putative zinc translocator [18–20]. These studies suggest that comparison of transcriptomes between resistant and sensitive cultivars could be used

to identify which genes are associated with resistance and which are associated with damage. Among the inducible genes, the expression level of resistance genes would be greater in resistant cultivars, while that of genes associated with damage would be greater in sensitive cultivars. This approach has been used in some plant species to identify genes associated with resistance to rhizotoxic ions. For example, transcriptomes were compared between Al-tolerant and Al-sensitive cultivars of wheat [21] and maize [22]. The results of these comparisons showed that some genes encoding enzymes in organic acid metabolism were greatly induced by rhizotoxic stress in resistant cultivars, but less so in sensitive cultivars. These changes may reflect the greater capacity of Al-resistant cultivars to release organic acids, which is a critical factor in Al resistance [23]. Comparison between wild-type and sensitive mutants is another powerful approach to identify genes causing the mutation and those associated with damage in the sensitive mutants. For example, Al treatments up-regulated expression of various genes in an *Arabidopsis* mutant that is sensitive to both Al and proton rhizotoxicities (STOP1: sensitive to proton rhizotoxicity) [24]. Many of the up-regulated genes were also up-regulated in a different Al-sensitive mutant (*i.e.,* the Al-activated malate transporter knock-out mutant). These genes may be associated with damage induced by Al.

BIOINFORMATICS AND INTEGRATIVE '-OMICS' APPROACHES IN RHIZOTOXIC STRESS RESEARCH

Rhizotoxic stress induces complex responses among many biological processes. In fact, rhizotoxic treatments modify transcriptome, while it is concomitant with the change of proteome (e.g. Al for tomato [25] and soybean roots [26], Cd for Arabidopsis roots [27]) and metabolome [28]. It is thus application of bioinformatics and integrated '-omics' would be useful to understand such complex systems.

The transcriptome of plants responding to a particular rhizotoxic stressor can be further analyzed to identify signal transduction pathways and gene-to-gene interactions. For example, the Microsoft-excel based free program KAGIANA (Kazusa, gene expression analysis tool) [29] has been used to identify co-expression of genes in *Arabidopsis* in response to various stressors. This tool was used to identify members of a gene cluster that was up-regulated in response to multiple rhizotoxins. Many of the up-regulated genes encode various ROS-scavenging enzymes and ROS-related stress-responsive transcription factors (*e.g., MYB15* and an unidentified ZAT zinc finger protein containing an EAR repressor domain) [8] were grouped as co-expression gene cluster in commonly up-regulated genes groups by various rhizotoxic ions. This could account for previous physiological studies on rhizotoxicity that have identified that ROS accumulation is a common response to various treatments. There are other web-tools that can be used to analyze gene co-expression, such as the ATTED-II database (details at ATTED-II website; http://www.atted.bio.titech.ac.jp/) [30]. In addition, GENEVESTIGATOR is useful for profiling genetic responses to various treatments, and is useful for identifying gene clusters and their responses to other stressors [31]. Various web-tools, *e.g.*, Melina (http://melina2.hgc.jp/public/index.html) [32] and others, are available to identify *cis*-elements that are responsive to particular stressors. Some databases, such as ppdb (plant promoter database) [33], provide potential sequences of *cis*-elements that may be responsive to particular stressors. Integrating bioinformatic analyses with these tools would be useful to identify gene-to-gene interactions and signal transduction pathways related to rhizotoxic stressors.

Some rhizotoxic stressors can alter metabolism, resulting in production of metabolites that protect plant tissues. For example, organic acid metabolism is linked to the Al-responsive organic acid release [34], while metabolic changes associated with NaCl resistance result in accumulation of spermidine [35]. These metabolic changes can be visualized by integrating the transcriptome or/and metabolome on a metabolic pathway map using web-available tools such as MapMan (http://gabi.rzpd.de/projects/MapMan/) [36] and KaPPA-View (Kazusa Plant Metabolic Pathway Viewer; http://kpv.kazusa.or.jp/kappa-view/) [37]. This approach was used to show that accumulation of trehalose, a ROS-protectant, is commonly induced by rhizotoxic stressors as well as proline accumulation [38], and that a gamma-amino butyric acid shunt is responsive to Al stress, possibly protecting cells from Al-induced acidosis [24, 39]. These reports showed that the combination of transcriptomics and other –omics analyses is powerful approach for understanding the complex nature of resistance to and toxicity of rhizotoxic stressors.

CONCLUSIONS

In conclusion, -omics approaches have become the most powerful approaches to understand resistance to rhizotoxins. Although these approaches are limited for some model plants, progress in genomic sequencing projects and development of the the next-generation sequencing technologies (e.g. Illumina sequencer) may allow application of these approaches to various plant species.

REFERENCES

[1]   Kochian LV, Hoekenga OA, Pineros MA. How do crop plants tolerate acid soils? mechanisms of aluminum tolerance and phosphorous efficiency. Annu Rev Plant Biol 2004; 55: 459-493.

[2]   Kinraide TB, Pedler JF, Parker DR. Relative effectiveness of calcium and magnesium in the alleviation of rhizotoxicity in wheat induced by copper, zinc, aluminum, sodium, and low pH. Plant Soil 2004; 259: 201-208.

[3]   Berthomieu P *et al.* Functional analysis of *AtHKT1* in *Aabidopsis* shows that Na$^+$ recirculation by the phloem is crucial for salt tolerance. EMBO J 2003; 22(9): 2004-2014.

[4]   Shi H, Ishitani M, Kim C, Zhu JK. The *Abidopsis thaliana* salt tolerance gene *SOS1* encodes a putative Na$^+$/H$^+$ antiporter. Proc Natl Acad Sci USA 2000; 97(12): 6896-6901.

[5]   Steffens J. The heavy metal-binding peptides of plants. Annual Review of Plant Biol 1990; 41: 553-575.

[6]   Kinraide TB. Three mechanisms for the calcium alleviation of mineral toxicities. Plant Physiol 1998; 118: 513-520.

[7]   Kinraide TB. Toxicity factors in acidic forest soils: attempts to evaluate separately the toxic effects of excessive Al^{3+} and H$^+$ and insufficient Ca^{2+} and Mg^{2+} upon root elongation. Eur J Soil Sci 2003; 54(2): 323-333.

[8]   Zhao CR, Ikka T, Sawaki Y *et al.* Comparative transcriptomic characterization of aluminum, sodium chloride, cadmium and copper rhizotoxicities in *Arabidopsis thaliana*. BMC Plant Biol 2009; 9: doi:10.1186/1471-2229-9-32.

[9]   Hoekenga OA *et al.* *AtALMT1*, which encodes a malate transporter, is identified as one of several genes critical for aluminum tolerance in *Arabidopsis*. Proc Natl Acad Sci USA 2006; 103(25): 9738-9743.

[10]   Kobayashi Y *et al.* Characterization of *AtALMT1* expression in aluminum-inducible malate release and its role for rhizotoxic stress tolerance in Arabidopsis. Plant Physiol 2007; 145: 843-852.

[11]   Seki M *et al.* Monitoring the expression profiles of 7000 *Arabidopsis* genes under drought, cold and high-salinity stresses using a full-length cDNA microarray. Plant J 2002; 31: 279-292.

[12]   Sharma SS, Dietz KJ. The relationship between metal toxicity and cellular redox imbalance. Trends Plant Sci 2009; 14: 43-50.

[13]   Heidenreich B, Mayer K, Sandermann Jr H, Ernst D. Mercury-induced genes in *Arabidopsis thaliana*: Identification of induced genes upon long-term mercuric ion exposure. Plant, Cell Environ 2001; 24: 1227-1234.

[14] Kawaura K, Mochida K, Yamazaki Y, Ogihara Y. Transcriptome analysis of salinity stress responses in common wheat using a 22k oligo-DNA microarray. Funct Integr Genomics 2006; 6: 132-142.

[15] Jiang Y, Yang B, Harris NS, Deyholos MK. Comparative proteomic analysis of NaCl stress-responsive proteins in *Arabidopsis* roots. J Exp Bot 2007; 58(13): 3591-3607.

[16] Jiang Y, Deyholos MK. Comprehensive transcriptional profiling of NaCl-stressed *Arabidopsis* roots reveals novel classes of responsive genes. BMC Plant Biol 2006; 6: 25 doi:10.1186/1471-2229-6-25

[17] El-Din El-Assal S, Alonso-Blanco C, Peeters AJ, Raz V, Koornneef M. A QTL for flowering time in *Arabidopsis* reveals a novel allele of *CRY2*. Nat Genet 2001; 29: 435-440.

[18] van Hoof N *et al.* Enhanced copper tolerance in *silene vulgaris* (moench) garcke populations from copper mines is associated with increased transcript levels of a 2b-type metallothionein gene. Plant Physiol 2001; 126: 1519-1526.

[19] Hanikenne M *et al.* Evolution of metal hyperaccumulation required *cis*-regulatory changes and triplication of *HMA4*. Nature 2008; 453: 391-395.

[20] Becher M, Talke IN, Krall L, Kramer U. Cross-species microarray transcript profiling reveals high constitutive expression of metal homeostasis genes in shoots of the zinc hyperaccumulator *Arabidopsis halleri*. Plant J 2004; 37: 251-268.

[21] Guo P, Bai G, Carver B, Li R, Bernardo A, Baum M. Transcriptional analysis between two wheat near-isogenic lines contrasting in aluminum tolerance under aluminum stress. Mol Genet Genom 2007; 277: 1-12.

[22] Maron LG, Kirst M, Mao C, Milner M, Menossi M, Kochian LV. Transcriptional profiling of aluminum toxicity and tolerance responses in maize roots. New Phytol 2008; 179: 116-128.

[23] Kochian LV. Cellular mechanisms of aluminum toxicity and resistance in plants. Ann Rev Plant Physiol Plant Mol Biol 1995; 46: 237-260.

[24] Sawaki Y, Iuchi S, Kobayashi Y *et al.* STOP1 regulates multiple genes that protect arabidopsis from proton and aluminum toxicities. Plant Physiol 2009; 150: 281-294.

[25] Zhou S, Sauve R, Thannhauser TW Proteome changes induced by aluminium stress in tomato roots. J Exp Bot 2009; 60: 1849-1857

[26] Zhen Y *et al.* Comparative proteome analysis of differentially expressed proteins induced by Al toxicity in soybean. Physiol Plant 2007; 131: 542-554

[27] Roth U, von Roepenack-Lahaye E, Clemens S. Proteome changes in Arabidopsis thaliana roots upon exposure to Cd^{2+}. J Exp Bot 2006; 57: 4003-4013.

[28] Jahangir M, Abdel-Farid IB, Choi YH, Verpoorte R Metal ion-inducing metabolite accumulation in Brassica rapa. J Plant Physiol 2008; 165: 1429-1437.

[29] Ogata Y *et al.* KAGIANA: An excel-based tool for retrieving summary information on *Arabidopsis* genes. Plant Cell Physiol 2009; 50: 173-177.

[30] Obayashi T *et al.* ATTED-II: a database of co-expressed genes and *cis* elements for identifying co-regulated gene groups in *Arabidopsis*. Nucleic Acids Res 2007; 35: D863-9.

[31] Zimmermann P, Hirsch-Hoffmann M, Hennig L, Gruissem W. GENEVESTIGATOR. Arabidopsis microarray database and analysis toolbox. Plant Physiol 2004; 136: 2621-2632.

[32] Poluliakh N, Takagi T, Nakai K. Melina: motif extraction from promoter regions of potentially co-regulated genes. Bioinformatics 2003; 19(3): 423-4.

[33] Yamamoto YY, Obokata J. ppdb: a plant promoter database. Nucleic Acids Res 2008; 36: D977-81.

[34] Ma JF, Ryan PR, Delhaize E. Aluminium tolerance in plants and the complexing role of organic acids. Trends Plant Sci 2001; 6: 273-278.

[35] Shevyakova N, Strogonov B, Kiryan I. Metabolism of polyamines in NaCl-resistant cell lines from *nicotiana sylvestris*. Plant Growth Regul 1985; 3: 365-369.

[36] Thimm O *et al.* MAPMAN: a user-driven tool to display genomics data sets onto diagrams of metabolic pathways and other biological processes. Plant J 2004; 37: 914-939.

[37] Tokimatsu T *et al.* KaPPA-View: a web-based analysis tool for integration of transcript and metabolite data on plant metabolic pathway maps. Plant Physiol 2005; 138: 1289-1300.

[38] Kaul S, Sharma SS, Mehta IK. Free radical scavenging potential of L-proline: evidence from *in vitro* assays. Amino Acids, 2008; 34: 315-320.

[39] Zhao CR, Sawaki Y, Sakurai N, Shibata D, Koyama H. Transcriptomic profiling of major carbon and amino acid metabolism in the roots of *Arabidopsis thaliana* treated with various rhizotoxic ions. Soil Sci Plant Nutr 2010; (in press)

CHAPTER 10

Nitric Oxide, S-Nitrosoproteome and Abiotic Stress Signaling in Plants

Jasmeet Kaur Abat and Renu Deswal*

Plant Molecular Physiology, Biochemistry and Proteomics Laboratory, Department of Botany, University of Delhi, Delhi-110007, India

Abstract: S-nitrosylation, is a PTM (post translational modification of NO (nitric oxide). Abiotic stress conditions lead to NO evolution and SNO (s-nitosothiols, the cellular NO pool) accumulation. Recently, research on NO and nitrosylation with respect to stress has gained momentum. Partial S-nirosoproteome of Arabidopsis, *K. pinata* and *B.juncea* are identified following BST, affinity purification and mass spectometric identification. In this chapter a snapshot of the relevance of NO and nitrosylation in abiotic stress is presented.

NITRIC OXIDE AS A SIGNALING MOLECULE IN BIOLOGICAL SYSTEMS

Use of Nitric Oxide (NO) dates back to late 1940's when nitroglycerine was used for controlling high blood pressure. Nitroglycerine derived NO was suggested as pharmacologically active agent by Farid Murad in 1977 [1]. Later, it was shown to be involved in smooth muscle relaxation [2]. This led to initiation of massive research subsequently proving it to be a signaling molecule in mammalian physiology. Till this time, NO was known as an atmospheric pollutant only to the plant scientists. In mid nineties, the importance of NO as a regulator of plant growth was shown as it showed enhanced leaf expansion, root growth and phytoalexin production [3, 4]. Plants were also shown to evolve NO [5]. These studies changed the status of NO from an atmospheric pollutant to a growth regulator. Since then considerable progress is made in understanding the role of NO in regulating plant [6, 7, 8, 9]. NO is, one of the three oxides of nitrogen present in air. It is a diatomic gas under atmospheric conditions. It is a gaseous free radical containing an unpaired electron in its π^2 orbital. Removal of this electron forms Nitrosonium cation (NO^+), while addition of an electron forms Nitroxyl anion (NO^-) [10]. Both of these forms are reactive [11]. Simple structure and high diffusibility make NO a good candidate as a signalling molecule. NO has both beneficial as well as detrimental effects depending on its concentration. It has been found that high concentration of NO cause cellular damage while at low concentration it acts as a signaling molecule [12].

Pharmacological studies using NO donors, scavengers, NO synthase (NOS) inhibitors and physiological studies as described below have established it as a "do it all molecule" regulating majority of plant processes right from seed germination to cell death. As mentioned above first time NO was shown as inducer of leaf expansion, root growth and phytoalexin production. Later, involvement of NO was reported in root elongation [13], cell wall metabolism [14], cell division [15], xylem differentiation [16], wound healing [17], stem elongation [18], seed germination [19] and reproductive processes like flowering and re-orientation of pollen tube [20]. Apart from these NO was also shown to regulate senescence [21] and cell death [22].

NITRIC OXIDE AND ABIOTIC STRESS

Stress whether biotic or abiotic compromise the genetic potential of crops leading to fall in agricultural productivity. NO evolution was observed as a generalized stress response, in response to high temperature, osmotic stress or UV-B in tobacco leaf peels and mesophyll cell suspensions [23, 24]. Application of NO donor, Sodium nitroprusside (SNP), was protective against salinity and heavy metal stress on root growth of *Lupinus luteus*. The protective role of NO was partially due to stimulation of Superoxide dismutase (SOD) activity and /or by direct scavenging of superoxide anion [19]. Similarly, in rice seedlings, NO donor treatment emaroliated salt and heat stress effects [25]. Salinity induced a transient increase in NO which in turn enhanced salt tolerance through increased activities of the proton pump and Na^+/H^+ antiport in the tonoplast in maize leaves [26].

*****Address correspondence to: Dr. Renu Deswal,** Associate Professor, Plant Molecular Physiology, Biochemistry and Proteomics Laboratory, Department of Botany, University of Delhi, Delhi-110007, India; Telefax: 91-011-27662273; E-mail: rdeswal@botany.du.ac.in

Narendra Tuteja, Sarvajeet Singh Gill and Renu Tuteja [Eds.]

The role of NO in imparting drought tolerance was shown in wheat, where treatment with NO donor induced Abscissic acid (ABA) mediated stomatal closure resulting in drought tolerance [27]. NO regulates K^+ and Cl^- channels by activating ryanodine-sensitive Ca^{2+} channels in guard cells of V*icia* [28]. It was reported that in tomato, wheat and corn, application of NO donor mediated chilling tolerance, probably via suppression of high levels of Reactive Oxygen Species (ROS) that accumulate during stress conditions.

Inspite its involvement in these diverse stress coditions, NO was not considered as a universal plant stress response as its production was not observed following mechanical or light stress up to 400 $\mu mol\ m^{-2}s^{-1}$ [24]. Although, the above compiled information strongly suggest participation of NO in abiotic stress conditions. Unfortunately, the mechanistic details and signaling pathways of NO mediated abiotic stress signaling are still obscure.

NITRIC OXIDE SIGNALLING

As reported in animal systems, NO signaling proceeds either through cyclic guanosine monophosphate (cGMP) dependent or independent pathways (Fig. **1**). In *Arabidopsis*, involvement of cGMP was shown during NO induced stomatal closure [29] and cell death [30]. cGMP and NO were shown to act as second messengers during Indole acetic acid (IAA) induced adventitious rooting in cucumber seedlings [13] and defense genes (PR-1 and PAL) in tobacco [31]. These reports show the relevance of cGMP dependent signaling in plants. To bring about physiological actions of NO cGMP only acts as a temporary signaling molecule and activates its downstream signaling components including cGMP induced protein kinases, cGMP gated ion channels, cyclic ADP ribose (cADPR) and calcium [32]. cGMP mediated signaling is still not convinsingly established as a major NO signaling apthway in plants as no NO sensitive GC has been found in plants till date. Presence of cGMP independent signaling mechanisms was indicated by the observation that effects of NO were not completely blocked by drugs that inhibit sGC

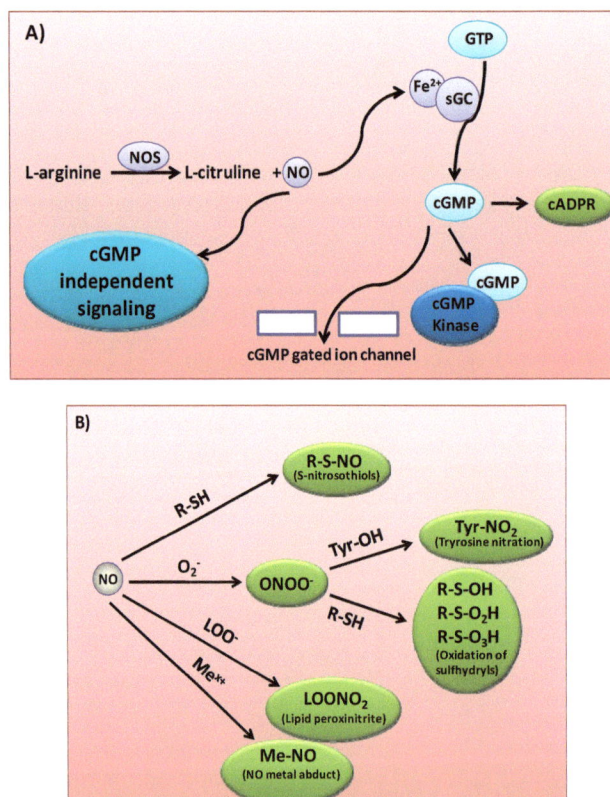

Figure 1: Schematic representation of Nitric oxide signaling mechanism. NO signalling is broadly classified into A) cGMP dependent and B) cGMP independent signalling. cGMP dependent signalling involves activation of soluble guanylyl cyclase (sGC) by NO, leading to production of cGMP and thus affecting its downstream targets including cADPR, cGMP dependent protein kinases and cGMP gated ion channels. Acting in a cGMP independent manner NO may also activate or inhibit proteins serving as signal transducer by promoting S-nitrosylation, tyrosine nitration and binding to metal centers.

NO signaling by direct interaction of NO with major cellular macromolecules like proteins, lipids and nucleic acids seems more apt at present. NO reacts with iron-sulfur, haem and free thiols of proteins. In addition, it induces oxidative and nitrative damage to nucleic acids and lipids [33]. NO itself is not very reactive with DNA but Reactive Nitrogen Species (RNS) are potent DNA-damaging agents which can cause alkylation, deamination, oxidation and nitration of bases and single and double strand breaks leading to increased mutations [34]. NO is also a potent inhibitor of lipid peroxidation and low-density lipoprotein oxidation. Nitrated membrane lipids and lipoproteins can transduce NO-signaling reactions and mediate pathways for anti-inflammatory lipid signalling [35].

Post-translational modifications (PTMs) of proteins constitute one of the major mechanisms of NO signaling. These could regulate protein activity, intracellular localization, stability, and protein-protein interactions. PTMs of proteins by NO include a) S-nitrosylation of free thiols, b) nitration of tyrosine, tryptophan residues and c) binding to metal centers [36]. Of these, S-nitrosylation is the major and most investigated route of NO signaling.

S-nitrosylation is the reversible attachment of NO to the free thiols of a protein, forming S-nitrosothiols (SNO). Chemically SNO are the thio-esters of nitrite, with the structure R-S-N=O [37]. S-nitrosylation can modulate protein function by blocking the free thiols (required for protein activity) and leading to induction of disulfide bond formation in the neighboring thiols. NO can be covalently incorporated into cysteine thiols, tryptophan indols and amines, but studies of free-SH groups by radioactive SH modifying reagents, ultraviolet visible spectrophotometry and electro spray ionization-mass spectrometry have demonstrated that cysteine residues are rapidly nitrosylated while reactions with other amino acids occur at much slower rates, therefore formation of S-nitrosothiol bond by cysteine thiol nitrosylation is considered most important due to its reactivity under physiological conditions and its influence on many protein functions. S-nitrosylation has been established as a key reversible, modification of proteins by NO, which regulates enzymatic activity, sub-cellular localization, complex formation and degradation of proteins [38]. SNO are biological metabolites of NO, as in comparison to NO these are stable compounds with longer half life. S-nitrosylation of glutathione forms S-nitrosoglutathione (GSNO) which functions as mobile reservoir of NO as well as endogenous NO donor. Under intense oxidative conditions, formation of sulphenic acid (SO⁻) and sulphonic acid (SO₃⁻), leads to irreversible oxidative modification of proteins [39].

S-nitrosylation has been established as an important redox-based modification which modulates multiple pathways via regulation of enzymes of these pathways. S-nitrosylation can be compared with phosphorylation (a PTM). It shares some similarities with phosphorylation in terms of being precisely targeted, reversible and spatio-temporally restricted. The only difference between the two PTM,s is that phosphorylation is enzyme driven while s-nitrosylation is a non-enzymatic reaction at least till date no enzyme has been discovered for nitrosylation [40].

Table 1: Routinely used methods for detection of S-nitrosoproteins

Method	Purpose
Saville-Griess	Spectrophotometric measurement of NO released
Immuno-histochemical	Immunohistochemistry, western blotting, semi-quantitative
Photolysis chemiluminescence	Gas phase measurement of NO released
Biotin Switch	Protein identification, western blotting, semi-quantitative, *insitu* detection*
SNOSID	Identification of redox-sensitive cysteine residues
AMCA Switch	Visualization and identification of proteins
DAF	Proteomics, semi-quantitative

Detection and quantification of SNO in biological systems faces a number of problems including their very low levels *in vivo*, labile nature (degradation in presence of metals and thiols), the presence of mixtures of iron-nitrosyls and S-nitrosothiols and even formation of these bonds during sample preparation. A number of methodologies are available as described in (Table **1**) for detection and quantification of S-nitrosothiols including Saville assay, Immuno-histochemical detection, Gas-phase Chemiluminescence and Biotin Switch Technique (BST). Out of these, BST is most commonly used [41]. It consists of 3 sequential steps which convert nitrosylated cysteines to biotinylated cysteines. In the first step (blocking), all free thiols in a protein mixture are blocked using either alkylating agent like N-ethylmaleamide (NEM) or thiol specific methyl thiolating agent like methyl methane thiosulfonate (MMTS). S-nitrosothiols and disulfides are not affected by these blocking agents. Presence of SDS while blocking ensures that blocking agent reaches to the buried cysteines. In the next step (labeling), SNO bonds are selectively decomposed using ascorbate, which specifically reduces SNO to thiols. In the final step, newly formed free thiols are then reacted with Biotin-HPDP (N-[6-(biotinamido)hexyl]-3-(2-pyridyldithio) propionamide), a sulfhydryl specific biotinylating agent. These biotinylated proteins are either detected using anti-biotin antibodies by western blotting or these are purified using avidin-biotin affinity chromatography and identified using mass spectrometry. Advantage of this technique include i) metal-NO adducts are not detected, ii) it allows general screening of S-NO. The disadvantages associated with this assay are i) it doesn't allow quantification of SNO, ii) it may lack specificity (if blocking is not adequate, it may identify reduced thiols as SNO) and iii) care is required during sample preparation steps to avoid loss of SNO. Also, endogenously biotinylated proteins (like decarboxylase) may interfere with the method. BST was widely used for detection and identification of S-nitrosylated proteins in animals [42-51] and plants [52, 53]. The identified s-nitrosylated proteins constitute the s-nitrosoproteome of a cell/ organ/ organism. Here, we are providing the snapshot of plant S-nitrosoproteoms as deciphered till date.

S-NITROSOPROTEOME IN PLANTS

S-nitrosylation is a well established redox-based signaling mechanism in animals. Over 200 S-nitrosylated proteins are known in animals [42-51]. These were associated with important functional categories including transcription factors, ion channels, membrane receptors, enzymes and structural proteins, indicating S-nitrosylation to be global regulator of metabolic processes. In plants, the first report on identification of S-nitrosylated proteins was published in 2005. Employing BST, 63 S-nitrosylated proteins were identified from GSNO, the NO donor, treated *Arabidopsis* cell cultures [52] (Table **2**). To know, whether S-nitrosylation occurs *in vivo* or not, the crude leaf extract from NO gas treated *Arabidopsis* seedlings was used. Out of 52 S-nitrosylated proteins, about 79% were similar to those identified in the *in vitro* treatment. These included metabolic enzymes, abiotic and biotic stress-related proteins, cytoskeletal, photosynthesis, redox-related and antioxidant enzymes. Of these, the majority were metabolic enzymes involved in carbon, nitrogen and sulfur metabolism. Target validation is necessary to understand physiological relevance of S-nitrosylation. In the above mentioned study Glyceraldehyde-3-phosphate dehydrogenase (GAPDH) was validated to be nitrosylated. Treatment of crude extract with GSNO reduced GAPDH activity up to 90% while glutathione (GSH, an inactive analogue of GSNO) had no effect on GAPDH activity [52]. Similar target validation studies have hypothesized functional relevance of S-nitrosylation in ethylene synthesis, cell death, biotic and abiotic stress conditions.

Table 2: Some of the putative stress associated s-nitrosylated proteins from Arabidopsis, *K. pinnata* and *B. juncea*

No	Protein	Arabidopsis	K. pinnata*	B.juncea*	Functional category
1	GST, putative	+	-	+	
2	Hsp 90, 81-3, 70	+	+	+	Stress-Related proteins
3	Cu/Zn-superoxide dismutase	+	-	-	
4	Glutathione peroxidase, putative	+	-	-	
5	Peroxiredoxin-related	+	-	+	Redox-Related proteins
6	L-ascorbate peroxidase	-	-	+	
7	Peroxiredoxin	+	-	+	
8	Type 2 peroxiredoxin-related	+	-	-	
9	Glutaredoxin, putative	+	-	-	
10	Elongation factor EF-2	+	-	+	
11	Elongation factor eEF-1α-chain, 1B α-subunit, 1B γ, putative	+	-	+	Signaling proteins
12	Initiation factor eIF-4A1, 5A-4-related	+	-	-	
13	60S Ribosomal protein L15-2	+	-	+	
					Tabel 2: cont....
14	Tubulin α 6 chain, β 4 chain	+	-	-	Cytoskeleton

15	Actin depolymerizing factor 3 like	+	-	-	Proteins
16	Annexin	+	-	-	
17	Kinesin-like protein	-	+	-	
18	Fructose 1,6 bisphosphate aldolase	+	+	-	
19	Triosephosphate isomerase	+	-	+	
20	GAPDH C-subunit	+	+	-	
21	Enolase	+	-	-	
22	Phosphoglycerate kinase	+	+	-	
23	Aconitase	+	-	-	
24	SAM synthetase, putative	+	-	-	
25	Adenosylhomocysteinase	+	-	-	
26	Methionine synthase	+	+	+	Metabolic enzymes
27	Cysteine synthase	+	-	-	
28	ATP synthase CF1 α-chain, β-chain	+	-	-	
29	UDP-glucose 4-epimerase	-	+	-	
30	Glutamate ammonia ligase	-	+	-	
31	Myrosinase	-	-	+	
32	Transketolase, putative	-	-	+	
33	ACC synthase	-	-	+	
34	Ascorbate peroxidase	-	-	+	
35	Rubisco large chain	+	+	+	
36	Rubisco small chain 1a precursor	+	+	+	
37	Rubisco activase, large subunit	+	-	-	
38	PSII P680 47-KD, D2 protein, fragment and PSII oxygen-evolving complex	+	-	-	
39	Rieske Fe-S protein	+	-	-	Proteins involved in photosynthesis
40	Carbonic anhydrase	-	+	-	
41	Glycolate oxidase	-	+	-	
42	PEP carboxylase	+	+	-	
43	Transketolase-like protein	+	-	+	
44	Sedoheptulose-bisphosphatase	-	-	+	
45	DNA topoisomerase II	-	+	-	Proteins involved in DNA replication
46	Reverse Transcriptase	-	-	+	

Following Arabidopsis, S-nitrosylated proteins were identified from *Kalanchoe pinnata* (a CAM plant) [54]. A total of 19 proteins were identified including enzymes for carbon, nitrogen and sulfur metabolism, cytoskeleton, stress related and proteins involved in photosynthesis (Table **2**). Out of these kinesin-like protein, glycolate oxidase, UDP glucose 4 epimerase and DNA topoisomerase II were the novel targets. Ribulose-1,5-bisphosphate carboxylase/oxygenase (Rubisco), which is the most abundant and also a major protein responsible for carbon fixation, was validated as target of S-nitrosylation. Treatment with S-nitrosoglutathione (a nitrosylating agent) reduced Rubisco carboxylase activity in a dose dependent manner. As expected, reduced glutathione did not change the activity suggesting that free thiols are required for Rubisco carboxylase activity.

Recently, 20 S-nitrosylated proteins were identified from *Brassica juncea* using BST and LC-MS/MS analysis [55] (Fig. **2**). Identified proteins included proteins related with primary and secondary metabolism, photosynthesis, DNA replication, abiotic, biotic stress responses, signaling, and a few unknowns. Of these photosynthesis and metabolism were the major categories followed by antioxidant, stress related, signaling, defense, hypothetical and DNA replication (Table **2**).

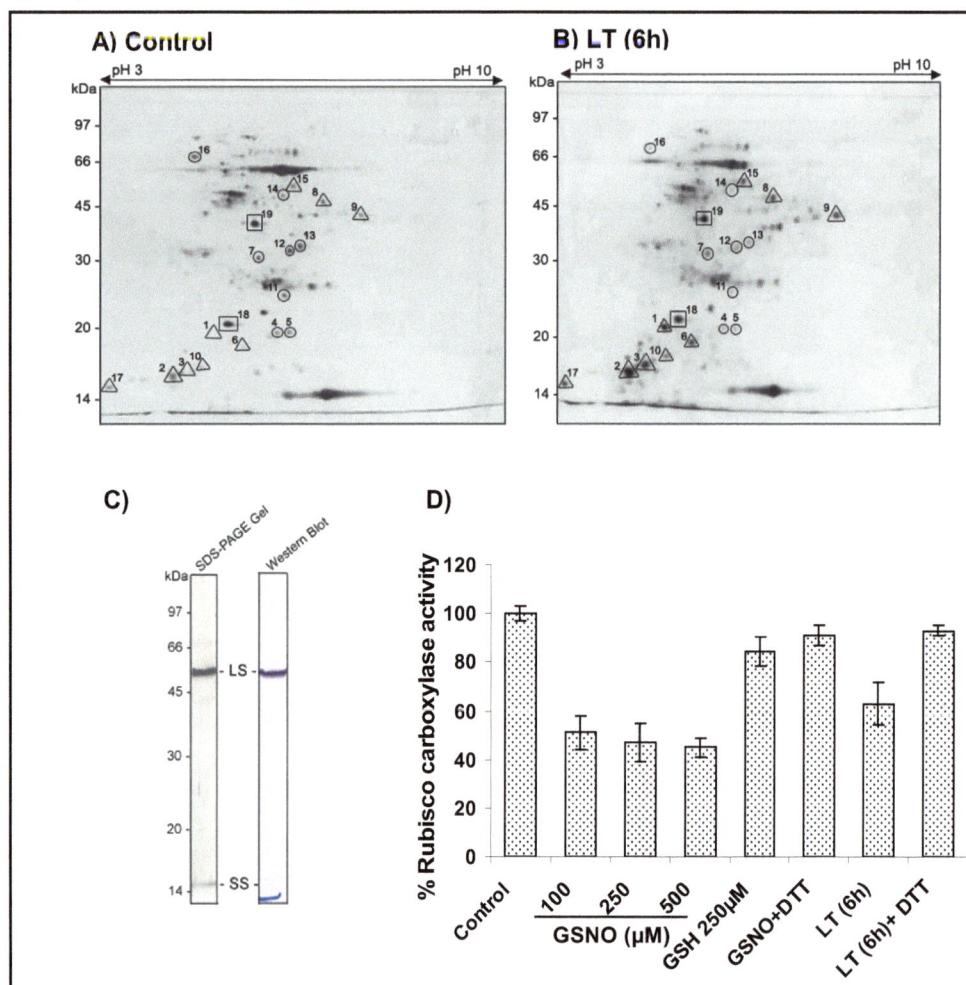

Figure 2: Detection of LT induced differentially S-nitrosylated proteins in *B. juncea*. Protein extracts (5mg) from A) untreated, B) LT (6h) treated seedlings were subjected to BST. Affinity purified S-nitrosylated proteins were resolved on 2-DE gels using linear 13cm IPG strips (pH 3-10) and 12% SDS-PAGE gels. Protein spots showing significant changes in intensity due to LT treatment are marked with triangle (showing increased S-nitrosylation) and circle (showing decreased S-nitrosylation). Landmark protein spots are marked by squares. Effect of S-nitrosylation on Rubisco carboxylase activity. (C) Rubisco was purified from *B. juncea* seedlings. SDS-PAGE profile and western blot of purified fraction. (D) Purified Rubisco was used either untreated (control) after treatment with GSNO or GSH (as indicted) prior to the measurement of enzyme activity. The Rubisco activity in the untreated control was taken as 100%. To see the effect of LT on Rubisco carboxylase activity, LT (6h) treated seedlings were extracted and the supernatant was used for activity assay.

STRESS RELEVANT TARGETS

Identification of stress associated proteins in S-nitrosylation target in Arabidopsis, *K. pinnata* and Brassica hinted that S-nitrosylation might have a significant role in stress tolerance. To find out, studies were carried out using both genomics as well as proteomics approach. Using genomics approach, mutants of GSNOR which modulates cellular SNO by utilizing GSNO as the substrate and removes it from the cellular pool were created in *Arabidopsis*. SNO were shown to regulate plant disease resistance [56]. Two T-DNA insertion lines lines Atgsnor1-1 and Atgsnor1-2 had increased levels of AtGSNOR1 transcripts and 189% and 165% GSNOR activity as compared to wild type. Loss of AtGSNOR1 function led to an increase in SNO levels and a decrease in plant defense responses to pathogen attack. AtGSNOR was also shown to positively regulate salicylic acid signalling. Recently an increase in SNO was shown during progression of HR in *Arabidopsis* using BST. To know the mechanism by which NO functions during HR, identification of targets in form of S-nitrosylated proteins is required. Combining BST with 2D and LC-MS/MS expression of 16 S-nitrosylated proteins was shown to vary during defense response. These proteins belonged to different protein families like metabolic enzymes, redox- related, stress-related and signaling proteins. Out of these

S-nitrosylation of Peroxidoredoxin IIE (PrxII E) and MDHAR was confirmed using western blot analysis [53]. In olive seedlings it was demonstrated that salt stress increased L-arginine-dependent NO production, total SNO and number of proteins that underwent tyrosine nitration, suggesting generation of RNS [57]. In *B. juncea* SNO content was quantified by Saville assay in stress treated seedlings. All the stresses increased total SNO. LT (1hr) showed the maximum (1.4 fold) increase followed by drought (1.2 fold), HT and salinity (1.1 fold each) supporting previous observations that stress induced NO contributes to higher SNO [55]. As LT showed highest SNO accumulation, it was analysed in detail.

COLD STRESS INDUCED DIFFERENTIAL NITROSYLATION

In *B. juncea* cold stress induced differential nitrosylation was identified using BST combined with 2D-PAGE and LC-MS/MS analysis. Initailly low temperature (LT) induced S-nitrosylated proteins were detected using chemiluminisence. LT (1-6hr) treated extracts showed maximum SNO at 6hr. 2D-PAGE gels were done for identification of differentailly S-nitrosylated proteins. S-nitrosylated proteins were purified from control and LT (6hr) treated seedlings and were resolved on 2D-PAGE gels. Seventeen spots showed statistically significant changes in intensity after LT stress. To confirm that the changes are due to differential S-nitrosylation and not differential protein expression, extracts from LT treated seedlings were incubated with GSH (250µM) and were subjected to BST followed by purification of S-nitrosylated proteins. Only 22 spots were visible in GSH treated sample with the major spots being Rubisco. These could constitute the stable SNO pool of *B. juncea*. Stable SNO are the SNO which are resistant to reduction with the reducing agent (GSH) due to stabilization conferred by S-nitrosylation induced conformational changes [58].

Of the seventeen LT induced differentially S-nitrosylated spots, nine spots showed an increase in intensity, suggesting higher S-nitrosylation while 8 spots showed a decrease. Major LT modulated S-nitrosylation targets were identified and these included photosynthetic, metabolic (glycolysis), pathogenesis related and signaling proteins. Rubisco was the major protein modulated by cold stress. Interestingly both up and down regulation of nitrosylation by LT was observed. To confirm if differential expression of Rubisco was contributing to differential nitrosylation, Rubisco expression was analyzed in S-nitrosylated proteins from control and LT (6h) treated seedlings by western blotting. cold stress also decreased the carboxylase activity by 39% in comparison to control, which was restored by DTT indicating cold induced reversible thiol modification of Rubisco. This study/analysis clearly emphasised the fact that stress (cold in this case) could change *in vivo* SNO content manifesting it in form of differential nitrosylation. Finally, measurement of electron transport chain and photosynthetic efficiency in stress and measurement of rubisco actiivity under these conditions further showed that stress induced inhibition in photosynthesis could be explained by nitrosylation mediated rubisco inhibition to some extent [55].

FUTURE PRESPECTIVES

Advances in NO plant research over last two decades have clearly established it as an important signaling molecule. PTM's of NO, namely nitrosylation and tyrosine nitration are important in abiotic stress conditions. Target of nitrosylation have been identified but a lot needs to be done in terms of (1) identifying more targets especially, the low abundant ones by abundant protein depletion, using gel free techniques and advance mass spectrometry; (2) validation of these targets needs to take up at large scale to show their physiological relevance; (3) metabolism of NO which includes its generation mechanism as well as its catabolism needs to be dissected; (4) Relationship between redox status of the cell/tissue/organism under different physiological/stress conditions needs to be known.

Analysis of thiol status of proteins in normal and stressed plants, using techniques like OxICAT (combination of ICAT (Isotope-coded-affinity tag) and differential thiol trapping technique), could complement the current information and provide functional significance. It is also important to understand the crosstalk between the oxidative and nitrosative stress *in vivo*. NO-mediated regulation of physiological processes and its contribution to important traits like crop yield is gradually being worked out. Current observations suggest that NO could modulate carbon, nitrogen and sulfur metabolism as well as important processes like photosynthesis and glycolysis, emphasizing that it could be identified as an important regulatory mechanism like phosphorylation. NO research in plants, at present, is just tip of the iceburg. The deep sea is still waiting to be discovered.

ACKNOWLEDGEMENTS

Work described in the chapter was partialy funded by a research grant by Govt. of India, DST (Department of Science and Technology) and CSIR (Council of Scientific and Industrial Research). JKA was supported by a CSIR, SRF fellowship.

REFERENCES

[1]   Katsuki S, Arnold W, Mittal C, Murad F. Stimulation of guanylate cyclase by sodium nitroprusside, nitroglycerin and nitric oxide in various tissue preparations and comparison to the effects of sodium azide and hydroxylamine. J Cyclic Nucleotide Res 1977; 3: 23-35.

[2]   Palmer RM, Ferrige AG, Moncada S. Nitric oxide release accounts for the biological activity of endothelium-derived relaxing factor. Nature 1987; 327: 524-526.

[3]   Leshmen YY. Nitric oxide in biological systems. Plant Growth Regul 1996; 18: 155-159.

[4]   Noritake T, Kawakita K, Doke N. Nitric oxide induces phytoalexin accumulation in potato tuber tissue. Plant Cell Physiol 1996; 37: 113-116.

[5]   Wildt J, Kley D, Rockel A, Rockel P, Segenschneider A. Emission of NO from several higher plant species. J Geophys Res 1997; 102: 5919-5927.

[6]   Besson-Bard A, Pugin A, Wendehenne D. New insights into nitric oxide signaling in plants. Annu Rev Plant Biol 2008; 59: 21-39.

[7]   Arasimowicz M, Floryszak-Wieczorek J. Nitric oxide as a bioactive signalling molecule in plant stress responses. Plant Sci. 2007; 172: 876-887.

[8]   Lamattina L, Garcia-Matta C, Graziano M, Pagnussat G. Nitric Oxide: the versatility of an extensive signal molecule. Ann Rev Plant Biol 2003; 54: 109-136.

[9]   Neill SJ, Desikan R, Hancock JT. Nitric oxide signalling in plants. New Phytol 2003; 159: 11- 35.

[10]   Stamler JS, David JS, Loscalzo J. Biochemistry of Nitric Oxide and its redox-activated forms. Science 1992; 258: 1898-1902.

[11]   Stamler J.S. Redox Signalling: nitrosylation and related target interactions of nitric oxide. Cell 1994; 78: 931-936.

[12]   del Río LA, Corpas FJ, Barroso JB. Nitric oxide and nitric oxide synthase activity in plants. Phytochem 2004; 65: 783-792.

[13]   Pagnussat GC, Lanteri ML, Lamattina L. Nitric Oxide and Cyclic GMP are messengers in the Indole Acetic Acid-induced adventitious rooting process. Plant Physiol 2003; 132: 1241-1248.

[14]   Pacoda D, Montefusco A, Piro G, Dalessandro G. Reactive oxygen species and nitric oxide affect cell wall metabolism in tobacco BY-2 cells. J Plant Physiol 2004; 161: 1143-1156.

[15]   Ötvös K *et al.* Nitric oxide is required for, and promotes auxin-mediated activation of, cell division and embryogenic cell formation but does not influence cell cycle progression in alfaalfa cell cultures. Plant J 2005; 43: 849-860.

[16]   Gabaldon C, Gomez RLV, Pedreno MA, Ros Barcelo A. Nitric oxide production by the differentiating xylem of *Zinnia elegans*. New Phytol 2005; 165: 121-130.

[17]   París R, Lamattina L, Casalongué CA. Nitric oxide promotes the wound-healing response of potato leaflets. Plant Physiol Biochem 2007; 45: 80-86.

[18]   Qu Y, Feng H, Wang Y, Zhang M, Cheng J, Wang X, Anv L. Nitric oxide functions as a signal in ultraviolet-B induced inhibition of pea stems elongation. Plant Sci 2006; 170: 994-1000.

[19]   Kopyra M, GwoŹdź EA. Nitric oxide stimulates seed germination and counteracts the inhibitory effect of heavy metals and salinity on root growth of *Lupinus luteus*. Plant Physiol Biochem 2003; 41: 1011-1017.

[20]   Prado AM, Porterfield DM, Feijo JA. Nitric oxide is involved in growth regulation and re-orientation of pollen tubes. Development 2004; 131: 2707-2714.

[21]   Cheng F, Hsu S, Kao CH. Nitric oxide counteracts the senescence of detached rice leaves induced by dehydration and polyethylene glycol but not by sorbitol. Plant Growth Regulon 2002; 38: 265-272.

[22]   Beligni MV, Fath A, Bethke PC, Lamattina L, Jones RL. Nitric oxide acts as an antioxidant and delays programmed cell death in barley aleurone layers. Plant Physiol 2002; 129: 1642-1650.

[23]   Beligni MV, Lamattina L. Nitric oxide: a non-traditional regulator of plant growth. Trends Plant Sci 2001; 6: 508-509.

[24]   Gould KS, Lamotte O, Klinguer A, Pugin A, Wendehenne D. Nitric oxide production in tobacco leaf cells: a generalized stress response? Plant Cell Environ 2003; 26: 1851-1862.

[25]   Uchida A, Jagendorf AT, Hibino T, Takabe T, Takabe T. Effect of hydrogen peroxide and nitric oxide on both salt and heat stress tolerance in rice. Plant Sci 2002; 163: 515-523.

[26]   Zhang Y, Wang L, Liu Y, Zhang Q, Wei Q, Zhang W. Nitric oxide enhances salt tolerance in maize seedlings through increasing activities of proton-pump and Na$^+$/H$^+$ antiport in the tonoplast. Planta 2006; 224: 545-555.

[27]   Garcia-Mata C, Lamattina L. Nitric oxide induces stomatal closure and enhances the adaptive plant responses against drought stress. Plant Physiol 2001; 126: 1196-1204.

[28]   Garcia-Mata C, Gay R, Sokolovski S, Hills A, Lamattina L, Blatt MR. Nitric oxide regulates K$^+$ and Cl$^-$ channels in guard cells through a subset of abscisic acid-evoked signaling pathways. Proc Natl Aacd Sci USA 2003; 100: 11116-11121.

[29]   Neill SJ, Desikan R, Clarke A, Hancock JT. NO is a novel component of ABA signaling in stomatal guard cells. Plant Physiol 2002; 128: 13-16.

[30]   Clarke A, Desikan R, Hurst RD, Hancock JT, Neill ST. NO way back: nitric oxide and programmed cell death in *Arabidopsis thaliana* suspension cultures. Plant J 2000; 4: 667-677.

[31]   Durner J, Wendehenne D, Klessig DF. Defense gene induction in tobacco by nitric oxide, cyclic GMP and cyclic ADP-ribose. Proc Natl Aacd Sci USA 1998; 95: 10328-10333.

[32]   Hancock JT. Cell Signaling. Harlow, UK: Longman. 1997.

[33]   Wendehenne D, Pugin A, Klessig DF, Durner J. Nitric oxide: comparative synthesis and signaling in animal and plant cells. Trends Plant Sci 2001; 6: 177-183.

[34]   Sawa T, Ohshima H. Nitrative DNA damage in inflammation and its possible role in carcinogenesis. Nitric Oxide 2006; 14: 91-100.

[35]   Kalyanaraman B. Nitrated lipids: a class of cell signaling molecules. Proc Natl Aacd Sci USA 2004; 101: 11527-11528.

[36]   Gow AJ, Farkouh CR, Munson DA, Posencheg MA, Ischiropoulos H. Biological significance of nitric oxide-mediated protein modifications. Am J Physiol Lung Cell Mol Physiol 2004; 287: L262-L268.

[37]   Broillet MC. S-nitrosylation of proteins. CMLS Cell Mol Life Sci 1999; 55: 1036-1042.

[38]   Hess DT, Matsumoto A, Kim SO, Marshall HE, Stamler JS. Protein S-nitrosylation: purview and parameters. Nat Rev Mol Cell Biol 2005; 6: 150-166.

[39]   López-Sánchez LM, Muntané J, de la Mata M, Rodríguez-Ariza A. Unraveling the S-nitrosoproteome: tools and strategies. Proteomics 2009; 9: 808-818.

[40]   Mannick JB, Schonhoff CM. Nitrosylation: the next phosphorylation? Arch Biochem Biophys 2002; 408: 1-6.

[41]   Jaffrey SR, Snyder SH. The Biotin Switch assay for the detection of S-nitrosylated proteins. Sci STKE 2001; PL1.

[42]   Jaffrey SR, Erdjument-Bromage H, Ferris CD, Tempst P, Snyder SH. Protein S-nitrosylation: a physiological signal for neuronal nitric oxide. Nat Cell Biol 2001; 3: 193-197.

[43]   Kuncewicz T, Sheta EA, Goldknopf IL, Kone BC. Proteomic analysis of S-nitrosylated proteins in mesangial cells. Mol Cell Proteom 2003; 2: 156-163.

[44]   Foster MW, Stamler JS. New insights into protein S-Nitrosylation. Mitochondria as a model system. J Biol Chem 2004; 279: 25891-25897.

[45]   Gao C, Guo H, Wei J, Mi Z, Wai PY, Kuo PC. Identification of S-nitrosylated proteins in endotoxin-stimulated RAW264.7 murine macrophages. Nitric Oxide 2005; 12: 121-126.

[46]   Rhee KY, Erdjument-Bromage H, Tempst P, Nathan CF. S-nitrosoproteome of *Mycobacterium tuberculosis*: Enzymes of intermediary metabolism and antioxidant defense. Proc Natl Aacd Sci USA 2005; 102: 467-472.

[47]   Yang Y, Loscalzo J. S-nitrosoprotein formation and localization in endothelial cells. Proc Natl Aacd Sci USA 2005; 102: 117-122.

[48]   Zhang Y, Keszler A, Broniowska KA, Hogg N Characterization and application of the biotin-switch assay for the identification of S-nitrosylated proteins. Free Radic Biol Med 2005; 38: 874-881.

[49]   Agnol MD, Bernstein C, Bernstein H, Garewal H, Payne CM. Identification of S-nitrosylated proteins after chronic exposure of colon epithelial cells to deoxycholate. Proteomics 2006; 6: 1654-1662.

[50]   Burwell LS, Nadtochiy SM, Tompkins AJ, Young S, Brookes PS. Direct evidence for S-nitrosation of mitochondrial complex I. Biochem J 2006; 394: 627-634.

[51]   Hao G, Derakhshan B, Shi L, Campagne F, Gross SS. SNOSID, a proteomics method for identification of cysteine S-nitrosylation sites in complex protein mixtures. Proc Natl Aacd Sci USA 2006; 103: 1012-1017.

[52]   Lindermayr C, Saalbach G, Durner J. Proteomic identification of S-nitrosylated proteins in *Arabidopsis*. Plant Physiol 2005; 137: 921-930.

[53]   Romero-Puertas MC, Campostrini N, Mattè A, Righetti PG, Perazzolli M, Zolla L., Roepstorff P, Delledonne M. Proteomic analysis of S-nitrosylated proteins in *Arabidopsis thaliana* undergoing hypersensitive response. Proteomics 2008; 8: 1459-69.

[54]   Abat JK, Mattoo, AK, Deswal R. S-Nitrosylated protein profile of a medicinal CAM Plant *Kalanchoe pinnata*: Ribulose-1, 5-bisphospate carboxylase/oxygenase activity targeted for inhibition. FEBS J 2008; 275: 2862-2872.

[55]   Abat JK, Deswal, R. Differential modulation of S-nitrosoproteome of *Brassica juncea* by low temperature: change in S-nitrosylation of Rubisco is responsible for the inactivation of its carboxylase activity. Proteomics 2009; 9: 4368-4380.

[56]   Feechan A, Kwon E, Yun BW, Wang Y, Pallas JA, Loake GJ. A central role for S-nitrosothiols in plant disease resistance. Proc Natl Aacd Sci USA 2005; 102: 8054-8059.

[57]   Valderrama R, Corpas FJ, Carreras A, Fernández-Ocaña A, Chaki M, Luque F, Gómez-Rodríguez MV, Colmenero-Varea P, Río LA, Barroso JB. Nitrosative stress in plants. FEBS Lett 2007; 581: 453-461.

[58]   Paige JS, Xu G, Stancevic B, Jaffrey SR. Nitrosothiol reactivity profiling identifies S-nitrosylated proteins with unexpected stability. Chem Biol 2008; 15: 1307-1316.

CHAPTER 11

Abscisic Acid in Abiotic Stress Tolerance: An 'Omics' Approach

Kailash C. Bansal[1]*, Sangram K. Lenka[1] and Narendra Tuteja[2]

[1]*National Research Centre on Plant Biotechnology, Indian Agricultural Research Institute, New Delhi – 110 012, India and* [2]*Plant Molecular Biology Group, International Centre for Genetic Engineering and Biotechnology, New Delhi- 110 067, India*

Abstract: Abiotic stresses impair crop production on irrigated land worldwide. Overall, the susceptibility or tolerance to the stress in plants is a coordinated action of multiple stress responsive genes, which also cross-talk with other components of stress signal transduction pathways. Plant responses to abiotic stress can be determined by the severity of the stress and by the metabolic status of the plant. Abscisic acid (ABA) is a phytohormone critical for plant growth and development and plays an important role in integrating various stress signals and controlling downstream stress responses. Plants have to adjust ABA levels constantly in response to changing physiological, metabolic and environmental conditions. To date, the mechanisms for fine-tuning of ABA levels remain elusive. The mechanisms by which plants respond to abiotic stresses include both ABA-dependent and ABA-independent processes. Various transcription factors such as DREB2A/2B, AREB1, RD22BP1 and MYC/MYB are known to regulate the ABA-responsive gene expression by interacting with their corresponding *cis*-acting elements such as DRE/CRT, ABRE, and MYCRS/MYBRS, respectively. Due to polygenic nature of the trait, it is becoming important to apply genome wide tools for precise understanding of the mechanisms and to ultimately improve stress tolerance in crops plants. This chapter describes the ABA-induced stress response pathway and the application of 'omics' technologies to unravel the complex mechanisms governing abiotic stress tolerance in plants.

INTRODUCTION

Plants experience different kinds of abiotic stresses including higher concentration of salt (salinity), extremes of temperature (low temperature i.e. cold [chilling or freezing], higher temperature [heat], and water shortage (drought or dehydration). These stresses are the principal cause of reducing the yield of crops significantly [1]. Overall, stress tolerance is a complex phenomenon because plants may go through multiple stresses at the same time during their development.

Plants perceive and respond adaptively to abiotic stresses imposed by salt, cold, high temperature and drought. These adaptive process are controlled mainly by the phytohormone abscisic acid (ABA), which acts as an endogenous messenger in the regulation of plant's water status [2]. ABA is generated as a signal during plant's life cycle to control seed germination and developmental processes. The action of ABA can target specifically guard cells for induction of stomatal closure but may also signal systemically for adjustment towards severe water shortage. Since the synthesis of ABA is induced by various stresses, it is now considered as a plant stress hormone [1, 2]. Various transcription factors are known to regulate the ABA-responsive gene expression [1, 3]. Stress responsive genes can be expressed either through an ABA-dependent or ABA-independent pathway [4]. Due to multigene involvement in the ABA responsive signaling pathway, it is becoming important to apply genome scale tools for precise understanding of the mechanisms of ABA-mediated stress tolerance in crops plants. A tremendous lead has been obtained in advancement of 'omics' technologies in recent years. The aim of this chapter is to provide a comprehensive overview of various 'omics' approaches potentially useful in understanding the complexity of ABA responsive genetic regulatory mechanisms in plants.

Generic Pathway for Plant Response to Stress

Plants can respond to stresses as individual cells and synergistically as a whole organism. The extracellular stress signal is first perceived at membrane level by the membrane receptors, ion channel,

*Address correspondence to: Dr. Kailash C. Bansal, National Research Centre on Plant Biotechnology, Indian Agricultural Research Institute, New Delhi - 110 012, India; Tel.: +91-11-25843554; Fax: +91-11-25843984; E-mail: kailashbansal@hotmail.com

receptor-like kinase (RLK) or histidine kinase (HK), which then activate large and complex signaling cascade intracellularly including the generation of secondary signal molecules such as Ca^{2+}, inositol phosphates (InsP), reactive oxygen species (ROS) and ABA. The stress signal is then transduced inside the nucleus to induce multiple stress responsive gene expression, the products of which ultimately lead to plant adaptation to stress tolerance directly or indirectly [1]. Overall, the stress response could be a coordinated action of many genes, which may cross-talk with each other. The stress-induced gene products are also involved in the generation of regulatory molecules like ABA, salicylic acid and ethylene, which can initiate the second round of signaling. Small molecules like ABA play important role in this process [1-4].

ABA and Abiotic Stress Signaling

ABA is an important phytohormone and plays a critical role in response to various stress signals. The application of ABA to plant mimics the effect of a stress condition. As many abiotic stresses ultimately results in desiccation of the cell and osmotic imbalance, there is an overlap in the expression pattern of stress genes after cold, drought, high salt or ABA application. This suggests that various stress signals and ABA share common elements in the signaling pathway and these common elements cross-talk with each other to maintain cellular homeostasis [5,6]. ABA also plays important roles in many other physiological processes such as seed dormancy, development of seed, promotion of stomatal closure, embryo morphogenesis, synthesis of storage proteins and lipids, leaf senescence and also defense against pathogens [2]. ABA levels are induced in response to various stress signals. ABA actually helps the seeds to surpass the stress conditions and geminate only when the conditions are conducive for seed germination and growth. It also prevents the precocious germination of premature embryos.

Main function of ABA seems to be the regulation of plant water balance and osmotic stress tolerance. Several ABA deficient mutants namely *aba1*, *aba2* and *aba3* have been reported for *Arabidopsis* [7]. ABA deficient mutants for tobacco, tomato and maize have also been reported [2]. Without any stress treatment, the growth of these mutants is comparable to wild type plants. However, under drought stress, ABA deficient mutants readily wilt and die if the stress persists. Under salt stress also ABA deficient mutants show poor growth. In addition, ABA is required for freezing tolerance, which also involves the induction of dehydration tolerance genes [8].

Studies suggest that osmotic stress imposed by high salt or drought is transmitted through at least two pathways; one is ABA-dependent and the other ABA-independent. Cold exerts its effects on gene expression largely through an ABA-independent pathway [5,6]. ABA induced expression often relies on the presence of *cis*-acting element called ABRE element (ABA-responsive element) [5, 6, 9, 10]. Genetic analysis indicates that there is no clear line of demarcation between ABA-dependent and ABA-independent pathways and the components involved may often cross talk or even converge in the signaling pathway. Calcium, which serves as a second messenger for various stresses, represents a strong candidate, which can mediate such cross talks. Several studies have demonstrated that ABA, drought, cold and high salt result in rapid increase in calcium levels in plant cells [1,2,3,4].

The transcript accumulation of *RD29A* gene is reported to be regulated in both ABA-dependent and ABA-independent manner [11]. Proline accumulation in plants can also be mediated by both ABA-dependent and ABA-independent signaling pathways [1]. The role of calcium in ABA-dependent induction of *P5CS* gene during salinity stress has been reported by Kinight *et al.* (1997) [12]. It is known that the expression of *RD29A*, *RD22*, *COR15A*, *COR47* and *P5CS* genes was reduced in the *los5* mutant [8]. The signaling mechanism behind the activation of these genes is not well known, but the transcriptional activation of few stress induced genes represented by *RD29A* is known to some extent [3]. It is also suggested that phospholipase D (*PLD*) along with ABA and calcium act as a negative regulator of proline biosynthesis in *Arabidopsis* [13]. The salinity stress-induced transcript of pea DNA helicase 45 (*PDH45*) followed ABA-dependent pathway while calcineurin B-like protein (*CBL*) and CBL-interacting protein kinase (*CIPK*) from pea followed the ABA-independent pathway [14-16]. Overall, the ABA-dependent pathways are involved essentially in osmotic stress-induced gene expression. Recently, Lee *et al.* (2006) have proposed that the activation of inactive ABA pools by polymerized AtBG1 (a beta-glucosidase, hydrolyzes glucose-conjugated) could be a mechanism by which plants rapidly adjust ABA levels and respond to changing environmental cues [17].

ABA Biosynthesis

ABA accumulation under stress conditions mainly occurs due to the induction of genes encoding enzymes responsible for ABA biosynthesis. Several ABA biosynthesis genes have been cloned which includes Zeathanxin epoxidase (Known as *ABA1* in *Arabidopsis*), 9-*cis*-epoxycarotenoid dioxygenase (*NCED*), ABA aldehyde oxidase and ABA3 also known as *LOS5* [3, 8]. The abiotic stress-induced activation of many ABA biosynthetic genes such as zeaxanthin oxidase (*ZEP*), 9-cis-epoxycarotenoid dioxygenase (*NCED*), ABA-aldehyde oxidase (*AAO*) and molybdenum cofactor sulphurase (*MCSU*) appeared to be regulated through calcium-dependent phosphorylation pathway [3,4,18]. The accumulation of ABA can also feedback stimulate the expression of ABA biosynthetic genes through calcium-signaling pathway and can also activate the ABA catabolic enzymes to degrade the ABA. The mechanisms by which the abiotic stresses up-regulate ABA biosynthesis genes are not well understood. Recently, Verslues *et al.* (2007) suggested that metabolic changes that alter hydrogen peroxide levels could also affect both ABA accumulation and ABA sensitivity [19].

ABA and Transcription Factors in Stress Tolerance

Transcriptional regulatory network of *cis*-acting elements and transcription factors involved in ABA and abiotic stress responsive gene expression is depicted in Fig. **1**. The promoters of the stress-induced genes contain *cis*-regulatory elements such as DRE/CRT (A/GCCGAC), ABRE (PyACGTGGC), MYC recognition sequence (MYCRS; CANNTG) and MYB recognition sequence (MYBRS; C/TAACNA/G), which are regulated by various upstream transcriptional factors (Fig. **1**) [1,18]. The ABA-dependent stress signaling activates basic leucine zipper transcription factors called AREB, which binds to ABRE element to induce the stress responsive gene (*RD29B*). In *Arabidopsis*, it is reported that two ABRE motifs are involved in the regulation of ABA-responsive expression of the RD29B gene, which encodes a LEA-like (late embryogenesis abundant) protein [9]. Transcription factors like DREB2A and DREB2B trans-activate the DRE *cis*-element of osmotic stress genes and thereby are involved in maintaining the osmotic equilibrium of the cell [1]. Some genes like *RD22* lack the typical CRT/DRE elements in their promoter suggesting their regulation by some other mechanism. *RD22* gene encodes a protein having a homology to an unidentified seed protein. The drought-inducible expression of *DREB1D* is regulated by ABA-dependent pathways, indicating that DREB1D protein may function in the slow response to drought that depends upon the accumulation of ABA (Fig. **1**). The MYC/MYB transcription factors, RD22BP1 and AtMYB2, could bind MYCRS and MYBRS elements, respectively, and help in activation of *RD22* gene (Fig. **1**). These MYC and MYB proteins are known to be synthesized only after endogenous levels of ABA accumulate, hence suggesting that their role is in late stages of the stress response [1]. The Clp protease regulatory subunit encoding gene, *ERD1* (early responsive to dehydration 1), responds to dehydration and salinity stress before the accumulation of ABA, which suggested that an ABA-independent pathway also exists in the dehydration stress response of *Arabidopsis* [20]. Overexpression of both ZF-HD and NAC proteins activate the expression of *ERD1* gene in normal growth conditions (non-stressed) in the transgenic *Arabidopsis* plants. Overall, these transcription factors may also cross-talk with each other for their maximal response to stress tolerance. Kim *et al.* (2004) reported that over-expression of a transcription factor regulating ABA-responsive gene expression conferred multiple stress tolerance [21].

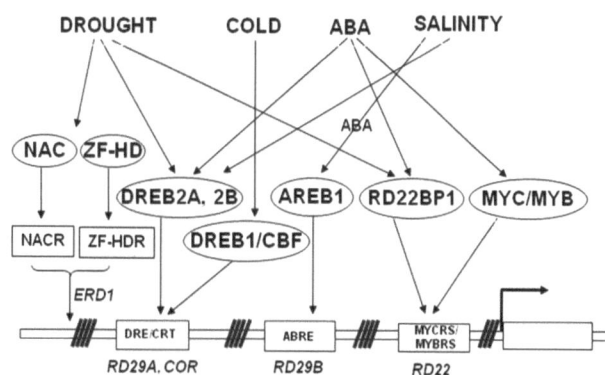

Figure 1: Transcriptional regulatory network of *cis*-acting elements and ABA-dependent transcription factors involved in drought, cold and salinity stress gene expression. Abiotic stress signaling seems to be mediated by transcription factors such as

NAC, ZF-HD, DREB2A/DREB2B and DREB1/CBF, AREB1, RD22BP1 and MYC/MYB transcription activators, which interact with NACR, ZF-HDR, DRE/CRT, ABRE and MYCRS/MYBRS elements in the promoter of the stress genes, respectively. AtMYC2 and AtMYB2 act cooperatively to activate the expression of ABA-inducible genes such as *RD22*. *Cis*-acting elements that are involved in transcription of stress-responsive gene are shown in boxes. Transcription factor that regulate stress-inducible gene expression are shown in ovals. ABA, abscisic acid; ABRE, ABA-responsive element; AREB, ABRE-binding protein; CBF, C-repeat-binding factor; *COR*, cold regulated genes; CRT, C-repeat; DRE, dehydration-responsive element; DREB, DRE-binding protein; *ERD*, early responsive to dehydration, MYB, myeloblastosis; MYBRS, MYB-recognition sequence; MYC, myelocytomatosis; MYCRS, MYC-recognition sequence; NACR, NAC-recognition site; *RD*, genes responsive to dehydration; ZF-HD; zinc-finger homeodomain.

The ABA level goes down during seed imbibition, which allow embryos to germinate and develop into seedlings, while ABA level remains high during abiotic stress conditions, which can arrest the growth and development. Several transcription factors, including abscisic acid-insensitive, ABI3 and ABI5, are known to control this developmental checkpoint. Recently, Reyes and Chua (2007) have shown that in germinating *Arabidopsis thaliana* seeds, ABA induces the accumulation of microRNA 159 (miR159) in an ABI3-dependent fashion and miRNA159 mediates cleavage of MYB101 and MYB33 transcripts *in vitro* and *in vivo* [22]. Here, they have shown that, the two MYB transcription factors function as positive regulators of ABA responses, as null mutants of myb33 and myb101 show hyposensitivity to the hormone. These results suggested that ABA-induced accumulation of miR159 controls transcript levels of two MYB factors during *Arabidopsis* seed germination. This is a homeostatic mechanism to direct MYB33 and MYB101 transcript degradation to desensitize hormone signaling during seedling stress responses [22].

ADVANCES IN 'OMICS' APPROACHES

Considerable advances have been made in understanding the plant's adaptation in stress environments, complex genetics involving multitude of genes and stress tolerance mechanisms. Application of an array of new innovative omics technologies are growing at a rapid pace, which facilitate researchers to systematically analyze the plant genomes to a greater extent. These sophisticated technologies have brought about paradigm shift in understanding the complexity of biology of an organism and its response to environmental disturbances. As ABA influences regulation of several stress responsive genes in the plant genome, application of high throughput techniques are essential to decipher ABA responsive molecular genetic network. Several 'omics' approaches have emerged over the past years due to the significant technological advancements, helping us to obtain an in-depth understanding in genome structure, function and evolution through studies on genomics, proteomics and metabolomics.

Application of Functional Genomics, Proteomics and Metabolomics in ABA Responsive Abiotic Stress Tolerance

Several components of regulatory elements responsible for mediating ABA responsive gene expression have been identified by application of genome-wide 'omics' approaches. 27 genes were found to be induced by combined sucrose and ABA treatment from rice cultured cells with cDNA-AFLP [23]. Almost 2000 drought-responsive genes were identified in *Arabidopsis thaliana* under progressive soil drought stress using whole-genome oligonucleotide microarrays. About two-thirds of drought-responsive genes (1310 out of 1969) were regulated by ABA and/or the ABA analogue (+)-8'-acetylene-ABA (PBI425) [24]. An oligonucleotide microarray was used for transcriptomic analysis of *Physcomitrella* treated with abscisic acid (ABA), or subjected to osmotic, salt and drought stress. This analysis identified 56 genes responsive to ABA. Among abiotic stress responsive genes, many ABA- and drought-responsive genes are homologues of angiosperm genes expressed during drought stress and seed development [25]. Gene expression patterns were analyzed using more than 10,000 seed-expressed sequences by microarray, which reveal tissue-specific signaling networks controlling programmed cell death and ABA- regulated maturation in developing barley seeds [26]. Similarly using c-DNA microarray it has been shown that abscisic acid (ABA) signaling plays a prominent role in the early response to darkness, although this effect is not mediated by an increase in the ABA level [27].

ABA plays important role in the induction and maintenance of seed dormancy. Expression analysis using whole-genome Affymetrix tiling arrays unraveled the differences in effects of exogenous ABA on ABA-mediated gene transcription from those of endogenous ABA-mediated gene transcription in dry and imbibed *Arabidopsis* seeds [28]. Genome-wide transcriptome analysis using gene chip revealed that ROP10 negatively regulates ABA responses by specifically and differentially modulating the ABA sensitivity of a subset of genes including protein kinases and zinc-finger family proteins [29]. Several other transcriptomics studies unraveled many novel features of ABA dependent signaling network in different model systems [30,31,32,33]. AREB1, AREB2 and ABF3 are the master regulator of ABRE dependent ABA signaling involved in drought stress tolerance. Large-scale transcriptome analysis revealed the overlapping cooperative function of these master regulators in ABRE dependent manner [34]. ABA-responsive elements (ABREs) were found to be over-represented as consensus *cis*-motifs among the co-expressed ABA responsive genes. The presence of ABREs within the promoters of many ABA upregulated genes suggests that transcription factors binding to ABREs may represent major master regulator of downstream targets of ABA signaling responses [10,35]. Transcriptome analysis of the moss *Physcomitrella patens* reveals evolutionarily conserved abscisic acid signaling components during embryophyte evolution [36]. Both structural, functional genomics and bioinformatics analysis revealed glucose- and ABA-regulated transcription networks in *Arabidopsis* [37]. *Arabidopsis* Affymetrix tiling arrays was used to study the whole genome transcriptome under drought, cold, high-salinity and ABA treatment conditions. It has been observed that 7,719 non-AGI transcriptional units (TUs) exist in the unannotated "intergenic" regions of *Arabidopsis* genome. The biogenesis mechanisms of the stress- or ABA-inducible antisense RNAs was studied and observed that the expression of sense TUs is necessary for the stress- or ABA-inducible expression of the antisense TUs in the fSATs (AGI code/non-AGI TU) [38]. Response of C4 cereal *Sorghum bicolor* towards different abiotic stresses and ABA was monitored using cDNA microarray having 12,982 unique gene clusters. Analysis of response profiles demonstrated the existence of a complex gene regulatory network that differentially modulates gene expression in a tissue- and kinetic-specific manner in response to ABA, high salinity and water deficit [39].

Using proteomics, such as two-dimensional protein profiles, it was shown that the mechanisms blocking germination of the non-dormant (ND) seeds by ABA application are different from those preventing germination of the dormant (D) seeds of *Arabidopsis thaliana* [40]. Seed dormancy breaking involves proteins of various processes but the proteasome proteins, S-adenosylmethionine synthetase, glycine-rich RNA binding protein, ABI3-interacting protein 1, EF-2 and adenosylhomocysteinase are of particular importance. This conclusion was drawn by proteome analysis (by mass spectrometry) in *Acer platanoides* influenced by cold stratification, and by the participation of the ABA and GA hormones [41]. Proteomic mechanism of ABA mediated seed dormancy breaking was also analyzed in *Fagus sylvatica* seeds [42]. Using 2DE approaches Lee *et al.* (2004) were able to identify role of AnnAt1 and AnnAt4 in osmotic stress and ABA signaling in a Ca^{2+}-dependent manner in *Arabidopsis thaliana* root microsomal fraction [43]. Embryonic proteome modulation studies using various proteomics techniques by applying GA and ABA to germinating rice seeds demonstrated a dramatic down-regulatory role of ABA for rice isoflavone reductase (OsIFR) and rice PR10 (OsPR10) proteins [44]. Applying proteomics approach, it has been shown that rice seedlings acquire salt tolerance upon ABA pretreatment [45]. Six ABA-regulated phosphoproteins were identified such as, G protein beta subunit-like protein, ascorbate peroxidase, manganese superoxide dismutase, triosephosphate isomerase, putative $Ca^{(2+)}/H^{(+)}$ antiporter regulator protein, and glyoxysomal malate dehydrogenase in rice using proteomics strategy [46]. The minimal set of core components of ABA signaling has been constituted *in vitro*. Using proteomics strategy, ABA-triggered phosphorylation of the transcription factor ABF2/AREB1 was achieved *in vitro* by combining ABA receptor PYR1 with the type 2C protein phosphatase (PP2C) ABI1, the serine/threonine protein kinase SnRK2.6/OST1 and the transcription factor ABF2/AREB1 [47]. By application of LC-MS/MS it was demonstrated that ABI1- a member of group-A PP2Cs can interact with several PYR/PYL/RCAR family members in *Arabidopsis in vivo*. PYR1-ABI1 interaction is rapidly stimulated by ABA in *Arabidopsis* establishing a new SnRK2 kinase-PYR/PYL/RCAR interaction for PYR/PYL/RCAR-mediated ABA signalling. Crystal structures of apo and ABA-bound receptors as well as a ternary PYL2-ABA-PP2C complex were resolved recently by identifying a conserved gate-latch-lock mechanism underlying ABA signaling [48]. The crystallographic structure of pyrabactin resistance 1 (PYR1), a prototypical PYR/PYR1-like (PYL)/regulatory component of ABA receptor (RCAR) protein was also illustrated. This study demonstrated an alpha/beta helix-grip fold and homodimeric assembly of PYR1, to which ABA binds and makes it more compact symmetric closed-lid dimer [49]. The crystal structures of the ABA receptor PYL1 bound with (+)-ABA, and the complex formed by the further binding of (+)-ABA-bound PYL1 with the PP2C protein ABI1 were also demonstrated by identifying the structural basis of the mechanism of (+)-ABA-dependent inhibition of ABI1 by PYL1 in ABA signalling [50].

The metabolic phenotypes of *Arabidopsis* wild-type and a knockout mutant of the NCED3 gene (nc3-2) under dehydration stress was analyzed using two types of mass spectrometry (MS) systems, gas chromatography/time-of-flight MS (GC/TOF-MS) and capillary electrophoresis MS (CE-MS). Using an integrated metabolome and transcriptome analysis, it was revealed that ABA-dependent transcriptional regulation of the biosynthesis of the branched-chain amino acids, saccharopine, proline and polyamine takes place in *Arabidopsis*, whereas the level of the oligosaccharide raffinose was regulated by ABA-independent route under dehydration stress [51]. Metabolomics analysis of the xylem sap of maize under drought stress revealed induction of ABA [52].

CONCLUSIONS AND PERSPECTIVES

Overall, abiotic stress signaling is an important area of research with respect to understanding mechanisms associated with sustenance of crops yields under sub-optimal conditions. The involvement of the stress responsive genes in various metabolic processes to enhance the stress tolerance in plant might have general implications. Different mechanisms of stress tolerance are just beginning to be understood. The overall progress of research on ABA regulated stress responsive genes and their products reflect their central role in plant growth and development under stress conditions. A lot of effort is still required to uncover in detail the genes and products thereof that are induced by ABA and their interacting partners to understand the complexity of the abiotic stress signal transduction pathways. The role of endogenous siRNA in regulating the ABA induced stress tolerance will be further helpful in enhancing our understanding of the mechanisms that impart stress tolerance. Determination of the upstream receptors or sensors that monitor the stimuli as well as the downstream effectors that regulate the responses is essential, which will also expedite our understanding of ABA mediated stress signaling mechanisms in plants. We still lack understanding of the mechanisms by which the abiotic stresses upregulate ABA biosynthesis genes. In addition, it is important to understand the function of all the different kinds of ABA-responsive genes to gain further insight in this complex trait of abiotic stress tolerance. Overall, it is becoming clear that ABA action enforces a sophisticated regulation at all levels.

ACKNOWLEDGEMENTS

This work was partially supported by ICAR-sponsored Network Project on Transgenics in Crops (NPTC), and the grants from the Department of Biotechnology, Department of Science and Technology and Defence Research and Development Organisation, Government of India. We apologize if some references could not be cited due to space constraint.

REFERENCES

[1]   Mahajan S, Tuteja N. Cold, salinity and drought stresses: an overview. Arch Biochem Biophys 2005; 444(2): 139-158.
[2]   Swamy PM, Smith B. Role of abscisic acid in plant stress tolerance, Curr Sci 1999; 76: 1220-1227.
[3]   Xiong L, Schumaker KS, Zhu JK. Cell signaling during cold, drought, and salt stress. Plant Cell 2002; 14 (Suppl): S165-183.
[4]   Chinnusamy V, Schumaker K, Zhu JK. Molecular genetic perspectives on cross-talk and specificity in abiotic stress signalling in plants. J Exp Bot 2004; 55(395): 225-236.
[5]   Thomashow MF. PLANT COLD ACCLIMATION: Freezing Tolerance Genes and Regulatory Mechanisms. Annu Rev Plant Physiol Plant Mol Biol 1999; 50: 571-599.
[6]   Shinozaki K, Yamaguchi-Shinozaki K. Molecular responses to dehydration and low temperature: differences and cross-talk between two stress signaling pathways. Curr Opin Plant Biol 2000; 3(3): 217-223.
[7]   Koornneef M, Leon-Kloosterziel KM, Schwartz SH, Zeevaart JAD. The genetic and molecular dissection of abscisic acid biosynthesis and signal transduction in Arabidopsis. Plant Physiol Biochem 1998; 36: 83-89.
[8]   Xiong L, Ishitani M, Lee H, Zhu JK. The Arabidopsis LOS5/ABA3 locus encodes a molybdenum cofactor sulfurase and modulates cold stress- and osmotic stress-responsive gene expression. Plant Cell 2001; 13(9): 2063-2083.
[9]   Uno Y, Furihata T, Abe H, Yoshida R, Shinozaki K, Yamaguchi-Shinozaki K. Arabidopsis basic leucine zipper transcription factors involved in an abscisic acid-dependent signal transduction pathway under drought and high-salinity conditions. Proc Natl Acad Sci USA 2000; 97(21): 11632-11637.
[10]   Lenka SK, Lohia B, Kumar A, Chinnusamy V, Bansal KC. Genome-wide targeted prediction of ABA responsive genes in rice based on over-represented cis-motif in co-expressed genes. Plant Mol Biol 2009; 69(3): 261-271.
[11]   Yamaguchi-Shinozaki K, Shinozaki K. Characterization of the expression of a desiccation-responsive rd29 gene of *Arabidopsis thaliana* and analysis of its promoter in transgenic plants. Mol Gen Genet 1993; 236(2-3): 331-340.

[12]   Knight H, Trewavas AJ, Knight MR. Calcium signalling in Arabidopsis thaliana responding to drought and salinity. Plant J 1997; 12(5): 1067-1078.

[13]   Thiery L, Leprince AS, Lefebvre D, Ghars MA, Debarbieux E, Savoure A. Phospholipase D is a negative regulator of proline biosynthesis in Arabidopsis thaliana. J Biol Chem 2004; 279(15): 14812-14818.

[14]   Mahajan S, Sopory SK, Tuteja N. Cloning and characterization of CBL-CIPK signalling components from a legume (*Pisum sativum*). FEBS J 2006; 273(5): 907-925.

[15]   Sanan-Mishra N, Pham XH, Sopory SK, Tuteja N. Pea DNA helicase 45 overexpression in tobacco confers high salinity tolerance without affecting yield. Proc Natl Acad Sci USA 2005; 102(2): 509-514.

[16]   Mahajan S, Sopory SK, Tuteja N. CBL-CIPK paradigm: role in calcium and stress signaling in plants. PINSA 2006; 72: 63-78.

[17]   Lee KH *et al.* Activation of glucosidase *via* stress-induced polymerization rapidly increases active pools of abscisic acid. Cell 2006; 126(6): 1109-1120.

[18]   Zhu JK. Salt and drought stress signal transduction in plants. Annu Rev Plant Biol 2002; 53: 247-273.

[19]   Verslues PE, Kim YS, Zhu JK. Altered ABA, proline and hydrogen peroxide in an Arabidopsis glutamate:glyoxylate minotransferase mutant. Plant Mol Biol 2007; 64(1-2): 205-217.

[20]   Nakashima K, Kiyosue T, Yamaguchi-Shinozaki K, Shinozaki K. A nuclear gene, erd1, encoding a chloroplast-targeted Clp protease regulatory subunit homolog is not only induced by water stress but also developmentally up-regulated during senescence in *Arabidopsis thaliana*. Plant J 1997; 12(4): 851-861.

[21]   Kim JB, Kang JY, Kim SY. Over-expression of a transcription factor regulating ABA-responsive gene expression confers multiple stress tolerance. Plant Biotechnol J 2004; 2(5): 459-466.

[22]   Reyes JL, Chua NH. ABA induction of miR159 controls transcript levels of two MYB factors during Arabidopsis seed germination. Plant J 2007; 49(4): 592-606.

[23]   Akihiro T *et al.* Genome wide cDNA-AFLP analysis of genes rapidly induced by combined sucrose and ABA treatment in rice cultured cells. FEBS Lett 2006; 580(25): 5947-5952.

[24]   Huang D, Wu W, Abrams SR, Cutler AJ. The relationship of drought-related gene expression in *Arabidopsis thaliana* to hormonal and environmental factors. J Exp Bot 2008; 59(11): 2991-3007.

[25]   Cuming AC, Cho SH, Kamisugi Y, Graham H, Quatrano RS. Microarray analysis of transcriptional responses to abscisic acid and osmotic, salt, and drought stress in the moss, Physcomitrella patens. New Phytol 2007; 176(2): 275-287.

[26]   Sreenivasulu N, Radchuk V, Strickert M, Miersch O, Weschke W, Wobus U. Gene expression patterns reveal tissue-specific signaling networks controlling programmed cell death and ABA- regulated maturation in developing barley seeds. Plant J 2006; 47(2): 310-327.

[27]   Kim BH, von Arnim AG. The early dark-response in Arabidopsis thaliana revealed by cDNA microarray analysis. Plant Mol Biol 2006; 60(3): 321-342.

[28]   Okamoto M *et al.* Genome-wide analysis of endogenous abscisic acid-mediated transcription in dry and imbibed seeds of Arabidopsis using tiling arrays. Plant J 2010; 62(1): 39-51.

[29]   Xin Z, Zhao Y, Zheng ZL. Transcriptome analysis reveals specific modulation of abscisic acid signaling by ROP10 small GTPase in Arabidopsis. Plant Physiol 2005; 139(3): 1350-1365.

[30]   Takahashi S *et al.* Monitoring the expression profiles of genes induced by hyperosmotic, high salinity, and oxidative stress and abscisic acid treatment in Arabidopsis cell culture using a full-length cDNA microarray. Plant Mol Biol 2004; 56(1): 29-55.

[31]   Yazaki J *et al.* Genomics approach to abscisic acid- and gibberellin-responsive genes in rice. DNA Res 2003; 10(6): 249-261.

[32]   Rabbani MA *et al.* Monitoring expression profiles of rice genes under cold, drought, and high-salinity stresses and abscisic acid application using cDNA microarray and RNA gel-blot analyses. Plant Physiol 2003; 133(4): 1755-1767.

[33]   Seki M *et al.* Monitoring the expression pattern of around 7,000 Arabidopsis genes under ABA treatments using a full-length cDNA microarray. Funct Integr Genomics 2002; 2(6): 282-291.

[34]   Yoshida T *et al.* AREB1, AREB2, and ABF3 are master transcription factors that cooperatively regulate ABRE-dependent ABA signaling involved in drought stress tolerance and require ABA for full activation. Plant J 2009; 61(4): 672-685.

[35]   Zhang W, Ruan J, Ho TH, You Y, Yu T, Quatrano RS. Cis-regulatory element based targeted gene finding: genome-wide identification of abscisic acid- and abiotic stress-responsive genes in *Arabidopsis thaliana*. Bioinformatics 2005; 21(14): 3074-3081.

[36]   Richardt S *et al.* Microarray analysis of the moss Physcomitrella patens reveals evolutionarily conserved transcriptional regulation of salt stress and abscisic acid signalling. Plant Mol Biol 2010; 72(1-2): 27-45.

[37]   Li Y *et al.* Establishing glucose- and ABA-regulated transcription networks in Arabidopsis by microarray analysis and promoter classification using a Relevance Vector Machine. Genome Res 2006; 16(3): 414-427.

[38] Matsui A *et al.* Arabidopsis transcriptome analysis under drought, cold, high-salinity and ABA treatment conditions using a tiling array. Plant Cell Physiol 2008; 49(8): 1135-1149.

[39] Buchanan CD *et al.* Sorghum bicolor's transcriptome response to dehydration, high salinity and ABA. Plant Mol Biol 2005; 58(5): 699-720.

[40] Chibani K, Ali-Rachedi S, Job C, Job D, Jullien M, Grappin P. Proteomic analysis of seed dormancy in Arabidopsis. Plant Physiol 2006; 142(4): 1493-1510.

[41] Pawlowski TA. Proteome analysis of Norway maple (*Acer platanoides* L.) seeds dormancy breaking and germination: influence of abscisic and gibberellic acids. BMC Plant Biol 2009; 9: 48.

[42] Pawlowski TA. Proteomics of European beech (*Fagus sylvatica* L.) seed dormancy breaking: influence of abscisic and gibberellic acids. Proteomics 2007; 7(13): 2246-2257.

[43] Lee S *et al.* Proteomic identification of annexins, calcium-dependent membrane binding proteins that mediate osmotic stress and abscisic acid signal transduction in Arabidopsis. Plant Cell 2004; 16(6): 1378-1391.

[44] Kim ST, Kang SY, Wang Y, Kim SG, Hwang du H, Kang KY. Analysis of embryonic proteome modulation by GA and ABA from germinating rice seeds. Proteomics 2008; 8(17): 3577-3587.

[45] Li XJ, Yang MF, Chen H, Qu LQ, Chen F, Shen SH. Abscisic acid pretreatment enhances salt tolerance of rice seedlings: Proteomic evidence. Biochim Biophys Acta 1804(4): 929-940.

[46] He H, Li J. Proteomic analysis of phosphoproteins regulated by abscisic acid in rice leaves. Biochem Biophys Res Commun 2008; 371(4): 883-888.

[47] Fujii H *et al. In vitro* reconstitution of an abscisic acid signalling pathway. Nature 2009; 462(7273): 660-664.

[48] Melcher K *et al.* A gate-latch-lock mechanism for hormone signalling by abscisic acid receptors. Nature 2009; 462(7273): 602-608.

[49] Nishimura N *et al.* Structural mechanism of abscisic acid binding and signaling by dimeric PYR1. Science 2009; 326(5958): 1373-1379.

[50] Miyazono K *et al.* Structural basis of abscisic acid signalling. Nature 2009; 462(7273): 609-614.

[51] Urano K *et al.* Characterization of the ABA-regulated global responses to dehydration in Arabidopsis by metabolomics. Plant J 2009; 57(6): 1065-1078.

[52] Alvarez S, Marsh EL, Schroeder SG, Schachtman DP. Metabolomic and proteomic changes in the xylem sap of maize under drought. Plant Cell Environ 2008; 31(3): 325-340.

CHAPTER 12

The Role of RNA Silencing in Plant Stress Responses

Ngoc Tuan Le and Ming-Bo Wang*

CSIRO Plant Industry, GPO Box 1600, Canberra, ACT 2601, Australia

Abstract: RNA silencing is an evolutionarily conserved mechanism in eukaryotes that control gene expression through small RNA-guided RNA degradation, translational repression and DNA methylation. Plants have evolved multiple small RNA pathways that have been demonstrated to play an essential role in developmental regulation and defence against invasive nucleic acids such as transposable elements and viruses. Recent studies have provided evidence that the different small RNA pathways play a more diverse role in plant defence against biotic and abiotic stresses. These findings are likely to result in new platforms for engineering stress tolerant crops in the future.

INTRODUCTION

RNA silencing is an eukaryotic mechanism for regulating the expression of endogenous or foreign genes in a sequence-specific manner. The RNA silencing pathways have been elucidated in a variety of eukaryotes ranging from fungi to humans [see [1] for a comprehensive review]. In plants, several distinct RNA silencing pathways have been characterized through molecular genetic studies with the model plant *Arabidopsis thaliana*. These silencing pathways share several common features (Fig. **1**). Firstly, the production of a double-stranded RNA (dsRNA) molecule is the trigger for RNA silencing (Fig. **1A**). dsRNA molecules may arise from self-complementary single-stranded RNAs (ssRNAs) which can fold on themselves to form dsRNA stem-loop hairpin structures or from ssRNAs that have been converted to dsRNAs through the action of one of six RNA-dependent RNA polymerase (RDR) enzymes. Secondly, the dsRNA molecules are processed and diced into small RNA (sRNA) fragments (~20-25nt in length) by one of four Dicer-like (DCL) enzymes (Fig. **1B**). DCL1 synthesises sRNAs of 20-22 nt in length, while the products of DCL2, DCL3 and DCL4 are 22 nt, 24 nt, and 21 nt respectively [2]. The dicing of long dsRNAs into sRNAs is facilitated by one of five double-stranded RNA binding (DRB) proteins (DRB1-5; DRB1 is also known as Hyponastic Leaves 1 or HYL1) (Fig. **1B**). Thirdly, the sRNAs are methylated at their 3' overhanging ends by the methyltransferase HUA Enhancer 1 (HEN1), which serves to stabilize the sRNAs and to protect them from degradation. The sRNAs are retained in the nucleus for transcriptional gene silencing (TGS) or they are exported to the cytoplasm, through the action of the exportin-5 homologue HASTY (HST), for post-transcriptional gene silencing (PTGS) (Fig. **1C**). Finally, one strand of the dsRNA molecule acts as a guide strand that is bound by one of ten Argonaute (AGO) proteins to form a ribonucleoprotein complex called an RNA-induced silencing complex (RISC). The guide strand provides the sequence specificity for AGOs to associate with partially or fully complementary target RNA (or DNA) for PTGS (or TGS). The target RNA/DNA is 'turned off' by one of three main mechanisms: by cleavage (degradation) of the target mRNA, by translational inhibition of the target mRNA, or by transcriptional silencing of the target gene through DNA methylation and/or histone (chromatin) modification (Fig. **1D** and **1E**).

Plant sRNAs have been classified according to their mode of biogenesis. MicroRNAs (miRNAs) originate from RNA polymerase II-transcribed RNAs that form imperfect hairpin structures which are cleaved in the nucleus by DCL1 into 20-24 nt dsRNA fragments. Accurate processing of miRNAs is provided by DRB proteins including HYL1 and the C_2H_2 Zn-finger protein SERRATE (SE) [3]. Trans-acting small interfering RNAs (ta-siRNAs) are 21-nt sRNAs generated from non-coding RNA precursors that are targeted for miRNA-mediated cleavage. The biogenesis of tasi-RNAs, which are found only in plants, requires DCL1 and DCL4, as well as HYL1, HEN1 and RDR6 [4]. Repeat-associated siRNAs (rasiRNAs), as their name suggests, are a unique class of 24-nt siRNAs that originate from transposons and other repetitive sequences in the genome. RasiRNAs are the predominant population

*Address correspondence to: Dr. Ming-Bo Wang, CSIRO Plant Industry, GPO Box 1600, Canberra, ACT 2601, Australia; E-mail: ming-bo.wang@csiro.au; ngoc.le@csiro.au

Narendra Tuteja, Sarvajeet Singh Gill and Renu Tuteja [Eds.]

of sRNAs in plants [5]. Genetic analyses of insertion mutants of *A. thaliana* have revealed that RDR2 is required for the production of all 24-nt endogenous rasiRNAs [2]. Natural antisense siRNAs (nat-siRNAs) are another class of endogenous siRNAs which are produced from transcription of gene pairs on opposite DNA strands [6]. The paired transcripts share regions of complementarity at their 3' ends and are able to form dsRNA transcripts that are cleaved by DCL2 into 24-nt nat-siRNAs. The resultant 24-nt nat-siRNAs subsequently target transcripts of one of the gene pairs for cleavage.

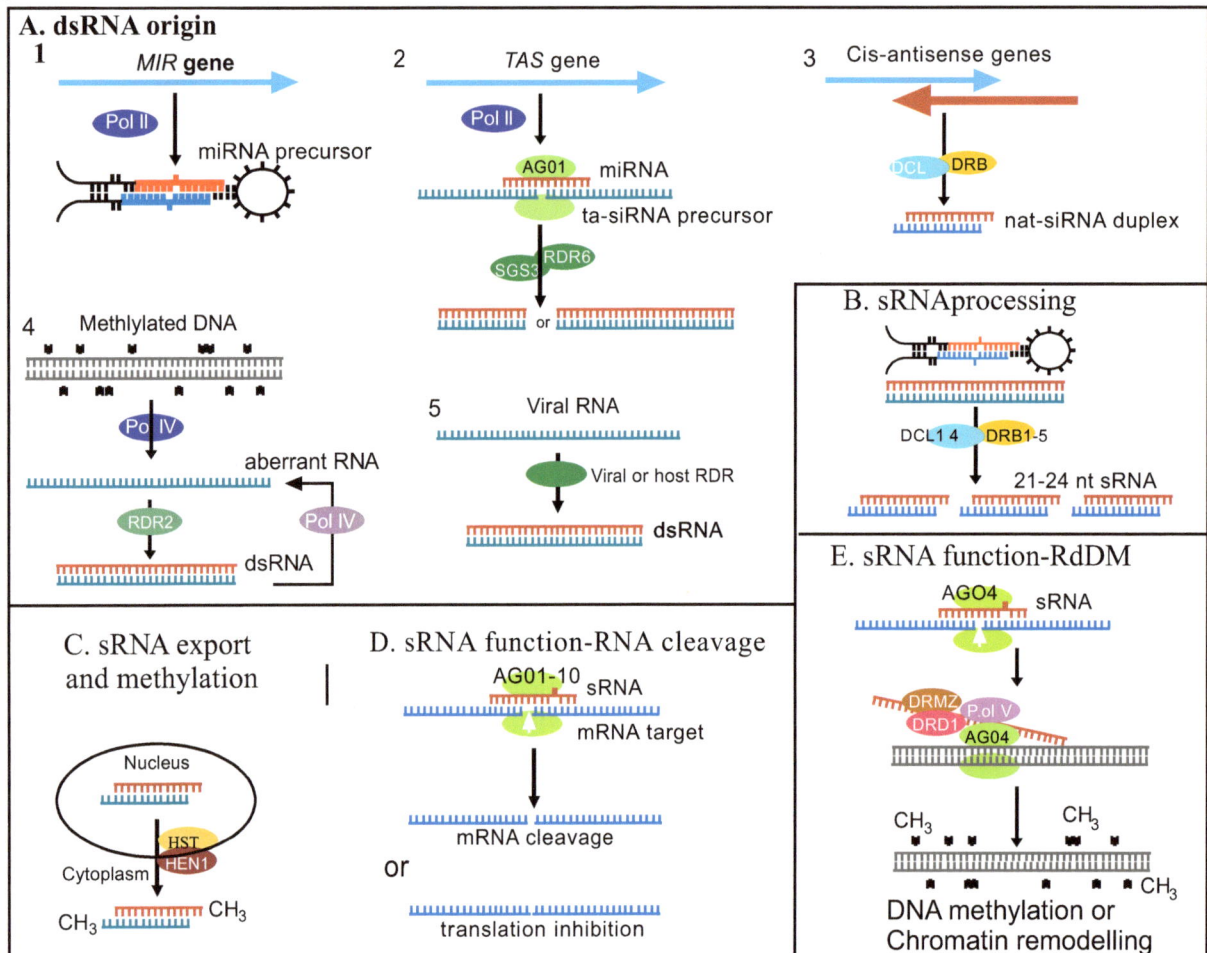

Figure 1: Common features of RNA silencing pathways in plants. A. The dsRNA trigger arises from different sources for the five classes of small RNAs discussed in this review. miRNA precursors are transcribed by RNA polymerase II (Pol II) from MIR genes that normally reside in intergenic regions, which fold back to itself to form an imperfect hairpin-loop structure (1). dsRNA for trans-acting siRNAs (ta-siRNAs) is synthesized by RNA-dependent RNA polymerase 6 (RDR6) from miRNA-cleaved TAS gene transcript (2). Natural antisense siRNAs (nat-siRNAs) are derived from dsRNA formed between complementary regions of two overlapping transcripts (3). For repeat-associated or hetechromatic siRNAs (ra-siRNAs), dsRNA is synthesized by RDR2 from aberrant ssRNA that is transcribed from methylated DNA or dsRNA by DNA-dependent RNA polymerase IV (Pol IV) (4). Viral dsRNA is derived from single-stranded viral RNA by viral- or host-encoded RDRs (5). B. Slicing of hairpin RNA (hpRNA) or dsRNA molecules into 21-24nt small RNAs by Dicer-like proteins (DCLs) involving double-stranded RNA binding proteins (DRBs). Four DCLs and five DRBs have been identified in Arabidopsis thaliana. DCL1 and DRB1 (or HYL1) are required for miRNA processing, while DCL4 and DRB4 for vsRNA and ta-siRNA processing. Ra-siRNAs are processed by DCL3. DCL2 is involved for the biogenesis of vsRNAs, transgene-associated siRNAs and a nat-siRNA in Arabidopsis. The specific function of the remaining three DRBs remains unclear. C. Methylation and stabilization of the small RNA duplex by HUA ENHANCER 1 (HEN1) and export to the cytoplasm through the action of HASTY (HST). D. Post-transcriptional RNA silencing via cleavage or translational inhibition of the target RNA. Single-stranded small RNA (sRNA) binds to one of ten Argonaute (AGO) proteins to form an RNA-induced silencing complex (RISC), and guide the RISC to bind and cleave homologous target RNA. E. Transcriptional gene silencing by RNA-directed DNA methylation (RdDM) and chromatin remodelling. Ra-siRNAs bind to AGO4, which in turn interacts with nascent RNA transcribed by DNA-dependent RNA polymerase V (Pol V) with the assistance of the chromatin remodelling factor DRD1. The de novo cytosine methyltransferase DRM2 is then recruited to adjacent DNA to cause cytosine methylation. (Figure is modified from ref [2])

The different classes of sRNAs identified in plants have been shown to negatively regulate the expression of endogenous genes or to repress the replication of viruses [7]. MiRNAs negatively regulate the expression of transcriptional factors and other regulatory genes and therefore play an essential role in control of plant development [8]. Ta-siRNAs also target regulatory genes and play a role in several aspects of plant development such as juvenile to reproductive phase transition and adaxial-abaxial patterning of lateral organs [9]. Ra-siRNAs guide sequence-specific cytosine methylation to target DNA leading to its transcriptional inactivation and play a key role in silencing of transposons and repetitive sequences [5]. Nat-siRNAs have not been associated with developmental control, but like miRNAs and ta-siRNAs, have recently been implicated in biotic and abiotic stress responses in plants [10]. While the diverse functions of plant small RNAs in developmental control have been extensively investigated in the past few years, this review focuses on the emerging evidence indicating an important role of small RNAs in plant responses to abiotic and biotic stress.

ABIOTIC STRESS

Plants are sessile organisms that must be able to respond to their constantly changing environment in a timely fashion. Throughout their life cycle, plants are potentially subjected to a variety of abiotic stresses including drought, cold, high salinity and hypoxia. In recent years, small non-coding RNAs such as miRNAs and siRNAs have emerged as important regulators of plant responses to abiotic stress [11]. Although the various types of abiotic stress will be discussed separately, many small RNAs are responsive to one or more form of stress (see below).

Drought

In *A. thaliana* seedlings exposed to dehydration, salinity, or cold stress or to the plant stress hormone abscisic acid (ABA), miR393 was strongly up-regulated by all stress treatments, while miR397b and miR402 were only slightly up-regulated [12]. In contrast, miR389a was down-regulated by all of the stress treatments. The expression of miR393 is also induced by drought stress in rice [13]. The predicted targets of miR393 include four closely related F-box auxin receptor proteins, including transport inhibitor response1 (TIR1), and a basic-helix-loop-helix (bHLH) protein [8,12]. TIR in turn targets AUX/IAA proteins for degradation by SCF E3 ubiquitin ligases in an auxin-dependent manner. Thus, the induction of miR393 under abiotic stress conditions leads to increased mRNA degradation or translational repression of TIR1 and this results in the repression of auxin signaling and seedling growth under stressful conditions. In a more recent study, Liu and colleagues identified 14 miRNAs which are induced by drought, salt and/or cold in *A. thaliana* [14]. Among these, miR168, miR171, and miR396 were responsive to all three types of stresses. These studies suggest that up-regulated miRNAs target negative regulators of stress responses or they target positive regulators of processes that are inhibited by stresses (e.g., cell division and expansion). In contrast, down-regulated miRNAs could repress the expression of stress-induced genes or positive regulators of stress responses [11].

Cold

As previously demonstrated, there is significant cross-talk among the salt, drought, and cold stress signaling pathways and several miRNAs are commonly induced by these stress treatments in both *Arabidopsis* [14] and rice [15]. However, the induction of some miRNAs is specific for low temperature stress. For example, miR319c was only induced by cold stress in *Arabidopsis* but not by drought, salt or ABA treatments [12]. In *Populus trichocarpa* (Ptc), Lu and co-workers identified 68 putative miRNA sequences (classified into 27 families, including 9 novel families, based on sequence homology) that were induced by various abiotic stresses. Among these, 19 Ptc-miRNAs were found to be cold stress-responsive [16]. Recently, deep sequencing of small RNAs in the model monocot plant, *Brachypodium distachyon*, led to the identification of 3 conserved miRNAs and 25 novel miRNAs which showed significant changes in expression during cold stress [17]. Two-thirds of the novel miRNAs were down-regulated by cold treatment, while the 3 conserved miRNAs (miR169e, miR172b and miR397) were up-regulated. These 3 miRNAs were previously found to be induced by low temperature in *Arabidopsis* [14] and poplar [16], suggesting that they are important regulators of the cold stress response in dicotyledonous and monocotyledonous plants.

Salt

In *A. thaliana*, salt stress induces the expression of a transcript (*At5g62520*) of unknown function (designated as *SRO5*), which is complementary to the constitutively expressed *P5CDH* (*pyrroline-5-carboxylate dehydrogenase*) transcript [6]. These two mRNAs are able to associate with each other and form a dsRNA molecule which is then processed by the RNA silencing machinery into a 24-nt nat-siRNA. The 24-nt nat-siRNA guides the cleavage of the *P5CDH* mRNA and leads to the suppression of proline degradation. This results in proline accumulation which is an important adaptive response to alleviate high salt stress in plants [6].

In rice, miR169g, miR169n and miR169o were identified as salt-responsive miRNAs [18]. All 3 miRNAs selectively targets a CCAAT-box binding transcription factor for mRNA cleavage. The promoter of *miR169o* contains an ABA-responsive element (ABRE), suggesting that it may also be regulated by ABA. On the other hand, *miR169g* is also induced by drought and its promoter contains a dehydration-responsive element (DRE). In contrast, miR169a and miR169c are down-regulated by drought stress in *A. thaliana* [19].

Osmotic Stress

Plants accumulate toxic reactive oxygen species (ROS) when they are exposed to stressful conditions such as salt, drought, cold, excessive light and heavy metals. ROS must be scavenged and detoxified by enzymes such as superoxide dismutases to limit damage to cellular membranes and organelles. In *Arabidopsis*, Cu-Zn superoxide dismutase1 (CSD1) and CSD2 play important roles in the scavenging of ROS in the cytosol and chloroplast, respectively. Both *CSD1* and *CSD2* are negatively regulated by miR398 [20]. Under oxidative stress conditions such as high light, heavy metals or treatment with the ROS-inducer methyl viologen, suppression of the *CSD1* and *CSD2* genes in *Arabidopsis* seedlings is released through the transcriptional down-regulation of miR398 [20]. Salt stress in two-week-old *A. thaliana* seedlings also resulted in the down-regulation of miR398 and the corresponding up-regulation of *CSD1* and *CSD2* transcripts [21]. However, in the woody plant *Populus tremula*, miR398 was induced upon salt or ABA treatment. This indicates that miR398 is differentially regulated by abiotic stress in a herbaceous annual plant (*Arabidopsis*) and a woody perennial plant (*Populus*) [21].

UV Stress

Continual depletion of the stratospheric ozone layer exposes plants to more harmful UV-B radiation. Recent evidence suggests that UV-induced changes in gene expression, mediated by miRNAs, are necessary for plants to cope with this adverse stress. Using computational methods, Zhou and co-workers identified 21 miRNA genes (belonging to 11 miRNA families) that are up-regulated under UV-B stress conditions in *A. thaliana* [22]. More recently, 24 UV-B stress-responsive miRNAs (13 up-regulated and 11 down-regulated) were identified in *P. tremula* [23]. Among these, miR156, miR160, miR165/166, miR167, and miR398 are up-regulated by UV-B radiation in both *Arabidopsis* [22] and *P. tremula* [23], indicating that these UV-B-responsive miRNAs may be conserved across plant species. In contrast, miR159 and miR393 were induced in *A. thaliana* [22], but reduced in *P. tremula* [23], suggesting that the expression patterns of some UV-B responsive miRNAs exhibit species-specific differences. In addition, some miRNAs are involved in multiple stress responses in plants. For example, miR156, miR160, miR164, miR168a/b, miR169, miR390, miR393, and miR395 were up-regulated in response to both UV-B radiation [23] and cold stress [16].

Hypoxia

Low-oxygen stress (hypoxia), commonly observed in natural events such as waterlogging, is a major limiting factor on plant productivity. In a recent study, Moldovan and colleagues identified 65 unique miRNA sequences from 46 families, and 14 tasiRNAs from three families in *Arabidopsis* plants under hypoxic conditions [24]. Some hypoxia-induced miRNAs have previously been shown to be expressed in *Arabidopsis* during viral infection while tasiRNAs are thought to play a role in regulating mitochondrial respiration and energy production [24]. In addition, a potential nat-siRNA implicated in the regulation of lipid signalling is also induced in *Arabidopsis* during low oxygen stress [25]. Several of the hypoxia-induced miRNAs, most notably miR166, miR167, and miR171, were also up-regulated in the roots of submerged maize plants [26]. Interestingly, miR166 also regulates root and nodule development in *M. truncatula* [27], suggesting that miR166 serves several root-specific functions.

Nutrients

Essential nutrients such as inorganic phosphate (P_i) and sulfur are limiting factors for plant growth. In *A. thaliana*, the expression of miR399 is strongly induced in P_i-starved plants, an induction that is mediated by PHR1 (Phosphate

Starvation Response1), a MYB-like transcriptional activator [28]. Interestingly, a substantial amount of miR399* (the sequence on the opposite strand of the miRNA399) also accumulated under P_i deficiency, and that, like miR399, miR399* can move across the graft junction from scions to rootstocks [29]. This result indicates that both miR399 and miR399* may serve important physiological functions. miR399 is a negative regulator of *UBC24*, a gene encoding a putative ubiquitin conjugating enzyme involved in protein degradation [12]. The protein targets of UBC24 are thought to include transcriptional activators of genes induced during P_i-starvation conditions [30].

P_i deficiency in *A. thaliana* also up-regulates the expression of the non-protein coding gene *IPS1* (*INDUCED BY PHOSPHATE STARVATION 1*) [31]. Interestingly, *IPS1* contains a 24-nt sequence with almost complete complementarity to miR399. However, there is a mismatch between bases 10 and 11, where mRNA cleavage is expected to occur. The authors show that *IPS1* mRNA is not cleaved but instead its function is to inhibit miR399 activity by sequestering the miRNA itself. In addition, *IPS1* over-expression results in increased accumulation of the *PHO2* mRNA, a target of miR399, and also in reduced shoot P_i content [31]. This mechanism of inhibition of miRNA activity by sequestration has been termed 'target mimicry'. Target mimicry may also occur in other plants such as *M. truncatula*, tomato and rice where non-protein coding transcripts possessing the 24-nt sequence complementary to miR399 have been shown to be expressed under conditions of P_i deficiency [see [32] and refs therein].

Several other miRNAs have also been discovered to be differentially expressed in response to P_i starvation. In *Arabidopsis*, Hsieh and co-workers found that the expression of miR156, miR399, miR778, miR827, and miR2111 was induced, whereas the expression of miR169, miR395, and miR398 was repressed, in response to P_i deficiency [29]. The authors also discovered significant cross-talk co-ordinated by these miRNAs under different nutrient deficiencies. Recently, in P_i-deficient white lupin (*Lupinus albus*) plants, Zhu and co-workers identified 167 miRNAs belonging to 35 miRNA families that show differential expression [33].

Another miRNA that plays a role in nutrient homeostasis is miR395, which has been shown to be strongly induced under conditions of sulfur starvation in *Arabidopsis* [8]. MiR395 targets several ATP sulfurylase genes, as well as a low-affinity sulfate transporter, and thus has been implicated to play a role in regulating sulfate translocation and assimilation. Interestingly, in the woody plant *P. tremula*, two miRNAs implicated in playing important regulatory roles in nutrient homeostasis, namely miR395 and miR398, are induced by salt stress or ABA treatment [21].

BIOTIC STRESS

At any point in their lifecycle, plants are potentially subjected to attack from many pathogens including viruses, bacteria, fungi, nematodes and insects. Work in the past decade has highlighted the various RNA silencing mechanisms, involving miRNAs and siRNAs, that plants have evolved to defend themselves against invading organisms [1]. Upon encountering a pathogen, plants recognize pathogen-associated molecular patterns (PAMPs) and in turn differentially regulate a subset of miRNAs or siRNAs that play important roles in plant defence.

Viruses

Most pathogenic plant viruses have single-stranded RNA genomes which, during an infection, are replicated into dsRNA by a viral- or host-encoded RDR. Among the six predicted *RDRs* in the *A. thaliana* genome, *RDR1*, *RDR2* and *RDR6* have been shown to play major and redundant roles in the production of dsRNAs from RNA viruses such as tobacco rattle virus [34] and cucumber mosaic virus [35] and possibly other RNA viruses. The viral-derived dsRNAs produced by RDRs are recognized by one or more of the host DCLs and processed into 21- to 24-nt viral siRNAs (vsRNAs). DCL4 primarily processes viral dsRNAs into 21-nt vsRNAs. When *DCL4* is mutated or its function is suppressed by viruses (see below), DCL2 can act as the anti-viral dicer and process viral dsRNAs into 22-nt vsRNAs [36,37]. Infection of plants by DNA viruses is also associated with the accumulation of 21-24 nt vsRNAs, presumably derived from DCL processing of dsRNAs synthesized by host RDRs from single-stranded viral transcripts [38]. All four *Arabidopsis* DCLs have been shown to be required for vsRNA production from DNA viruses [39,40]. Additionally, optimal processing of vsRNAs from DNA, as well as RNA, viruses also requires DRB4 [41]. These vsRNAs are bound by AGOs to guide the RISC to target viral genomic RNAs for degradation or viral genomic DNA for methylation in infected plants, providing a sequence-specific defence mechanism against viral infection. Among the ten *Arabidopsis* AGOs, anti-viral activities have been shown for AGO1 and AGO7 [41].

To evade the host-encoded RNA silencing mechanisms, viruses have evolved silencing suppressor proteins which inhibit diverse steps within the various RNA silencing pathways. The helper component-proteinase (HC-Pro) protein, a viral suppressor of gene silencing (VSR) from potyviruses such as the turnip mosaic virus (TuMV), interferes with the function of RISC and therefore inhibits the activity of miRNAs (e.g. miR171) and siRNAs which could result in developmental defects in vegetative and reproductive organs of plants [42]. The turnip yellow mosaic virus VSR, p69, suppresses the siRNA pathway but induces the expression of the miRNA-processing *DCL1* and several miRNAs as well as a corresponding enhanced cleavage of the target host mRNAs [43]. Other VSRs have been shown to interfere with the antiviral activities of DCL4, DRB4, RDR6/SGS3, AGO1 and HEN1 [see [1] for a recent review]. Other viral strategies to suppress or avoid host-directed RNA silencing include: sequestration of viral small RNAs (vsRNAs) of particular lengths (e.g. 21-nt) so as to prevent assembly of RISC complexes, viral inhibition of miR162 which targets *DCL1* (a negative regulator of the antiviral *DCL4*), and formation of stable secondary structures within single-stranded viral RNAs that are resistant to small RNA-directed cleavage.

Bacteria

Navarro and co-workers showed that miR393 is induced in *Arabidopsis* plants treated with a bacterial flagellin-derived peptide [44]. Similarly, in an independent study, enhanced miR393 accumulation was observed in *Arabidopsis* plants challenged with *Pseudomonas syringae* pv. tomato (*Pst*) DC3000 *hrcC*, which lacks a functional type III-secretion system required for virulence [45]. In addition, transgenic plants over-expressing miR393a exhibited a higher level of resistance to virulent *Pst* DC3000 [44]. Similar to the abiotic stress response (see above), induction of miR393 results in repression of auxin signaling and plant growth which, in turn, restricts the growth of the pathogen. Thus, auxin promotes disease susceptibility and miRNA-mediated suppression of auxin signaling appears to be an important mechanism in pathogen resistance [44].

In contrast to miR393, infection of *Arabidopsis* plants with *Pst* DC3000 *hrcC* also led to the down-regulation of several miRNAs, most notably miR825 which may target at least three positive regulators of PAMP-triggered immunity (PTI) [45,46]. In addition, non-pathogenic *Pseudomonas fluorescens* and *Escherichia coli* strains and non-virulent *Pst* DC3000 *hrcC* were able to grow in some miRNA-deficient mutants of *Arabidopsis* (e.g. *dcl1* and *hen1*) [47]. Thus, these results implicate miRNAs as key components of plant basal defence against pathogenic bacteria.

The role of siRNAs and long siRNAs (lsiRNAs), which are a recently discovered new class of siRNAs that are typically 30-40nt in length, in plant defence against bacterial pathogens have also been identified recently [10,46]. Infection of *Arabidopsis* plants with *P. syringae* carrying the avirulence (avr) gene, *avrRpt2*, but not with other *Ps* strains carrying different *avr* genes, induces the expression of a 22-nt nat-siRNA called nat-siRNAATGB2 [10]. This induction requires several genes involved in RNA silencing (among them, *DCL1* and *RDR6*), as well as the host disease resistance (R) gene, *RPS2*. The nat-siRNAATGB2 is thought to repress a negative regulator of the *RPS2*-mediated host resistance pathway. This siRNA is also complementary to the antisense transcript of a pentatricopeptide-repeat protein-like (*PPRL*) gene, and may potentially down-regulate *PPRL* expression [10]. Infection of *Arabidopsis* plants with *Ps avrRpt2* also induces AtlsiRNA-1, a lsiRNA derived from the overlapping region of a natural antisense transcript (NAT) pair between a putative leucine-rich repeat receptor-like protein kinase (SSRLK) gene and a putative RNA-binding domain abundant in Apicomplexans (RAP) gene [46]. AtlsiRNA-1 is complementary to the 3' UTR of the *AtRAP* gene, a negative regulator of plant resistance to bacterial pathogens. It is hypothesized that AtlsiRNA-1 enhances plant disease resistance by down-regulating *AtRAP* through mRNA decapping and XRN4-mediated 5'-to-3' degradation [46].

Like viruses, pathogenic bacteria are known to secrete effector proteins that inhibit the host RNAi-mediated defence mechanisms to cause disease. When infecting *A. thaliana* plants, pathogenic *P. syringae* bacteria secrete effectors that suppress transcriptional activation of some PAMP-responsive miRNAs or inhibit miRNA biogenesis, stability, or activity [47]. The bacterial effector HOPT1-1 inhibits mRNA cleavage or translational repression of several host miRNA targets, possibly by direct interference with AGO1 or with a component of AGO1-mediated RISC [47]. Another Argonaute protein, AGO4, is required for resistance to *Pseudomonas* in *Arabidopsis* [48]. Recessive *ago4* mutants exhibit enhanced disease susceptibility to virulent *Pst* DC3000 as well as to the avirulent *Pst* DC3000 (*avrRpm1*) and the non-host bacterium *P. syringae* pv *tabaci*. AGO4 is one of the critical components of the transcriptional silencing process mediated via the RNA-directed DNA methylation (RdDM) pathway (Fig. **1E**). This involvement of an RdDM factor in bacterial resistance suggests a potential role of epigenetic control of disease resistance gene expression in plants. Consistent with this possibility, recent studies have shown that epigenetic changes, due to alteration in DNA methylation or histone modification, affect the expression of many host disease resistance (R) genes and have been shown to be important regulators of plant defence against bacterial [49] and fungal [50] pathogens.

Interactions between plants and bacteria can also be symbiotic and are also regulated by miRNAs. In the legume *Phaseolus vulgaris*, miR169b was isolated from seedlings infected with *Rhizobium tropici* [51]. MiR169 also accumulates in the developing nodules of *Medicago truncatula* and has been shown to regulate the expression of *MtHAP2-1*, a transcriptional regulator of symbiotic nodule development [52]. Several other miRNAs, including some novel and species-specific miRNAs, have been shown to play a role in the symbiosis between rhizobia and their host plants [53,54]. Interestingly, miR169 is also induced by abiotic stresses (see above) and by ABA treatment in *P. vulgaris* seedlings [51]. This suggests that, apart from its role in plant response against different osmotic stresses, miR169 also plays a role in symbiotic nodule development.

Fungi

Fungi are known to possess fully functional RNA silencing machinery [55]. In order to facilitate disease, like bacteria and viruses, fungi also secrete effectors (fungal proteins and small molecules) into the host to alter cell structure or function [56]. However, it is currently unknown whether these effectors, if any, specifically inhibit the host RNA silencing machinery to cause disease symptoms. Several recent studies have shown that RNA silencing is also important for plant defence against fungal pathogens. Mutation in *SGS1* (*Suppressor of Gene Silencing 1*), *SGS2* (also known as *RDR6* or *Silencing Defective1/SDE1*) or *SGS3* results in *Arabidopsis* plants becoming more susceptible to the wilt fungus *Verticillium dahliae* [57]. The *sgs1* mutant is impaired in PTGS [58], while *SGS2* and *SGS3* are essential for dsRNA synthesis in different RNA silencing pathways [59]. Recently, it has been shown that 26 miRNAs (corresponding to 11 miRNA families) from stem xylem of loblolly pine (*Pinus taeda*) were differentially expressed during infection by the rust fungus *Cronartium quercuum f. sp. fusiforme* [60]. Four of the 11 miRNA families (miR156, miR159, miR160, and miR319) are conserved, while the other 7 families (numbered from miR946 to miR952) are novel and have so far been found only in *P. taeda*. Members of these 11 miRNA families are predicted to target dozens of disease-related genes in *P. taeda*. Interestingly, expression of 10 of these 11 miRNA families (the notable exception being miR947) was significantly repressed in infected stems compared to healthy stems of loblolly pine [60].

Nematodes

Small RNAs have recently been shown to be involved in interactions between plants and parasitic cyst nematodes. Several siRNAs and miRNAs, including miR160, miR164, miR167, miR171, miR396, and miR398, were differentially expressed in *Arabidopsis* roots infected with the sugar beet cyst nematode *Heterodera schachtii* [61]. These miRNAs accumulated during infection while, concomitantly, transcripts of their corresponding target genes were reduced in abundance. In addition, wildtype *Arabidopsis* plants are more susceptible to *H. schachtii* infection than *dcl* and *rdr* mutants, suggesting a negative regulation of the small RNA silencing pathways during the plant–nematode interaction [61].

Insects

The role of RNA silencing in herbivore resistance was recently identified in the native tobacco *Nicotiana attenuata*. Pandey and colleagues demonstrated that silencing of the RNA-dependent RNA polymerases, *NaRDR1* and *NaRDR3*, makes *N. attenuata* plants more susceptible to herbivorous insects and reduces their fitness in nature [62,63]. The authors also identified NaRDR1-specific siRNAs by comparing the small RNA profiles of wildtype *N. attenuata* and *NaRDR1*-silenced transgenic plants before and after elicitation from oral secretions of the herbivorous *Manduca sexta* insect larvae [64]. These siRNAs are predicted to target genes involved in signal transduction of jasmonic acid (JA) and salicylic acid. These phytohormones are important mediators of plant defence against herbivore attack, as demonstrated by the complete restoration of insect resistance in *NaRDR1*-knockdown plants by exogenous application of JA [64].

Table 1: Small RNAs responsive to abiotic and biotic stress in plants.

Small RNA	Stress condition	Response	Plant	Function of small RNA	Ref.
miR156 miR159	*Cronartium quercuum* (rust fungus)	Down	Pta	Regulation of SPL and MYB TFs respectively involved in phase transition and flowering	[45]
miR160	(a) *Cronartium quercuum* (b) *Heterodera schachtii* (cyst nematode)	Down Up	Pta At	Regulation of Auxin Response Factors and growth	[45] [46]
miR164	*Heterodera schachtii* (cyst nematode)	Up	At	Regulation of NAC TFs involved in shoot apical meristem formation	[46]
miR166	(a) Hypoxia (b) Nodulation	Up Up	At Zm Mt	Regulation of HD-Zip TFs involved in axillary meristem initiation and leaf development	[17] [19] [20]
miR167	(a) Hypoxia (b) *Heterodera schachtii*	Up Up	At Zm At	Regulation of Auxin Response Factors and growth	[17] [19] [46]
miR168	Cold, drought, salt	Up	At	Regulation of ARGONAUTE1	[8]
miR169a miR169c	Drought	Down	At	Regulation of CCAAT binding factors	[14]
miR169b	*Rhizobium tropici* ABA	Up	Pv	Regulation of CCAAT binding factors involved in nodule development	[36]
miR169g miR169n miR169o	Salt	Up	Os	Regulation of CCAAT binding factors	[13]
miR169e	Cold	Up	At Bd Ptc	Regulation of CCAAT binding factors	[8] [11] [10]
miR171	(a) Cold, drought, salt (b) Hypoxia (c) *Heterodera schachtii*	Up Up Up	At At Zm At	Regulation of SCARECROW-like TFs involved in light/GA signaling and radial patterning in roots	[8] [17] [19] [46]
miR172b	Cold	Up	At Bd Ptc	Regulation of AP2 and AP2-like TFs involved in flowering	[8] [11] [10]
miR319	(a) Cold (b) *Cronartium quercuum*	Up Down	At Pta	Regulation of MYB and TCP TFs	[6] [45]
miR389a	Cold, drought, salt, ABA	Down	At	Unknown	[6]
miR393	(a) Cold, drought, salt, ABA (b) drought (c) Bacterial flagellin (d) *Pseudomonas syringae*	Up Up Up Up	At Os At At	Repression of auxin signaling and growth	[6] [7] [28] [29]
miR395	(a) Sulfur deficiency (b) Salt, ABA	Up Up	At Ptm	Regulation of sulfate translocation and assimilation	[3] [16]
miR396	(a) Cold, drought, salt (b) *Heterodera schachtii*	Up Up	At At	Regulation of Growth Regulating Factors	[8] [46]
miR397	Cold	Up	At Bd Ptc	Regulation of genes encoding laccases and beta-6 tubulin	[8] [11] [10]
miR398	(a) Oxidative stress (high light, heavy metals, methyl viologen) (b) Salt (c) Salt, ABA (d) *Heterodera schachtii*	Down Down Up Up	At At Ptm At	Suppression of superoxide dismutases (CSD1 and CSD2)	[15] [16] [16] [46]
miR399	Phosphate deficiency	Up	At	Suppression of *UBC24*, a gene encoding a putative ubiquitin conjugating enzyme involved in protein degradation	[21], [22]
SRO5-P5CDH nat-siRNA	Salt	Up	At	Activation of proline accumulation	[12]
nat-siRNA ATGB2	*Pseudomonas syringae*	Up	At	Activates *RPS2*-mediated host resistance	[32]
At-lsiRNA-1	*Pseudomonas syringae*	Up	At	Down-regulates *AtRAP* (a negative regulator of host resistance)	[30]

Abbreviations: At Arabidopsis thaliana, Bd Brachypodium distachyon, Mt Medicago truncatula, Os Oryza sativa, Pv Phaseolus vulgaris, Pta Pinus taeda, Ptc Populus trichocarpa, Ptm Populus tremula, Zm Zea mays, TFs transcription factors.

CONCLUSION

This review has discussed the role of RNA silencing in plant responses against biotic and abiotic stresses as separate entities. However, this situation may be more complex in the natural world. Several studies have shown that some abiotic stress-induced miRNAs are also regulated in response to biotic stress (see Table 1, e.g. miR393). Lu and co-workers [16] noted that several cold-induced miRNAs in poplar, including miR156 and miR160, were also induced in response to viral infection in both tobacco [65] and *Arabidopsis* [43,66]. Interestingly, these two miRNAs were significantly repressed in the stem of loblolly pine (*Pinus taeda*) infected with the rust fungus *C. quercuum* [60]. These studies indicate that cross-talk exists between miRNA pathways for both biotic and abiotic stress responses. These complex regulatory networks of miRNAs contribute to the ability of plants to survive in their rapidly changing and stressful environment [16].

Many miRNAs are conserved across plant species, suggesting that these miRNAs share conserved biological functions. However, some miRNAs are differentially regulated in response to stress among different plant species. For example, salt stress induces the expression of *miR398* in poplar but down-regulates its expression in *Arabidopsis* [21]. MiRNA genes, like most RNA polymerase II transcripts, have been shown to be transcriptionally regulated. Zhao and colleagues recently showed that the promoters of several rice *miR169* genes contain stress-responsive elements such as ABRE and DRE [18]. Other regulatory elements involved in plant stress responses will no doubt be discovered as the promoters of more miRNA genes are analysed. It is expected that miRNAs will also be regulated at the post-transcriptional level by processes such as RNA processing and stability. However, little is known about how siRNAs, for example the nat-siRNAs, are regulated in response to stress. It has been estimated that >30% of the *A. thaliana* genome can produce transcripts from both sense and antisense strands to form nat-siRNAs [67], some of which may be transcriptionally induced by different biotic and abiotic stresses. As more nat-siRNAs are discovered and analysed, the regulation of these small RNAs during stress will also become better characterized. It is also becoming clearer that small RNA-directed epigenetic variation, e.g. DNA methylation and/or histone modification, can affect the expression of many genes that mediate defence responses against abiotic and biotic stress [50,68].

Elucidation of the regulatory roles of miRNAs/siRNAs in plant stress responses has provided scientists with a novel strategy for genetically engineering plants for improved stress tolerance. Sunkar and co-workers showed that transgenic *Arabidopsis* plants over-expressing a form of the *CSD2* gene that is resistant to miR398-mediated cleavage actually accumulate more *CSD2* mRNA than plants over-expressing 'normal' *CSD2* and, compared to wildtype plants, are much more tolerant to heavy metal and excessive light stress [20]. Thus, attenuating miR398-guided suppression of *CSD2* in transgenic plants is an effective new approach to improving plant productivity under oxidative stress conditions [20]. RNAi technology has also been successfully used to control insect pests of corn and cotton respectively [69,70]. The advantage of RNAi technology, over the use of single genes, is that it is possible to manipulate the expression of many target genes by over-expressing or silencing only one miRNA/siRNA molecule.

Recent discovery of the role of small RNAs in stress responses has facilitated the diagnosis and screening of various diseases that inflict both animals and plants. In animals, miRNA expression data in various types of cancers demonstrate that cancer cells have different miRNA profiles compared with normal cells, thus highlighting the remarkable diagnostic and therapeutic potential of miRNAs in cancer [71]. In plants, specific miRNAs are responsive to specific stress conditions and these may be potentially useful as biomarkers for diagnosing infection or disease in natural habitats. For example, in *Arabidopsis* plants, infection with TuMV down-regulated the expression of miR171 [42] whereas infection with tobacco (or oilseed rape) mosaic virus up-regulates miR164a expression [72]. Interestingly, the level of miR164a induction was positively correlated with the severity of disease [72]. As deep sequencing technology becomes more affordable, it will be possible to use miRNA profiling in whole plants (or even whole field populations) for the diagnosis of specific diseases.

ACKNOWLEDGMENT

We wish to thank Dr Iain Wilson for critical reading of the manuscript.

REFERENCES

[1]   Ruiz-Ferrer V, Voinnet O. Roles of plant small RNAs in biotic stress responses. Ann Rev Plant Biol 2009; 60: 485-510.

[2]   Xie Z, Johansen LK, Gustafson AMGenetic and functional diversification of small RNA pathways in plants. PLoS Biol 2004; 2: e104.

[3]   Dong Z, Han M-H, Fedoroff N. The RNA-binding proteins HYL1 and SE promote accurateprocessing of pri-miRNA by DCL1. Proc Natl Acad Sci USA 2008; 105: 9970-9975.

[4]   Gasciolli V, Mallory AC, Bartel DP, Vaucheret H. Partially Redundant Functions of Arabidopsis DICER-like Enzymes and a Role for DCL4 in Producing trans-Acting siRNAs. Curr Biol 2005; 15: 1494-1500.

[5]   Chan SWL, Henderson IR, Jacobsen SE. Gardening the genome: DNA methylation in *Arabidopsis thaliana*. Nat Rev Genet 2005; 6: 351-360.

[6]   Borsani O, Zhu J, Verslues PE, Sunkar R, Zhu J-K. Endogenous siRNAs derived from a pair of natural cis-antisense transcripts regulate salt tolerance in Arabidopsis. Cell 2005; 123: 1279-1291.

[7]   Eamens A, Wang M-B, Smith NA, Waterhouse PM. RNA silencing in plants: yesterday, today, and tomorrow. Plant Physiol 2008; 147: 456-468.

[8]   Jones-Rhoades MW, Bartel DP. Computational identification of plant microRNAs and their targets, including a stress-induced miRNA. Mol Cell 2004; 14: 787-799.

[9]   Fahlgren NRegulation of AUXIN RESPONSE FACTOR3 by TAS3 ta-siRNA affects developmental timing and patterning in Arabidopsis. Curr Biol 2006; 16: 939-944.

[10] Katiyar-Agarwal SA pathogen-inducible endogenous siRNA in plant immunity. Proc Natl Acad Sci USA 2006; 103: 18002-18007.

[11] Phillips JR, Dalmay T, Bartels D. The role of small RNAs in abiotic stress. FEBS lett 2007; 581: 3592.

[12] Sunkar R, Zhu JK. Novel and stress-regulated microRNAs and other small RNAs from Arabidopsis. Plant Cell 2004; 16: 2001-2019.

[13] Zhao BIdentification of drought-induced microRNAs in rice. Biochem Biophys Res Commun 2007; 354: 585-590.

[14] Liu H-H, Tian X, Li Y-J, Wu C-A, Zheng C-C. Microarray-based analysis of stress-regulated microRNAs in *Arabidopsis thaliana*. RNA 2008; 14: 836-843.

[15] Jian XIdentification of novel stress-regulated microRNAs from Oryza sativa L. Genomics 2010; 95: 47-55.

[16] Lu S, Sun Y-H, Chiang VL. Stress-responsive microRNAs in Populus. The Plant J 2008; 55: 131-151.

[17] Zhang J, Xu Y, Huan Q, Chong K. Deep sequencing of Brachypodium small RNAs at the global genome level identifies microRNAs involved in cold stress response. BMC Genomics 2009; 10: 449 doi:410.1186/1471-2164-1110-1449.

[18] Zhao BMembers of miR-169 family are induced by high salinity and transiently inhibit the NF-YA transcription factor. BMC Mol Biol 2009; 10: 29 doi:10.1186/1471-2199-1110-1129.

[19] Li W-XThe Arabidopsis NFYA5 transcription factor is regulated transcriptionally and posttranscriptionally to promote drought resistance. Plant Cell 2008; 20: 2238-2251.

[20] Sunkar R, Kapoor A, Zhu J-K. Posttranscriptional induction of two Cu/Zn Superoxide dismutase genes in Arabidopsis is mediated by downregulation of miR398 and important for oxidative stress tolerance. Plant Cell 2006; 18: 2051-2065.

[21] Jia X, Wang W-X, Ren LDifferential and dynamic regulation of miR398 in response to ABA and salt stress in *Populus tremula* and *Arabidopsis thaliana*. Plant Mol Biol 2009; 71: 51-59.

[22] Zhou X, Wang G, Zhang W. UV-B responsive microRNA genes in *Arabidopsis thaliana*. Mol Syst Biol 2007; 3.

[23] Jia X, Ren L, Chen Q-J, Li R, Tang G. UV-B-responsive microRNAs in *Populus tremula*. J Plant Physiol 2009; 166: 2046-2057.

[24] Moldovan D, Spriggs A, Yang J, Pogson BJ, Dennis ES, Wilson IW. Hypoxia-responsive microRNAs and trans-acting small interfering RNAs in Arabidopsis. J Exp Bot 2010; 61: 165-177.

[25] Moldovan D, Spriggs A, Dennis ES, Wilson IW. The hunt for hypoxia responsive natural antisense short interfering RNAs. Plant Signal Behav 2010; 5: 1-5.

[26] Zhang Z, Wei L, Zou X, Tao Y, Liu Z, Zheng Y. Submergence-responsive microRNAs are potentially involved in the regulation of morphological and metabolic adaptations in maize root cells. Ann Bot 2008; 102: 509-519.

[27] Boualem AMicroRNA166 controls root and nodule development in Medicago truncatula. The Plant J 2008; 54: 876-887.

[28] Bari R, Datt Pant B, Stitt M, Scheible W-R. PHO2, MicroRNA399, and PHR1 define a phosphate-signaling pathway in plants. Plant Physiol 2006; 141: 988-999.

[29] Hsieh L-CUncovering small RNA-mediated responses to phosphate deficiency in Arabidopsis by deep sequencing. Plant Physiol 2009; 151: 2120-2132.

[30] Sunkar R, Chinnusamy V, Zhu J, Zhu J-K. Small RNAs as big players in plant abiotic stress responses and nutrient deprivation. Trends Plant Sci 2007; 12: 301-309.

[31] Franco-Zorrilla JMTarget mimicry provides a new mechanism for regulation of microRNA activity. Nat Genet 2007; 39: 1033-1037.

[32] Shukla LI, Chinnusamy V, Sunkar R. The role of microRNAs and other endogenous small RNAs in plant stress responses. Biochimica et Biophysica Acta (BBA)-Gene Regul Mech 2008; 1779: 743-748.

[33] Zhu YY, Zeng HQ, Dong CX, Yin XM, Shen QR, Yang ZM: microRNA expression profiles associated with phosphorus deficiency in white lupin (*Lupinus albus* L.). Plant Sci 2010; 178: 23-29.

[34] Donaire L, Barajas D, Martinez-Garcia B, Martinez-Priego L, Pagan I, Llave C. Structural and genetic requirements for the biogenesis of tobacco rattle virus-derived small interfering RNAs. J Virol 2008; 82: 5167-5177.

[35] Wang X-BRNAi-mediated viral immunity requires amplification of virus-derived siRNAs in *Arabidopsis thaliana*. Proc Natl Acad Sci USA 2010; 107: 484-489.

[36] Deleris A, Gallego-Bartolome J, Bao J, Kasschau KD, Carrington JC, Voinnet O. Hierarchical action and inhibition of plant Dicer-like proteins in antiviral defense. Science 2006; 313: 68-71.

[37] Diaz-Pendon JA, Li F, Li WX, Ding SW. Suppression of antiviral silencing by cucumber mosaic virus 2b protein in Arabidopsis is associated with drastically reduced accumulation of three classes of viral small interfering RNAs. Plant Cell 2007; 19: 2053-2063.

[38] Wang M-B, Metzlaff M. RNA silencing and antiviral defense in plants. Curr Opin Plant Biol 2005; 8: 216-222.

[39] Blevins TFour plant Dicers mediate viral small RNA biogenesis and DNA virus induced silencing. Nucl Acids Res 2006; 34: 6233-6246.

[40] Moissiard G, Voinnet O. RNA silencing of host transcripts by cauliflower mosaic virus requires coordinated action of the four Arabidopsis Dicer-like proteins. Proc Natl Acad Sci USA 2006; 103: 19593-19598.

[41] Qu F, Ye X, Morris TJ. Arabidopsis DRB4, AGO1, AGO7, and RDR6 participate in a DCL4-initiated antiviral RNA silencing pathway negatively regulated by DCL1. PNAS USA 2008; 105: 14732-14737.

[42] Kasschau KD, Xie Z, Allen EP1/HC-Pro, a viral suppressor of RNA silencing, interferes with Arabidopsis development and miRNA function. Developmental Cell 2003; 4: 205-217.

[43] Chen J, Li WX, Xie D, Peng JR, Ding SW. Viral virulence protein suppresses RNA silencing-mediated defense but upregulates the role of microRNA in host gene expression. Plant Cell 2004; 16: 1302-1313.

[44] Navarro L, Dunoyer P, Jay FA plant miRNA contributes to antibacterial resistance by repressing auxin signaling. Science 2006; 312: 436-439.

[45] Fahlgren N, Howell MD, Kasschau KDHigh-throughput sequencing of Arabidopsis microRNAs: Evidence for frequent birth and death of MIRNA Genes. PLoS ONE 2007; 2: e219.

[46] Katiyar-Agarwal S, Gao S, Vivian-Smith A, Jin H. A novel class of bacteria-induced small RNAs in Arabidopsis. Genes Dev 2007; 21: 3123-3134.

[47] Navarro L, Jay F, Nomura K, He SY, Voinnet O. Suppression of the microRNA pathway by bacterial effector proteins. Science 2008; 321: 964-967.

[48] Agorio A, Vera P. ARGONAUTE4 is required for resistance to Pseudomonas syringae in Arabidopsis. Plant Cell 2007; 19: 3778-3790.

[49] Stokes TL, Kunkel BN, Richards EJ. Epigenetic variation in Arabidopsis disease resistance. Genes Dev 2002; 16: 171-182.

[50] Yi H, Richards EJ. A cluster of disease resistance genes in Arabidopsis is coordinately regulated by transcriptional activation and RNA silencing. Plant Cell 2007; 19: 2929-2939.

[51] Arenas-Huertero C, Pérez B, Rabanal FConserved and novel miRNAs in the legume Phaseolus vulgaris in response to stress. Plant Mol Biol 2009; 70: 385-401.

[52] Combier J-PMtHAP2-1 is a key transcriptional regulator of symbiotic nodule development regulated by microRNA169 in Medicago truncatula. Genes Dev 2006; 20: 3084-3088.

[53] Lelandais-Briere CGenome-wide Medicago truncatula small RNA analysis revealed novel microRNAs and isoforms differentially regulated in roots and nodules. Plant Cell 2009; 21: 2780-2796.

[54] Simon SA, Meyers BC, Sherrier DJ. MicroRNAs in the rhizobia-legume symbiosis. Plant Physiol 2009; 151: 1002-1008.

[55] Nakayashiki H, Nguyen QB. RNA interference: roles in fungal biology. Curr Opin Microbiol 2008; 11: 494-502.

[56] Ellis JG, Rafiqi M, Gan P, Chakrabarti A, Dodds PN. Recent progress in discovery and functional analysis of effector proteins of fungal and oomycete plant pathogens. Curr Opin Plant Biol 2009; 12: 399-405.

[57] Ellendorff U, Fradin EF, de Jonge R, Thomma BPHJ. RNA silencing is required for Arabidopsis defence against Verticillium wilt disease. J Exp Bot 2009; 60: 591-602.

[58] Elmayan TArabidopsis mutants impaired in co-suppression. Plant Cell 1998; 10: 1747-1758.

[59] Brodersen P, Voinnet O. The diversity of RNA silencing pathways in plants. Trends Genetics 2006; 22: 268-280.

[60] Lu S, Sun Y-H, Amerson H, Chiang VL. MicroRNAs in loblolly pine (*Pinus taeda*) and their association with fusiform rust gall development. Plant J 2007; 51: 1077-1098.

[61]   Hewezi T, Howe P, Maier TR, Baum TJ. Arabidopsis small RNAs and their targets during cyst nematode parasitism. Mol Plant-Microbe Interac 2008; 21: 1622-1634.

[62]   Pandey SP, Baldwin IT. RNA-directed RNA polymerase1 (RdR1) mediates the resistance of Nicotiana attenuata to herbivore attack in nature. Plant J 2007; 50: 40-53.

[63]   Pandey SP, Gaquerel E, Gase K, Baldwin IT. RNA-directed RNA polymerase3 from *Nicotiana attenuata* is required for competitive growth in natural environments. Plant Physiol 2008; 147: 1212-1224.

[64]   Pandey SP, Shahi P, Gase K, Baldwin IT. Herbivory-induced changes in the small-RNA transcriptome and phytohormone signaling in Nicotiana attenuata. Proc Natl Acad Sci USA 2008; 105: 4559-4564.

[65]   Bazzini AA, Hopp HE, Beachy RN, Asurmendi S. Infection and coaccumulation of tobacco mosaic virus proteins alter microRNA levels, correlating with symptom and plant development. Proc Natl Acad Sci USA 2007; 104: 12157-12162.

[66]   Tagami Y, Inaba N, Kutsuna N, Kurihara Y, Watanabe Y. Specific enrichment of miRNAs in Arabidopsis thaliana infected with tobacco mosaic virus. DNA Res 2007; 14: 227-233.

[67]   Yamada K, Lim J, Dale JMEmpirical analysis of transcriptional activity in the Arabidopsis genome. Science 2003; 302: 842-846.

[68]   Zhai J, Liu J, Liu BSmall RNA-directed epigenetic natural variation in *Arabidopsis thaliana.* PLoS Genet 2008; 4: e1000056.

[69]   Baum JA, Bogaert T, Clinton WControl of coleopteran insect pests through RNA interference. Nat Biotech 2007; 25: 1322-1326.

[70]   Mao Y-B, Cai W-J, Wang J-WSilencing a cotton bollworm P450 monooxygenase gene by plant-mediated RNAi impairs larval tolerance of gossypol. Nat Biotech 2007; 25: 1307-1313.

[71]   Paranjape T, Slack FJ, Weidhaas JB. MicroRNAs: tools for cancer diagnostics. Gut 2009; 58: 1546-1554.

[72]   Bazzini AA, Almasia NI, Manacorda CAVirus infection elevates transcriptional activity of miR164a promoter in plants. BMC Plant Biol 2009; 9: 12.

CHAPTER 13

Systems Biology: A Promising Tool to Study Abiotic Stress Responses

Konika Chawla[+], Pankaj Barah[+], Martin Kuiper and Atle M. Bones*

Department of Biology, Norwegian University of Science and Technology, Høgskoleringen 5, NO-7491 Trondheim, Norway

Abstract: Plant abiotic stress responses are a major yield-limiting factor in agriculture and thereby in the production of food, feed and fibre. Recent technology developments allow studies of such stress responses at a global molecular scale using omics data (metabolome, proteome, transcriptome and more). Significant progress has been made in statistical, mathematical and informatics driven analysis of omics data. Genes, proteins and metabolites can now be classified, categorized and linked at a genomic scale, and network-based analysis of various biological processes is becoming reality. However, in order to gain a complete overview of all processes and active networks in each cell type of the plant at all developmental stages and under all types of environmental variation, data production needs to become feasible at a significantly more massive scale. Systems biology studies the organization of system components and their networks, with the idea that unique properties of a system can only be observed through study of the system as a whole. A system-based analysis can involve multiple scales, ranging from single cells, tissues, organs to whole organisms. One of the foundations of systems biology is the analysis of networks of interacting and interdependent components that produce the system's unique properties. Network analysis provides intuitive ways for omics data visualization, as it reduces the intrinsic complexity of such data. In this chapter we discuss systems biology as a promising tool to study plant stress responses. We list various network and visualization tools available to biologists, to help them analyse high throughput omics data sets.

OMICS AND DATA INTEGRATION

The success of technology development able to study the diversity and quantities of molecules at a genomic scale has resulted in a wide proliferation of terms ending with the suffix –ome, referring to a comprehensive number of observations of a certain type of data. The availability of entirely sequenced genomes (which can be used to predict genes, mRNAs and proteins) has subsequently spawned technology development to enable proteomics ("all" proteins), transcriptomics ("all" transcripts), interactomics ("all" interactions between biomolecules) and metabolomics ("all" metabolites). The term subsequently was introduced to cover other data types like phenomics (all the phenotypes related to gene mutants) or bibliomics (the study of complete literature on a biological topic), just to name a few. Omics technologies strive for completeness in high throughput mode, producing vast amounts of data. The advent of high throughput techniques and high throughput sequencing and analysis methods has improved our understanding of the structure and function of many components participating at the cellular level. It will soon be possible to get a whole eukaryotic plant genome of moderate size (<1 Gb) sequenced for less than one thousand Euros [1]. Developments in techniques like mass spectrometry, microarrays, NMR, and next generation sequencing technologies have helped in discovering new proteins and transcripts and have resulted in generatation of copious amounts of data. These data can be in the form of large amounts of numbers reflecting colour intensity readings in microarrays, or character strings reflecting formatted or unformatted text describing biological annotations, their functional annotations or experimental observations. It is difficult for a biologist to make sense of all these data and relate them to a particular component or interaction. This situation has created a great need for data integration and management [2] (Fig. **1**).

WHY SYSTEMS BIOLOGY?

A system can be broadly defined as a group of independent but interconnected components that function together as a unified whole [3]. For the stability and functional robustness of a system, the components have to work together in a coherent manner. In the biological world systems can be identified at different hierarchical levels like ecosystems, organisms, organs, tissues, cells, molecules (genes, proteins, metabolites) and interactomic level. The boundaries of

*Address correspondence to: Dr. Atle M. Bones,** Department of Biology, Norwegian University of Science and Technology, Høgskoleringen 5, NO-7491 Trondheim, Norway; E-mail: atle.bones@bio.ntnu.no
+Contributed Equally

Narendra Tuteja, Sarvajeet Singh Gill and Renu Tuteja [Eds.]

the systems may not be clearly defined or definable if multi-cellular organisms are being investigated, as many systems have connections at multiscale levels. The interactions within a single hierarchy itself can already be tremendously complex, let alone the overall functionality of a system comprising multiple hierarchical levels.

In the last decade, systems biology has emerged as a promising field which integrates vast amounts of data from genome-scale technologies and builds computational models to help understand the topology and dynamical function of the molecular systems that build and sustain an organism [4]. With a systems model in place, systems biology research proceeds with both data-driven and hypothesis-driven approaches (Fig. **1**). It combines data integration based descriptive approaches and computational simulation based predictive approaches, thereby helping biologists to derive information from data obtained with the multitude of omics technologies, synthesise network models and observe their behaviour in an understandable and intuitive manner. It is an approach that brings down boundaries between biological studies and merges it with mathematics, physics, chemistry, and computer science. Nevertheless, a single definition of systems biology is still contentious; some emphasise the role of dynamic modelling, whereas others stress multidimensional data analysis. Considering the early age of the field, this range of opinions is not surprising [5-8]. Perhaps a pragmatic approach in defining systems biology is to follow the definitions put forward by granting bodies like EU framework programmes or UK's BBSRC, who put great emphasis on mathematical models as the core of systems biology.

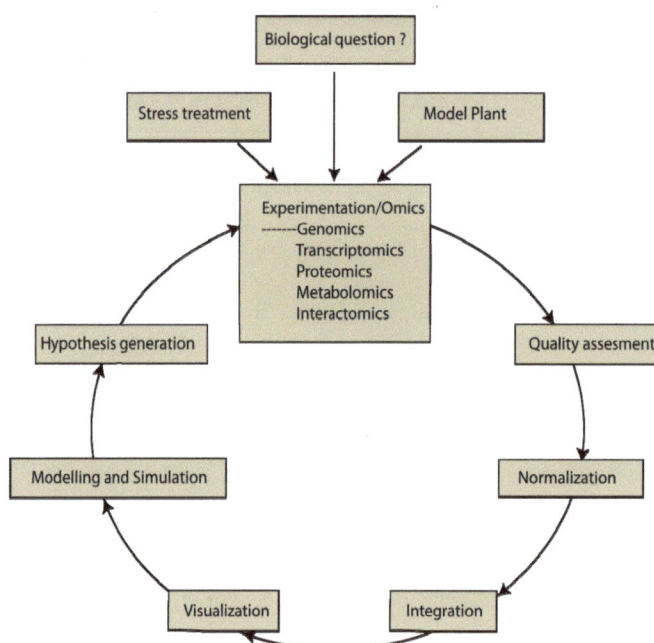

Figure 1: Systems biology cyclic approach for plant stress responses. A biological hypothesis is addressed and tested applying one or multiple stresses, using one or more omics technologies. Following integration of data, modelling and simulation, the hypothesis is validated, adjusted or new hypotheses generated.

PLANT SYSTEMS BIOLOGY

Plant research has the potential to contribute significantly to solve several of the most daunting problems that our planet and we face in the 21st century. These problems include an increasing shortage of available food, depletion of global fossil fuel reserves, and a growing scarcity of freshwater. Plants are sessile systems unable to escape from harmful environmental conditions. As a result, they have evolved a dazzling level of flexibility in their responses to environmental challenges caused by abiotic and biotic factors like sunlight/UV, drought, temperature, nutritional stress and pathogen and pest attacks. This flexibility of plants is known as a genotype by environment response, reflecting a genetic potential optimised for survival in a range of environmental conditions. Many plant genomes have undergone duplication events providing vast diversification of duplicated genes and pathways. Analysis of this interplay is complicated, as variable environmental conditions will elicit complex responses often resulting from different sets of pathways and components. Further complications arise from the fact that specific geno-phenotype relationships also depend on the growth stage and nutritional status. Owing to this complexity, following systems responses at the molecular level using transcript and protein profiling (for instance, when a single plant gene is manipulated) is difficult. Plant tissues consist of heterogeneous cell populations with multiple cellular compartments. Consequently, multiple gene and pathway types can be expressed in complex molecular concentration gradients in tissue samples. These plant-specific properties of highly compartmentalised intercellular networks and the resulting complexity arising from the environmental interactions pose great challenges for plant biologists [9].

The practice of integrating physiological, morphological, molecular, biochemical and genetic information pieces has long been applied in biological research, and in diverse fields such as plant breeding and ecology [3]. Systems biology has made this practice genome-scale, offering solutions to a more complete analysis by allowing virtual experimentation and hypothesis testing [10]. Plant systems biology helps in predicting functions of many genes and find missing links and crosstalk between various pathways and their effect on the plant [4]. This is also evident in the analysis of plant stress responses. Efforts are being taken to define and model the stress response mechanism of plants and its effect on other plant processes [11]. These efforts are aided by a wide variety of software [12], developed to get an improved and more intuitive visualisation of the experimental data, and for building networks and models that help to generate new hypotheses for further research. Much of the development of data and supporting software is targeted on the reference plant *Arabidopsis thaliana* (Table **1**).

ABIOTIC STRESS TOLERANCE

Plants have to tolerate, adapt to or defend against environmental stresses to survive. Plants have developed elaborate networks of defense mechanisms against stress, but the level of network crosstalk makes it challenging to correlate various types of network responses to a particular stress [13]. Various studies have increased our knowledge about general responses to changes in e.g. light, pH, water, salt and temperature. The effect of various combinations of stress in a particular crop variety and the interplay with other signaling systems that are activated or deactivated in response to stress remain difficult to predict [14,15]. This has inspired efforts to build virtual plant [10, 16] or plant part models that support the simulation of various stress conditions to analyze signaling network behaviour and predict for instance changes in plant metabolic pathways. At the Centre of Plant Integrated Biology in Nottingham, UK (CPIB) they aim to create a "virtual root" of *Arabidopsis thaliana* [17,18]. Various aspects of shoot and leaf development are the subject of systems biology studies in e.g. the European FP6 project Agron-omics (http://www.agron-omics.eu/; http://rico-coen.jic.ac.uk/index.php/Researchprogramme); and research at the Coen laboratory / John Innes Centre, UK (http://www.cpib.ac.uk /2009/virtual-root/). These plant and plant part models will be of outmost value for stress research in plants and provide unique modelling opportunities. Even though the data are generated from modelling of Arabidopsis thaliana, a large part of this information will most likely be transferable to other plant species including crop species. The virtual plants project [19]: (http://www.virtualplant.org) aims at constructing a dynamic model of plant that can mimic the responses of the model plant. In this direction, a virtual peach fruit model was designed to predict the growth and development of fruit under various environmental conditions. It thus acts as a tool to generate new hypotheses. Model-based predictions and hypothesis generation will help to rationalise the efforts and resources spent on wet lab research. Many databases are dedicated to annotate stress responsive genes, their products and their transcription factor binding sites [20] (Table **1**). Stress and stress responses are hot topics in plant systems biology, as it is so closely connected with productivity and yield. Understanding plant stress responses is essential if we want to offer solutions to the increasing demand of food, feed and biomass.

KINETIC MODELLING OF METABOLIC PATHWAYS AND GENE CIRCUITS

Ideally a model should include the possibility to exploit the dynamic processes in plant cells and tissues. Such a model would be based on vast amounts of quantitative data from all tissues and developmental stages and from plants cultured under all kinds of conditions and stresses. Although such data are largely missing today, we might expect that this will be gradually available over the next decade or so. The building of topologically correct models, aided by information from metabolic pathway databases, has allowed the generation of mathematical models [21] that represent the metabolic capabilities (as far as they are currently predictable) of various organisms. This information can be used in stoichiometric models (e.g. flux balance analysis [22, 23] that assume physicochemical constraints that have to be satisfied when an organism reaches a steady state (e.g. mass and energy balance, system boundaries, and flux limitations). These constraints are not known for every organism and every environmental condition, and so the possible steady-state solutions have to be further restricted by optimization with regard to an assumed objective (e.g. optimal growth, minimization of damage by excess light, or maximization of nutrient uptake).

NETWORKS AS A PROMISING TOOL FOR PLANT SYSTEMS BIOLOGY

Linus Pauling said – "Life is a relationship among molecules and not a property of any molecule" [24]. In systems biology, this relationship is referred to as 'emergent behaviour': a function that derives from the interaction of network components [3]. One of the main targets in this post genomics era is to understand how molecules interact with each other, how this enables signaling and how information processing is done to connect specific signals to specific molecular responses. Molecular interaction networks can be compared to functional modules that resemble attenuators, enhancers, adaptors, switches and feedback loops as in electronic circuits. These circuits are designed to maintain the stability of the system (homeostasis) by adaptive mechanisms. Several common themes are emerging from analysis of these circuits, one of which is that specific network based topologies appears to underlie the operation of many biological systems [25]. Networks have proven to be a powerful tool to deal with extremely complex systems. Several studies have shown that the underlying networks in biological systems have a similar behaviour as other physical systems [26].

STRATEGIES TO BUILD NETWORKS

It remains a significant challenge to infer biologically relevant network structures directly from biological data like sequences, transcripts, proteins or metabolites. Additional information like interactions or process of regulations is necessary to build up a network. These can be made and visualised with the help of software like Cytoscape [27], Cell Designer [28] and Pajek [29]. In addition, the discovery of network motifs including feedback and feed-forward loops is helpful to identify biologically relevant network structures in networks generated automatically from data. Building biological networks needs integration of heterogeneous data into a single model. Many different methods and algorithms have been developed [30-33], some of these are general in nature and others are specific to single species or particular processes. Here we list some of the available tools and databases potentially useful for systems biology related research in plants.

DIFFERENT LEVELS OF NETWORKS

As described above, biological systems exhibit multiple levels of complex hierarchical organisation. Depending on the level of hierarchical organisation and interactions the biological networks can be divided in two major groups [34]:

- Microscopic: molecular level (genes, proteins, metabolites), process level, tissue level, and organ level.

- Macroscopic: organism level (biotic) and environmental ecosystem level (abiotic).

Plants have evolved intricate mechanisms to perceive external signals, allowing optimal response to environmental conditions. In plants studies of crosstalk between biotic and abiotic stress and promise of network based studies will aim to uncover the cross talking signals. Under natural conditions plants encounter combinations of stresses and therefore have to mount an integrated response towards these stresses via synergistic and antagonistic actions. Various stress signaling pathways are likely able to crosstalk and adjust their contribution to the total response. Recent studies have revealed that several hormone signaling pathways regulated by abscisic acid, salicylic acid, jasmonic acid and ethylene and reactive oxygen species (ROS) signaling pathways are involved in crosstalk. Transcription factors like MYB, MYC, NAB, ZF as well as MAP kinases and metabolites like glutamate have been identified as promising candidates for common players that are involved in crosstalk between stress signaling pathways [35,36]. Network based approaches offer extremely promising tools to study such complex multi-level cross talking events.

TYPES OF NETWORKS AND THEIR BUILDING STRATEGIES

Four major types [37] of networks can be identified:

Gene-Metabolite Networks

These networks comprise various genes, proteins and metabolites in a system that can be highly complex because of the number of interacting metabolites and molecules. In correlation studies using multivaried data, similar behaviour or clustering of genes suggests their participation in the same biological process (this generally holds true for all kind of networks). Clustering approaches have been applied to study sulphur and nitrogen scarcity response in Arabidopsis [38,39] where networks were derived by using PCA (Principal Component Analysis) and SOM (Self Organising Maps) clustering. Clustering was also applied to study other nutrient deficiency stresses in Arabidopsis. It allowed the detection of signal transduction pathways in response to certain stress or novel candidate gene or metabolite involved in glucosinolate biosynthetic pathway [40]. Genes related to common metabolites in these pathways are often co-regulated. Many signaling pathways are included in this category and help understand the role of various enzymes in stress response [41]. These pathways and metabolic networks are stored in databases like KEGG [42], and ARACYC [43].

Protein-Protein Interaction (PPI) Networks

These networks comprise interactions between various proteins at a physical level. *In vitro* techniques like Y2H (yeast two hybrid) systems, and *in vivo* methods like FRET (fluorescence resonance energy transfer), BiFC (bimolecular fluorescence complementation) and TAP-tagging (tandem affinity purification protein-tagging) are used to characterize protein interactions in plants. In wheat plants, a protein-protein interaction network of 73 proteins and 97 interactions has been made to study the functions of up- and down- regulated stress responsive genes [44]. These studies provide network scaffolds that help to build signaling networks for understanding systemic regulation of biological processes [45]. These networks are stored in various databases like e.g. IntAct [46], MIPS [47] (Table **1** and **2**).

Table 1: Plant specific databases and tools

Arabidopsis Reactome [60]	http://arabidopsisreactome.org/
AraCyc [43]	http://www.arabidopsis.org/biocyc/index.jsp
AraNet [59]	http://www.functionalnet.org/aranet/about.html
ArrayXpath [61]	http://www.snubi.org/software/ArrayXPath/
AtPID [62]	http://atpid.biosino.org/
ATTED [63]	http://atted.jp/
GENEVESTIGATOR [64]	https://www.genevestigator.com/
MapMan [65]	http://mapman.gabipd.org/web/guest/mapman/
PathoPlant [58]	http://www.pathoplant.de/
STIFDB [20]	http://caps.ncbs.res.in/stifdb/
TAIR [66]	http://www.arabidopsis.org/
The Bio-Array Resource for Plant Functional Genomics [67]	http://bar.utoronto.ca/
Virtual Plant [16]	http://virtualplant.bio.nyu.edu/

Transcriptional Regulatory Network

These networks are also known as transcriptional regulatory networks. They result from the combination of various interactions taking place between different regulatory genes and genes located downstream of them. These networks have been used to describe interconnections between various genes involved for instance in abscisic acid biosynthesis [48]. Gene interactions are determined from gene expression data, protein and DNA interaction from gel shift assays and data mining from literature and public databases [37]. This approach has among others been used to decipher mRNA biogenesis processes [49]. In a recent study reverse engineering network modelling and analysis has been done to understand the transcription control of *Arabidopsis thaliana* [50]. Microarray data obtained from samples exposed to various abiotic stresses and at various stages of development were used to build the regulatory network. This model helps to predict the transcriptome of the microarray data under different conditions. It also suggested high connectivity between pathways related to stress responses.

Table 2: General databases and tools

Arena 3D [68]	*http://Arena3d.org/*
BiNGO [69]	http://www.psb.ugent.be/cbd/papers/BiNGO/
BiNoM [70]	*http://bioinfo-out.curie.fr/projects/binom/*
BioModules [71]	*http://galitski.systemsbiology.net/projs/biomodules/*
Cell Designer [72]	*http://www.celldesigner.org/*
ClueGO [73]	http://www.ici.upmc.fr/cluego/
Cytoscape [27]	*http://www.cytoscape.org/*
GeneTools [74]	*http://www.genetools.microarray.ntnu.no/*
GO [75]	*http://www.geneontology.org/*
IntAct [46]	*http://www.ebi.ac.uk/intact*
KaPPA-View [76]	http://kpv.kazusa.or.jp/kappa-view/
KEGG [77]	http://www.genome.jp/kegg/
MCODE [78]	*http://Baderlab.org/Software/MCODE*
MIPS [47]	http://mips.helmholtz-muenchen.de/proj/ppi/
NAViGaTOR [79]	*http://ophid.utoronto.ca/navigator/*
Pajek [29]	http://vlado.fmf.uni-lj.si/pub/networks/pajek/
PathwayExplorer [80]	https://pathwayexplorer.genome.tugraz.at/
Semantic Systems Biology [81]	http://www.semantic-systems-biology.org/
VANTED [82]	http://vanted.ipk-gatersleben.de/
VisANT [83]	http://visant.bu.edu/
VistaClara [84]	http://chianti.ucsd.edu/cyto_web/plugins/index.php/
WGCNA [85]	http://www.genetics.ucla.edu/labs/horvath/CoexpressionNetwork/Rpackages/WGCNA/

Gene Regulatory Networks

They constitute a general form for studying interaction between various genes, with genes depicted as nodes and their interactions as links. These links represent either activation or repression of the expression of a gene. These networks are built mainly on information curated from literature, and from gene expression data [51]. The resulting networks can be further validated by carrying out specific gene perturbation experiments (knock-out, knock-down, knock-in). When time-series data is taken into account these networks may provide insight into the dynamics of processes. A gene regulatory network represents a first stage approach to explain gene function in biological processes in a dynamic fashion. Equipping these networks with qualitative parameters has allowed the modelling of the isoprenoid biosynthesis pathway [52]; it has been useful to dynamically model flower development in *Arabidopsis thaliana* [53], or model a gene network controlling stomata and guard cell closure in response to abscisic acid treatment [54].

Co-expression Networks

In global co-expression analysis it has been observed that the network consists of many modules where nodes are tightly connected within the module, whereas they are only sparsely connected between modules. These networks exhibit a modular structure. Co-expressed or co-regulated genes can indicate their involvement in similar biological processes, meaning that individual modules can be attributed to specific biological processes. Using this basic concept, modular network topology based analysis has been proven to be useful in identifying functional modules [55]. Recently Mao *et al.* have provided new insight into the topological properties of biological networks using an Arabidopsis gene co-expression network (AGCN) generated from 1094 ATH1 microarrays [56]. In an another co-expression study, Weston *et al.* have shown how a mechanistic understanding of adaptive physiological responses to abiotic stress can provide plant researchers with a tool of great predictive value in understanding species and population level adaptation to climate change [57]. Similarly PathoPlant is likely to be a useful online database to analyze co-regulated genes involved in plant defence responses [58]. It displays signal perception and signal transduction pathways at a molecular level during plant pathogenesis as well as the corresponding interactions between plants and pathogens on the organismal level. A systems biology approach was also used to detect metabolite-protein co-regulation in a systemic response to temperature stress in *Arabidopsis thaliana* wild-type and a starch-deficient mutant (phosphoglucomutase-deficient) variety [2]. The study has also been used to identify functional marker candidates for abiotic stress tolerance and yield new insights for crop improvement. Another example is the AraNet, which is a probabilistic network of around 19,000 genes from Arabidopsis, which represents around 73 % of all genes of the Arabidopsis genome. This was built by first classifying the abiotic stress phenotypes and then linking with gene coexpression network analyses to find out genes responsible for a particular phenotype [59].

CHALLENGES AND FUTURE PROSPECTS

Systems biology of plant stress responses requires large amount of genome information for microarray studies from samples in various stress conditions. Methods like mass spectroscopy (MS) and nuclear magnetic resonance (NMR) need to be developed further for detecting a larger number of proteins and metabolites in plant samples and measure their abundance. Computational methods to make more robust models that capture the dynamics of plant biological processes are also required. There is a need to develop expertise and systems to integrate them all into data models consisting of complex networks and pathways. There is also a need for studying RNA:RNA interaction and small RNAs that are important for post transcriptional regulatory networks and the epigenome (http://neomorph.salk.edu/epigenome/epigenome.html)

Systems Biology is in a developing stage and yet the future seems very bright. It offers a platform to support the global research efforts dedicated to collect information about each and every component of a given biological system. Many strategies and technologies to integrate this information have been developed, including the semantic integration of knowledge. This has for instance resulted in the new paradigm of semantic systems biology, where hypothesis based research is not driven by a computational model of a process, but by integrative querying and automated reasoning [81]. The combination of data integration and modelling may help to make the "virtual plant" become a reality. The ability to carry out *in silico* experiments on this plant will revolutionise Plant Systems Biology.

ACKNOWLEDGEMENT

We want to thank the Norwegian Research Council (grant # 184146/S10 and 182897/S10), the Human frontier science program (HFSP) and Norwegian University of Science and Technology (NTNU) for financial support.

REFERENCES

[1]   Marco CAM, Bink TS, David Marshall. Statistical Challenges On The 1000 Euro Genome Sequences In Plants (EU COST Action TD0801), in Plant & Animal Genomes XVIII Conference. 2010.
[2]   Wienkoop S *et al.* Integration of Metabolomic and Proteomic Phenotypes. Mol Cell Proteom 2008; 7(9): 1725-1736.
[3]   Trewavas AA. Brief History of Systems Biology: "Every object that biology studies is a system of systems." Francois Jacob (1974). Plant Cell 2006; 18(10): 2420-2430.
[4]   Hammer GL *et al.* On systems thinking, systems biology, and the in silico plant. Plant Physiol 2004; 134(3): 909-911.
[5]   Kirschner MW. The meaning of systems biology. Cell 2005; 121(4): 503-504.
[6]   Kitano H. Computational systems biology. Nature 2002; 420(6912): 206-210.
[7]   Ideker T, Galitski T, Hood L. A NEW APPROACH TO DECODING LIFE: Systems Biology. Ann Rev Genom Hum Genet 2001; 2(1): 343-372.
[8]   Snoep J. Westerhoff H. From isolation to integration, a systems biology approach for building the Silicon Cell 2005; 13-30.
[9]   Sumner LW, Mendes P, Dixon RA. Plant Metabolomics: Large-Scale Phytochemistry in the Functional Genomics Era. ChemInform 2003; 34(19).
[10]   Gutierrez RA, Shasha DE, Coruzzi GM. Systems biology for the virtual plant. Plant Physiol 2005; 138(2): 550-554.
[11]   Minorsky PV. Achieving the in Silico Plant. Systems Biology and the Future of Plant Biological Research. Plant Physiol 2003; 132(2): 404-409.
[12]   Gehlenborg N *et al.* Visualization of omics data for systems biology. Nat Methods 2010; 7(3 Suppl): S56-68.

[13] Knight H, Knight MR. Abiotic stress signaling pathways: specificity and cross-talk. Trends Plant Sci 2001; 6(6): 262-267.

[14] Bowler C, Fluhr R. The role of calcium and activated oxygens as signals for controlling cross-tolerance. Trends Plant Sci 2000; 5(6): 241-246.

[15] Mittler R *et al.* Reactive oxygen gene network of plants. Trends Plant Sci 2004; 9(10): 490-498.

[16] Katari MS et al. Virtual Plant: a software platform to support systems biology research. Plant Physiol 2010; 152(2): 500-515.

[17] Benfey PN, Bennett M, Schiefelbein J. Getting to the root of plant biology: impact of the Arabidopsis genome sequence on root research. Plant J 2010; 61(6): 992-1000.

[18] Péret B *et al.* Arabidopsis lateral root development: an emerging story. 2009; 14(7): 399-408.

[19] Tardieu F. Virtual plants: modelling as a tool for the genomics of tolerance to water deficit. Trends Plant Sci 2003; 8(1): 9-14.

[20] Shameer K *et al.* STIFDB-Arabidopsis Stress Responsive Transcription Factor DataBase. Int J Plant Genom 2009; 2009: 583429.

[21] Poolman MG, Fell DA, Raines CA. Elementary modes analysis of photosynthate metabolism in the chloroplast stroma. Eur J Biochem 2003; 270(3): 430-439.

[22] Vahala E et al. Registration in interventional procedures with optical navigator. J Magn Reson Imaging 2001; 13(1): 93-98.

[23] Schuster S, Dandekar T, Fell DA. Detection of elementary flux modes in biochemical networks: a promising tool for pathway analysis and metabolic engineering. Trends Biotechnol 1999; 17(2): 53-60.

[24] Zuckerkandl E, Pauling LB. In: Kasha M, Pullman B, Eds. Horizons in Biochemistry, Molecular disease, evolution, and genetic heterogeneity. Academic Press. 1962; pp. 189-225.

[25] Schöner D *et al.* Network analysis of systems elements. 2007; 331-351.

[26] Barabasi A-L, Oltvai ZN. Network biology: understanding the cell's functional organization. Nat Rev Genet 2004; 5(2): 101-113.

[27] Shannon P *et al.* Cytoscape: a software environment for integrated models of biomolecular interaction networks. Genome Res 2003; 13(11): 2498-504.

[28] Funahashi A. Celldesigner: A Modeling Tool for Biochemical Networks. 2006.

[29] Batagelj V, Mrvar A. Pajek-Analysis and Visualization of Large Networks. 2002; 8-11.

[30] Breitling R *et al.* A structured approach for the engineering of biochemical network models, illustrated for signaling pathways. Brief Bioinform 2008; 9(5): 404-421.

[31] Gilbert D *et al.* Computational methodologies for modelling, analysis and simulation of signaling networks. Brief Bioinform 2006; 7(4): 339-353.

[32] van Riel NAW. Dynamic modelling and analysis of biochemical networks: mechanism-based models and model-based experiments. Brief Bioinform 2006; 7(4): 364-374.

[33] Kell DB. Metabolomics and systems biology: making sense of the soup. Curr Opin Microbiol 2004; 7(3): 296-307.

[34] Björn H, Junker FS. In: Pan Y, Ed. Wiley Series on Bioinformatics: Computational Techniques and Engineering. Analysis of Biological Networks. P.A.Y.Z. 2007.

[35] Kissen R *et al.* Transcriptional profiling of an Fd-GOGAT1 mutant in Arabidopsis thaliana reveals a multiple stress response and extensive reprogramming of the transcriptome. BMC Genom 2010; 11(1): 190.

[36] Fujita M *et al.* Crosstalk between abiotic and biotic stress responses: a current view from the points of convergence in the stress signaling networks. Curr Opin Plant Biol 2006; 9(4): 436-442.

[37] Yuan JS *et al.* Plant systems biology comes of age. Trends Plant Sci 2008; 13(4): 165-171.

[38] Hirai MY *et al.* Integration of transcriptomics and metabolomics for understanding of global responses to nutritional stresses in *Arabidopsis thaliana.* Proc Natl Acad Sci USA 2004; 101(27): 10205-10210.

[39] Nikiforova VJ *et al.* Integrative gene-metabolite network with implemented causality deciphers informational fluxes of sulphur stress response. J Exp Bot 2005; 56(417): 1887-1896.

[40] Hirai MY *et al.* Elucidation of Gene-to-Gene and Metabolite-to-Gene Networks in Arabidopsis by Integration of Metabolomics and Transcriptomics. J Biol Chem 2005; 280(27): 25590-25595.

[41] Cardinale F *et al.* Convergence and Divergence of Stress-Induced Mitogen-Activated Protein Kinase Signaling Pathways at the Level of Two Distinct Mitogen-Activated Protein Kinase Kinases. Plant Cell 2002; 14(3): 703-711.

[42] Kanehisa M The KEGG database. Novartis Found Symp, 2002; 247: 91-101; discussion 101-3, 119-28, 244-52.

[43] Mueller LA, Zhang P, Rhee SY, AraCyc: a biochemical pathway database for Arabidopsis. Plant Physiol 2003; 132(2): 453-460.

[44] Tardif G *et al.* Interaction network of proteins associated with abiotic stress response and development in wheat. Plant Mol Biol 2007; 63(5): 703-718.

[45] Uhrig JF. Protein interaction networks in plants. Planta 2006; 224(4): 771-781.

[46]   Bader GD, Betel D, Hogue CW. BIND: the Biomolecular Interaction Network Database. Nucleic Acids Res 2003; 31(1): 248-250.

[47]   Pagel P *et al.* The MIPS mammalian protein-protein interaction database. Bioinform 2005; 21(6): 832-834.

[48]   Nambara E, Marion-Poll A, ABSCISIC ACID BIOSYNTHESIS AND CATABOLISM. Ann Rev Plant Biol 2005; 56(1): 165-185.

[49]   Belostotsky DA, Rose AB. Plant gene expression in the age of systems biology: integrating transcriptional and post-transcriptional events. Trends Plant Sci 2005; 10(7): 347-353.

[50]   Carrera J *et al.* Reverse-engineering the Arabidopsis thaliana transcriptional network under changing environmental conditions. Genome Biol 2009; 10(9): R96.

[51]   Needham CJ et al. From gene expression to gene regulatory networks in Arabidopsis thaliana. BMC Syst Biol 2009; 3: 85.

[52]   Wille A *et al.* Sparse graphical Gaussian modeling of the isoprenoid gene network in *Arabidopsis thaliana*. Genome Biol 2004; 5(11): R92.

[53]   Espinosa-Soto C, Padilla-Longoria P, Alvarez-Buylla ER. A gene regulatory network model for cell-fate determination during Arabidopsis thaliana flower development that is robust and recovers experimental gene expression profiles. Plant Cell 2004; 16(11): 2923-39.

[54]   Li S, Assmann SM, Albert R. Predicting Essential Components of Signal Transduction Networks: A Dynamic Model of Guard Cell Abscisic Acid Signaling. PLoS Biol 2006; 4(10): e312.

[55]   Williams EJ, Bowles DJ. Coexpression of neighboring genes in the genome of Arabidopsis thaliana. Genome Res 2004; 14(6): 1060-1067.

[56]   Mao L *et al.* Arabidopsis gene co-expression network and its functional modules. BMC Bioinform 2009; 10(1): 346.

[57]   Weston D *et al.* Connecting genes, coexpression modules, and molecular signatures to environmental stress phenotypes in plants. BMC Systems Biol 2008; 2(1): 16.

[58]   Bulow L, Schindler M, Hehl R, PathoPlant(R): a platform for microarray expression data to analyze co-regulated genes involved in plant defense responses. Nucl Acids Res 2007; 35(suppl_1): D841-845.

[59]   Lee I *et al.* Rational association of genes with traits using a genome-scale gene network for *Arabidopsis thaliana*. Nat Biotechnol 2010; 28(2): 149-56.

[60]   Tsesmetzis N *et al.* Arabidopsis reactome: a foundation knowledgebase for plant systems biology. Plant Cell 2008; 20(6): 1426-1436.

[61]   Chung HJ *et al.* ArrayXPath: mapping and visualizing microarray gene-expression data with integrated biological pathway resources using Scalable Vector Graphics. Nucleic Acids Res 2004; 32(Web Server issue): W460-4.

[62]   Cui J *et al.* AtPID: Arabidopsis thaliana protein interactome database--an integrative platform for plant systems biology. Nucleic Acids Res 2008; 36(Database issue): D999-1008.

[63]   Obayashi T *et al.* ATTED-II provides coexpressed gene networks for Arabidopsis. Nucleic Acids Res 2009; 37(Database issue): D987-991.

[64]   Zimmermann P *et al.* GENEVESTIGATOR. Arabidopsis microarray database and analysis toolbox. Plant Physiol 2004; 136(1): 2621-32.

[65]   Thimm O *et al.* MAPMAN: a user-driven tool to display genomics data sets onto diagrams of metabolic pathways and other biological processes. Plant J 2004; 37(6): 914-39.

[66]   Rhee S *et al.* The Arabidopsis Information Resource (TAIR): a model organism database providing a centralized, curated gateway to Arabidopsis biology, research materials and community. Nucl Acids Res 2003; 31(1): 224-228.

[67]   Toufighi K *et al.* The Botany Array Resource: e-Northerns, Expression Angling, and promoter analyses. Plant J 2005; 43(1): 153-163.

[68]   Pavlopoulos GA *et al.* Arena3D: visualization of biological networks in 3D. BMC Syst Biol 2008; 2: 104.

[69]   Maere S, Heymans K, Kuiper M. BiNGO: a Cytoscape plugin to assess overrepresentation of gene ontology categories in biological networks. Bioinformatics 2005; 21(16): 3448-9.

[70]   Zinovyev A *et al.* BiNoM: a Cytoscape plugin for manipulating and analyzing biological networks. Bioinformatics 2008; 24(6): 876-877.

[71]   Prinz S *et al.* Control of yeast filamentous-form growth by modules in an integrated molecular network. Genome Res 2004; 14(3): 380-390.

[72]   Kitano H *et al.* Using process diagrams for the graphical representation of biological networks. Nat Biotech 2005; 23(8): 961-966.

[73]   Bindea G *et al.* ClueGO: a Cytoscape plug-in to decipher functionally grouped gene ontology and pathway annotation networks. Bioinformatics 2009; 25(8): 1091-1093.

[74]   Beisvag V et al. GeneTools--application for functional annotation and statistical hypothesis testing. BMC Bioinfo 2006; 7: 470.

[75]   Harris MA *et al.* The Gene Ontology (GO) database and informatics resource. Nucleic Acids Res 2004; 32(Database issue): D258-61.

[76]   Tokimatsu T *et al.* KaPPA-view: a web-based analysis tool for integration of transcript and metabolite data on plant metabolic pathway maps. Plant Physiol 2005; 138(3): 1289-1300.

[77]   Ogata H *et al.* KEGG: Kyoto Encyclopedia of Genes and Genomes. Nucleic Acids Res 1999; 27(1): 29-34.

[78]   Bader GD, Hogue CW. An automated method for finding molecular complexes in large protein interaction networks. BMC Bioinformatics 2003; 4: 2.

[79]   Brown KR *et al.* NAViGaTOR: Network Analysis, Visualization and Graphing Toronto. Bioinformatics 2009; 25(24): 3327-3329.

[80]   Mlecnik B *et al.* PathwayExplorer: web service for visualizing high-throughput expression data on biological pathways. Nucleic Acids Res 2005; 33(Web Server issue): W633-637.

[81]   Antezana E *et al.* BioGateway: a semantic systems biology tool for the life sciences. BMC Bioinformatics 2009; 10 Suppl 10: S11.

[82]   Junker BH, Klukas C, Schreiber F. VANTED: a system for advanced data analysis and visualization in the context of biological networks. BMC Bioinformatics 2006; 7: 109.

[83]   Hu Z *et al.* VisANT 3.5: multi-scale network visualization, analysis and inference based on the gene ontology. Nucleic Acids Res 2009; 37(Web Server issue): W115-121.

[84]   Kincaid R, Kuchinsky A, Creech M. VistaClara: an expression browser plug-in for Cytoscape. Bioinformatics 2008; 24(18): 2112-2114.

[85]   Langfelder P, Horvath S. WGCNA: an R package for weighted correlation network analysis. BMC Bioinformatics 2008; 9: 559.

Subject Index

A

ABA biosynthesis	145
Abiotic stress responses	163
Abiotic stress signaling	133,144
Abiotic stress tolerance	15,121,128,143,146,165
Abiotic stresses	10,40
Abscisic acid	45,123,143
Abscisic-acid-responsive element	95
Alkaloids	43
Allele mining	100
Aluminum	128
Amplified fragment length polymorphism	16
Antioxidant defense machinery	55
Arabidopsis	2,69,11,103,122,123,129
Ascorbate peroxidase	43,44
Ascorbic acid	43

B

Bacteria	156
Bioinformatics	130

C

Cadmium	128
Calcineurin B-like protein	45
cAMP-responsive element binding protein	95
Catalase	43,44
CBL-interacting protein kinase	46
Chemical genetics	78
Chilling stress	76,143
Chromatin remodeling	124
Co-expression networks	168
Cold acclimation	51
Cold stress	39,51,139,153
Cold tolerance	76
Combined stress	110
Comparative transcriptomics	129
Computer-aided technologies	106
Copper	128
Crop productivity	10,39

D

Dehydration	143
Dehydration-responsive element binding protein	95
Dehydroascorbate reductase	43,44
DNA methylation	121
DNA microarrays	17,19
Drought stress	10,39,48, 143,153

E

Epigenetics	121

Narendra Tuteja, Sarvajeet Singh Gill and Renu Tuteja [Eds.]

www.ingramcontent.com/pod-product-compliance
Lightning Source LLC
Chambersburg PA
CBHW041705210326
41598CB00007B/532